Animal Learning and Cognition

Handbook of Perception and Cognition
2nd Edition

Series Editors
Edward C. Carterette
and **Morton P. Friedman**

Animal Learning and Cognition

Edited by
N. J. Mackintosh

Department of Experimental Psychology
University of Cambridge
Cambridge, United Kingdom

Academic Press
San Diego New York Boston
London Sydney Tokyo Toronto

This book is printed on acid-free paper. ∞

Academic Press, Inc.
A Division of Harcourt Brace & Company
525 B Street, Suite 1900, San Diego, California 92101-4495

United Kingdom Edition published by
Academic Press Limited
24-28 Oval Road, London NW1 7DX

Library of Congress Cataloging-in-Publication Data

Animal learning and cognition / edited by N. J. Mackintosh.
        p.   cm. -- (Handbook of perception and cognition series : 3)
    Includes bibliographical references and index.
    ISBN 0-12-161953-2
    1. Learning in animals.   2. Cognition in animals.   I. Mackintosh,
N. J. (Nicholas John), date.   II. Series.
QL785.A728   1994
591-.51--dc20   94-7473
                                                             CIP

PRINTED IN THE UNITED STATES OF AMERICA
94   95   96   97   98   99   BC   9   8   7   6   5   4   3   2   1

Contents

3   Instrumental Conditioning
Anthony Dickinson

4   Reinforcement and Choice
Ben A. Williams

5 *Discrimination and Categorization*
John M. Pearce

6 *The Neural Basis of Learning with Particular Reference to the Role of Synaptic Plasticity: Where Are We a Century after Cajal's Speculations?*
Richard G. M. Morris

7 Biological Approaches to the Study of Learning
Sara J. Shettleworth

12 *Human Associative Learning*
David R. Shanks

Contributors

Numbers in parentheses indicate the pages on which the authors' contributions begin.

Anthony Dickinson (45)
Department of Experimental
  Psychology
University of Cambridge
Cambridge CB2 3EB, United
  Kingdom

C. R. Gallistel (221)
Department of Psychology
University of California, Los Angeles
Los Angeles, California 90024

William C. Gordon (255)
Department of Psychology
University of New Mexico
Albuquerque, New Mexico 87131

Geoffrey Hall (15)
Department of Psychology
University of York
York Y01 5DD, United Kingdom

C. M. Heyes (281)
Department of Psychology
University College London
London WC1E 6BT, United Kingdom

Rodney L. Klein (255)
Department of Psychology
University of New Mexico
Albuquerque, New Mexico 87131

N. J. Mackintosh (1)
Department of Experimental
  Psychology
University of Cambridge
Cambridge CB2 3EB, United
  Kingdom

Richard G. M. Morris (135)
Department of Pharmacology and
    Centre for Neuroscience
University of Edinburgh Medical
    School
Edinburgh, EH8 9LE, United
    Kingdom

John M. Pearce (110)
School of Psychology
University of Wales
College of Cardiff
Cardiff CF1 3YG, United Kingdom

Duane M. Rumbaugh (307)
Departments of Psychology and
    Biology
Language Research Center
Georgia State University
Atlanta, Georgia 30304

E. Sue Savage-Rumbaugh (307)
Departments of Biology and
    Psychology
Language Research Center
Georgia State University
Atlanta, Georgia 30304

David R. Shanks (335)
Department of Psychology
University College London
London WC1E 6BT, United Kingdom

Sara J. Shettleworth (185)
Department of Psychology
University of Toronto
Toronto, Ontario, Canada M5S 1A1

Ben A. Williams (81)
Department of Psychology
University of California, San Diego
La Jolla, California 92093

Foreword

The problem of perception and cognition is in understanding how the organism transforms, organizes, stores, and uses information arising from the world in sense data or memory. With this definition of perception and cognition in mind, this handbook is designed to bring together the essential aspects of this very large, diverse, and scattered literature to give a précis of the state of knowledge in every area of perception and cognition. The work is aimed at the psychologist and the cognitive scientist in particular, and at the natural scientist in general. Topics are covered in comprehensive surveys in which fundamental facts and concepts are presented, and important leads to journals and monographs of the specialized literature are provided. Perception and cognition are considered in the widest sense. Therefore, the work will treat a wide range of experimental and theoretical work.

The *Handbook of Perception and Cognition* should serve as a basic source and reference work for all in the arts or sciences, indeed for all who are interested in human perception, action, and cognition.

Edward C. Carterette and Morton P. Friedman

Preface

That this edition of the *Handbook of Perception and Cognition* should contain a volume on animal learning and cognition is both gratifying and appropriate. It marks a welcome and modest return swing of the pendulum in the often uneasy relationship between experimental psychologists studying human perception, learning, memory and cognition, and those who study the behavior of other animals. John Watson's behaviorist call to arms ushered in a long era when much of North American experimental psychology was dominated, to an unhealthy extent, by the classic learning theories of Hull, Tolman, and others—an era that was followed by, perhaps, an equally unhealthy split between human and animal psychology. I hope that some cognitive psychologists who browse through these pages will agree that both of these earlier attitudes were extremist, and that there is something here of interest and relevance for them. They will surely discover that the study of animal learning and cognition is richer and more diverse than anything they learned from their introductory psychology course.

I am grateful to the editors of the *Handbook of Perception and Cognition* for the opportunity to edit this volume and, thus, provide a sample of our wares to a wider community. They have given me a free hand in the choice of topics and authors, and I accept complete responsibility for that choice. North American readers, in particular, may note a certain British bias in the choice of authors. If this needs further defense than that provided by the chapters they have written, I plead that it is often easier to twist a friend's arm. The choice of topics may also seem idiosyncratic to some; another editor might have produced a volume with a rather different look—but surely not wholly different. I believe that the following chapters cover most of the central issues in the field of animal learning and cognition. It is probably my own interests that explain what some may see, with the exception of Chapter 4, as a certain bias against the operant tradition or the experimental analysis of behavior. This may seem odd, given that B. F. Skinner and many of his followers held more strongly than most to the faith that the experimental analysis of animal behavior was essential to understanding the central issues in human psychology. Here, perhaps, my reasoning must be that they have tended to question whether those central issues include the study of perception, learning, or cognition.

CHAPTER 1

Introduction

N. J. Mackintosh

I. THE LEGACY OF PAVLOV AND THORNDIKE

People have shown an interest in the mental life and behavior of other animals for most of recorded history. Myths and legends attributed cunning, sagacity, fortitude, or nobility now to one animal, now to another. Few of the world's religions have implied such a sharp contrast between ourselves and other animals as has Christianity, and even in Christian Europe, Descartes's argument that they are mere inanimate machines was rejected by many writers and philosophers in the seventeenth and eighteenth centuries (Lovejoy, 1936; Thomas, 1983). Before Darwin, people may have resisted the idea that we are part of the animal kingdom, but they did not doubt that other animals shared many human attributes, that they could be brave, cowardly, playful, or long suffering, and that they could be trained to perform useful tasks. But by giving scientific respectability to the idea that we really are part of the animal kingdom, Darwinian theory also made the systematic study of the behavior of other animals a legitimate branch of scientific enquiry.

That study was eventually brought into the laboratory by Pavlov and by Thorndike at the end of the nineteenth century. Today, these two men are commonly, and rightly, credited with initiating the experimental analysis of associative learning in animals, but it was not always so. Neither had an immediate, major impact on the laboratory study of animal learning; in the

Animal Learning and Cognition

early years of the twentieth century, following the work of Small (1900), that study was dominated by experiments on maze learning. Watson himself, who did so much to place the study of animal behavior at the center of all psychology, had contributed to these experiments, and in his first book, *Behavior: An Introduction to Comparative Psychology,* there were no more than half a dozen or so brief references to Thorndike and only a single mention of Pavlov, in which the main use of his method was said to be "to determine the efficiency of animals' receptors" (Watson, 1914, p. 65).

A. Learning as Conditioning

In due course, however, the two experimental procedures that Pavlov and Thorndike pioneered, classical or Pavlovian conditioning and instrumental or operant conditioning, came to dominate psychological research on animal learning. In part, this was because Watson later endorsed Pavlov's argument that the conditioned reflex provided the mechanism underlying the formation of all habits (Watson, 1924). And in part, it was because, as Dickinson recounts in Chapter 3, by the 1930s the distinction between Pavlovian and instrumental conditioning became clearer and provided the focus for much subsequent research. Skinner (1932) was one of the first to draw attention to the operational distinction between the two, in terms of the experimental contingencies specifying the occurrence of reinforcers; in Pavlovian conditioning, the experimenter arranges a contingency between a stimulus (the conditioned stimulus or CS) and a reinforcer (the unconditioned stimulus or US), while in instrumental conditioning, the contingency is between the subject's behavior and a reinforcer. But Miller and Konorski (1928) had earlier argued that this operational distinction was of more fundamental theoretical importance: Pavlov's principle of reinforcement, while sufficient to account for the emergence of conditioned responding in his procedure, was unable to account for the appearance of instrumental responding in Thorndike's. According to Pavlov, an association between CS and US was sufficient to endow the CS with the power to elicit responses normally elicited by the US. Application of this principle to an instrumental conditioning experiment, for example, a hungry rat pressing a lever for food reinforcers, predicts that the sight or feel of the lever might in due course cause the rat to salivate (the unconditioned response to food), but would not explain why the rat chose to press the lever.

There have always been those who have questioned the importance of the distinction between Pavlovian and instrumental conditioning, and it is true that the circumstances that promote successful conditioning, a temporal relationship between an antecedent event and a reinforcer, with the added requirement that the reinforcer not be otherwise predicted, apply equally to both forms (see Chapters 2, 3, and 12 by Hall, Dickinson, and Shanks, respectively). But the fact remains that Pavlov's principle of reinforcement and Thorndike's law of effect make quite different predictions about the outcome of certain experimental manipulations. If a CS signals the delivery of food, but only if the animal refrains from performing a particular response in anticipation of food when the light comes on, the Pavlovian principle predicts that if the response is elicited by food, the animal will fail to learn this omission contingency. But the law of effect predicts successful learning regardless of the nature of the response. There is sufficient evidence that some responses are more likely to obey the Pavlovian principle, others Thorndike's, to justify maintaining the distinction between the two varieties of conditioning. None of this is to deny, of course, that both principles usually interact to determine an animal's behavior in most situations—even in the laboratory. That has long been a familiar proposition to proponents of two-factor theory. Less familiar will be its application to the theory of choice, discussed by Williams in Chapter 4.

B. Comparative Psychology and Physiology

Pavlov and Thorndike differed, of course, in many other ways, not least in the intellectual background to their work. Pavlov, a physiologist studying the digestive system, regarded his conditioned reflex method as an ideal procedure for investigating the organization of the nervous system—the neural substrate of learning and memory. There is an element of paradox here, for although Pavlov is rightly remembered for his behavioral analysis of conditioning, and although virtually all his behavioral observations have stood the test of time, his account of the physiology of higher nervous activity has not. The reason, as Konorski noted, is that although "the very notion of the conditioned reflex is, of course, purely connectionistic . . . Pavlov's concepts of the dynamics of inhibitory processes and of the interplay between excitation and inhibition go far beyond the connectionistic outlook and consider the brain as a sort of continuum in which the nervous processes do not travel along predetermined pathways but spread diffusely" (Konorski, 1968, p. xii). It was left to Konorski (1948) and Hebb (1949) to show how a connectionist neurophysiology was in principle capable of explaining the phenomena of associative learning that Pavlov had studied. Half a century later, work on simple systems such as precocial imprinting in chicks (Horn, 1985), conditioning in *Aplysia* (Hawkins, 1989), or the rabbit nictitating membrane CR (Thompson, 1986) has gone some way to justify Pavlov's original faith, while Morris's review, in Chapter 6, of our understanding of synaptic plasticity documents the impact of Konorski's and Hebb's insistence on applying a connectionist neuroscientific theory to the connectionism of conditioning experiments.

Thorndike, on the other hand, although equally concerned to provide "an explanation of the processes of association" (Thorndike, 1898, p. 1), located those associations "in the animal mind" rather than in the brain, and his work derived from a quite different tradition—one that stemmed direct- ly from the publication of Darwin's *Descent of Man* (1871). Darwin wanted proof of mental, as well as physical, continuity between humans and other animals, and the immediate impetus for Thorndike's work was the steady flow of books published in the last decades of the nineteenth century con- taining amazing tales of the feats of intellect of which monkeys, dogs, cats, and other animals were capable. What was required, Thorndike believed, was less reliance on unsubstantiated anecdote, and more experimental anal- ysis of the ways in which animals actually set about solving problems. The results of his own analyses persuaded Thorndike that these processes were actually rather simple, not only in cats and chickens, but also in monkeys. In brief, they were stimulus–response connections strengthened or weak- ened in accordance with the law of effect.

Thorndike's conclusions were vigorously challenged by those who insis- ted that some mammals, and surely most primates, solved problems in ways that transcended mere trial and error learning (or instrumental condi- tioning). Kohler (1924) and Yerkes (1916), in particular, argued that chim- panzees were capable of working out the solutions to problems without going through a series of random attempts until they hit on the correct solution by chance. They had insight into the relationship between means and ends. But the concept of insight proved unproductive: no one ever succeeded in providing operational criteria for the distinction between in- sight and trial and error, in the way that Tolman (1932) sought operational definitions of purposive behavior, expectancies, or means–end readiness. To advance beyond Thorndike's skeptical position, research on animal in- telligence and the mental continuity between humans and other animals required different concepts.

No such unifying concept has ever emerged—indeed, there is no good reason why one should; it seems rather more plausible to suppose that there is a wide variety of ways in which we and other animals differ from one another. But the idea that something called intelligence has evolved and that some animals, usually primates, have more of it than others, exerts a tena- cious hold on our imagination, and there has been no shortage of compara- tive psychological research addressing these issues. By and large, it has had a bad press. Some have questioned the simplistic notion that living animals can be rank ordered along a notional phylogenetic scale from less to more intelligent (Hodos & Campbell, 1969). Others, such as Macphail (1982), have roundly asserted that there is no evidence of any difference in anything we should want to call intelligence between one nonhuman vertebrate and another. One should not underestimate the difficulty of disproving Mac- phail's null hypothesis, but Pearce (Chapter 5) points to some apparent differences in the type of discrimination problem soluble by different ani- mals, and it is hard to believe that a pigeon or a frog (fairy tales notwithstanding) would provide as good evidence of comprehension of artificial, let alone natural, language as can some chimpanzees (Rumbaugh & Savage-Rumbaugh, Chapter 11). The final charge leveled against much traditional comparative psychology has been that laboratory studies of animal intelligence are guilty of incurable anthropocentrism, that they have taken scant account of the problems that animals actually face in their natural environment, and have necessarily given a wholly false picture of their real intelligence.

II. BIOLOGICAL AND ANTHROPOCENTRIC APPROACHES TO ANIMAL BEHAVIOR

Neither Pavlov nor Thorndike expressed great concern for the concept of ecological validity, and both were unashamedly anthropocentric in the sense that they believed that the study of learning in animals would tell them something about the human condition, and thought it important for precisely that reason. It should hardly be necessary for a psychologist to apologize for holding such beliefs, nor for studying, in animals, problems sometimes chosen for the light they might shed on human concerns and on the human case rather than because they are particularly relevant to that animal in its natural habitat. Psychologists are entitled to ask questions about memory of their animal subjects, suggested by theories of human memory, because it may be easier to answer *some* of those questions in the animal case, and such work may suggest new questions to ask of human subjects (Gordon & Klein, Chapter 9). To take just one other example, it may be obvious enough that only humans communicate with one another by anything approximating human language, but that does not mean that there has been nothing to learn from the search for rudiments of language learning in apes, whose vicissitudes are recounted by Rumbaugh and Savage-Rumbaugh (Chapter 11). With the usual advantage of hindsight, it is easy enough to see the false trails and dead ends that have marked much of this research. But it is also surely true that it has illuminated some of the senses in which human language *is* unique, as well as some of the senses in which it is not. Thus, it has not only had something to say about the larger question of "man's place in nature," but has also told the psycholinguist something about the process of language acquisition.

Nevertheless, it would be equally wrong to deny the force of the ethologists' and ecologists' criticism of what Shettleworth (Chapter 7) calls the anthropocentric tradition in comparative psychology. The proper study of any particular animal's behavior, learning abilities, or intelligence must benefit from some prior understanding of that animal's natural life (what it needs to learn and when), even if such a study will almost certainly require experimental analysis, often undertaken in the laboratory, to elucidate the nature of the underlying mechanisms. Shettleworth illustrates how well this lesson has been learned, and to the benefit of all concerned. Where once there was mutual suspicion, incomprehension, and, occasionally, vulgar abuse, now there is at least some recognition that comparative psychology has been enriched by broadening the range of paradigms it studies, and that the behavior studied by ethologists may, upon analysis, often reveal processes not unlike those postulated by experimental psychologists to account for Pavlovian or instrumental conditioning. For example, animals seeking food in the world must make decisions between alternative sources of food, and laboratory experiments illustrate some of the complexities of the processes underlying such choice (see also Williams, Chapter 4).

Two other chapters reinforce this message. Experiments on maze learning by rats have once again become as popular as they were fifty years ago, but now largely because they are thought to require specialized perceptual or learning abilities—those involved in the construction of a spatial map of an animal's environment. This work has become of great interest to neuroscientists, largely stemming from O'Keefe and Nadel's (1978) suggestion that the mammalian hippocampus is the locus where such a map is built (Morris, Chapter 6). But Gallistel's review of spatial learning (Chapter 8) relies mostly on work with insects. Here is a case that shows that it is often possible to undertake well-controlled field experiments, and that they may reveal mechanisms not readily studied in the laboratory: the behavior of a rat in a radial maze or in Morris's swimming pool may be controlled by complex configurations of spatial landmarks, but it is less likely than that of a foraging honey bee to reveal evidence of navigation by dead reckoning or the use of the sun's azimuth position as a compass.

Heyes (Chapter 10) reviews evidence on the "social intelligence" hypothesis. According to Jolly (1966) and Humphrey (1976), the intelligence of monkeys and apes evolved not in order to solve the sorts of problems traditionally given to them by comparative psychologists (problems bearing more than a passing resemblance to those found in human IQ tests), but in order to cope with the complexities of social life, engendered by living in relatively large, often fluid groups, where it becomes important to understand relationships between other members of the group and to predict their behavior. Since 1980 or so, this hypothesis has generated much enthusiasm and a large body of research. Although Heyes is, quite rightly in my judgment, critical of many of the conclusions so confidently drawn from some of this work, no one would wish to dismiss it as of no value. It has done much to reinvigorate primate research, not only by emphasizing the need for interplay between field observation and laboratory experiment, but also by looking for links with the study of cognitive development in children.

III. THE COGNITIVE REVOLUTION IN PSYCHOLOGY

A. Changing Views of the Learning Process

The new title of this handbook serves as one reminder of the "cognitive revolution" that swept through experimental psychology in the 1960s. That the handbook should contain a volume on animal behavior, which includes the term "cognition" in its title, says something about the distance that revolution has traveled. Among its first casualties, of course, was the claim that animal learning theory took a position at center stage of human experimental psychology. The next consequence, however, had much less to recommend it—the erection of a nearly impermeable barrier between the study of conditioning and learning in animals and that of learning, memory, and cognition in humans. Having freed themselves of the chains of old-fashioned behaviorism, few human cognitive psychologists took the trouble to discover that, in a final step, the cognitive revolution had permeated the study of animal conditioning and learning itself.

This much is now commonplace—at least among those studying animal learning and behavior. Books on animal cognition, comparative cognition, or cognitive ethology now abound (Flaherty, 1985; Pearce, 1987; Roitblatt, 1985; Roitblatt, Bever, & Terrace, 1985). Popular conceptions, however, although commonly containing an element of truth, even more commonly simplify that truth to the point of being misleading.

Chapters 2 and 3, by Hall and Dickinson, respectively, show how far modern conditioning theory has progressed from the simple S–R theory of Watson and Guthrie, or even from the more sophisticated, neo-behaviorist theories of Hull (1952) and Spence (1956). Few conditioning theorists today find anything strange or dangerous in talking of the formation of associations between central representations of CS and US, or response and outcome, rather than just between stimulus and response. Hall, indeed, gives expression to a widely shared view when he says that his concern is not with the behavioral consequences of Pavlovian conditioning, but with the underlying processes: "That the conditioning procedure frequently produces some change in behavior will be regarded as a happy accident—one that allows the experimenter, by monitoring the vigor of some CR, to make inferences about the changes going on within the organism as a consequence of the training to which it is being subjected."

But it is seriously misleading to suggest, as is sometimes done, that the cognitive revolution has merely replaced Hull's S–R theory with some version of Tolman's S–S theory. What has happened is rather more interesting than that, and represents significant progress rather than a mere restatement of old positions and a refighting of old battles. For example, contemporary analyses of Pavlovian conditioning allow a place for both S–R and

S–S processes. The first explicit account of Pavlovian conditioning in these terms was proposed by Konorski (1967), who drew a distinction between what he termed consummatory and preparatory conditioning, the former dependent on an association between the CS and a representation of the stimulus attributes of the US, the latter on an association between the CS and the affective reactions elicited by the US. Consummatory (S–S) conditioning, if one likes, provides the organism with information about the particular event predicted by the CS; preparatory conditioning is a form of S–R learning where the response in question is not the discrete, overt CR, but rather the internal, affective response to the US. It gives the relevant emotional tone to the information provided by consummatory conditioning. The distinction has been elaborated by Wagner and Brandon (1989), and incorporated into an extension of Wagner's earlier SOP model (Wagner, 1981).

Dickinson (Chapter 3) similarly argues that there is a place both for an S–R "habit" mechanism, and for an action–outcome "expectancy" theory in the analysis of instrumental conditioning. We now have some reasonable idea how to identify each of these processes, and what factors determine which will be more important in producing any particular instance of instrumental behavior. Dickinson even outlines a theory that provides one account of how the two processes interact to produce observed changes in behavior, while Williams (Chapter 4) shows that experiments using standard operant schedules of reinforcement may be used to analyze the fundamental question of how animals distribute their choices between alternative sources of reinforcement. The search for an adequate theory of choice may not have come up with a final answer (so much so that at times one suspects that we are asking the wrong questions of the data, or at least need a radically new perspective), but it has at least advanced some way beyond earlier, neo-behaviorist theories.

A second dispute between earlier theories of learning centered on the role of reinforcement in conditioning. This dispute appeared to cut across that between S–S and S–R theorists, since at least some of the latter, such as Watson and Guthrie, implied that temporal contiguity between a stimulus and a response was sufficient to establish an association between them, while others, such as Thorndike and Hall, argued that a reinforcer was needed to strengthen the S–R connection. The one point they all agreed on, the importance of strict temporal contiguity, is the one that has been overturned by the work of Garcia, Kamin, and others in the late 1960s (see Hall, Chapter 2). Garcia showed that successful conditioning could occur across intervals of several hours between CS and US—provided that these were chosen appropriately. A novel flavor was readily associated with gastric distress across such an interval, but an aversion could not easily be conditioned to the same flavor when the US was shock, nor were other, exteroceptive CSs readily associated with illness. Kamin's influential experiments established that temporal contiguity between a CS and a US was not sufficient to ensure successful conditioning to that CS: an additional requirement was that the US not be already predicted by some other event occurring at the same time as the target CS. As is documented in Hall's chapter, the discovery that the conditions responsible for successful conditioning are notably more complex than those envisaged by any traditional learning theory has prompted a renewed interest in attempts to unravel the "laws of association." Even if there is little consensus on what those laws should be, let alone how they might be explained at a more fundamental level, few would question that the search has invigorated the study of conditioning and led to a better understanding of the nature of conditioning and associative learning.

Associationism, as Hall recounts, has a long, philosophical history, and the relationship between philosophy and experimental psychology ensured that it would continue to figure prominently in psychological analyses of learning and memory. That has been equally ensured by psychology's relationship to neuroscience. As Morris (Chapter 6) makes clear, for most of the twentieth century, neuroscientists have rarely questioned the assumption that the physical substrate of learning and memory in the brain is to be found in the establishment of functional connections between neurons. But associationism has always had its critics, even as an account of supposedly simple associative learning. As I have already noted, Thorndike's S–R theory was soon challenged by Köhler and other Gestalt psychologists, who argued that learning was more profitably viewed as a matter of perceptual reorganization. The more recent discovery of the importance for successful conditioning of the contingency between CS and US has encouraged a number of theorists to propose that conditioning is better viewed as a matter of detecting the contingency between CS and US rather than associating the two events. Gallistel (Chapter 8) and Shanks (Chapter 12) discuss such contingency theories of conditioning in animals and people. Dickinson (Chapter 3) and Williams (Chapter 4) both note problems for associative analyses of instrumental performance. Such accounts typically assume that the strength of the association between response and outcome will be a function of the probability of the second given the first. Contingency theory finds support in the observation that a rat's instrumental lever pressing is sensitive to the correlation between its rate of pressing and the rate at which reinforcers are delivered, rather than to the probability of a lever press being reinforced, and equally that its choice of one lever over another is not determined in any simple way by the probability of a response to one rather than to the other having been reinforced.

B. Animal Learning and Human Behavior

In 1938, Tolman wrote:

> I believe that everything important in psychology (except such matters as the building up of a super-ego, that is everything save such matters as involve society and words) can be investigated in essence through the continued experimental and theoretical analysis of the determiners of rat behavior at a choice point in a maze. Herein I believe I agree with Professor Hull and also with Professor Thorndike. (p. 34)

Substitute rats pressing levers in operant chambers for rats turning left or right in a T-maze, and Tolman could equally have counted on the support of Professor Skinner. As I remarked when I quoted these words in the opening chapter of a book on conditioning and associative learning in 1983, "it is hard now to believe that anyone should seriously have thought that such experiments could tell us all we need to know about human development, perception, or intelligence" (Mackintosh, 1983, p. 1). But they certainly did. Learning theorists' arrogance received its just desserts; by about 1960, their claim to a central place in psychology would have seemed laughable to anyone interested in human learning, memory, decision making, thinking, or problem solving. Indeed, with the exception of Skinner and some of his followers, few of those studying animal learning and behavior themselves would have cared to defend their predecessors' views. In retrospect, those views can have seemed tenable only because so little was understood of the true complexity of human cognition: one need only recollect that in 1950 the study of language or cognitive development was still untouched by the work of Chomsky or Piaget. From 1960 until relatively recently, many of those engaged in the study of human cognition confidently dismissed animal learning theory as the last outpost of an outmoded behaviorist paradigm. Such traffic as there was between the two areas was more often a matter of importing research areas and theoretical ideas from the human to the animal case: Gordon and Klein's account of work on memory provides one example (Chapter 9). Others abound; here was one way in which the cognitive revolution was supposed to have penetrated this last outpost.

There has also, however, been a more interesting coming together of animal and human work, stemming from the recognition of a much wider range of common concern and related theorizing. The rise of connectionism may take some of the credit for this: not only is it possible to see connectionist theorizing as associationism writ large, but the error-correcting rules for changing weights in connectionist networks, as Sutton and Barto (1981) were the first to point out, are formally equivalent to the Rescorla–Wagner (1972) equation. That simple equation has given us a remarkably powerful account of a wide variety of experiments on Pavlovian conditioning in animals. Put another way, therefore, such experiments provide a powerful way to observe the operation of error-correcting rules in simple situations, untrammeled by other complexities. Among other consequences of this, such experiments have also revealed the inadequacy of such rules: animal learning theorists know quite well that the Rescorla–Wagner model, taken alone, is quite incomplete as an account of the formation of associations even in simple conditioning experiments. Perhaps the additional processes implicated in such experiments will provide more powerful accounts of human behavior also. To take just one example: the Rescorla–Wagner equation contains no mechanism to explain the phenomenon of latent inhibition—that conditioning proceeds more rapidly to a novel CS than to an already familiar CS (see Hall, Chapter 2). The incorporation of such a mechanism into a connectionist model provides a more powerful account of some aspects of categorization by humans than does many of its rivals (McLaren, Leevers, & Mackintosh, 1994).

The parallels discernible between animal and human work extend further than this. For example, Pearce's account of elemental and configural theories of generalization and discrimination in animals (Chapter 5) deals with many of the same issues discussed by Shanks (Chapter 12) in his review of models of categorization in people, where the distinction between instance or exemplar theories on one hand and connectionist models on the other precisely mirrors the distinction drawn by Pearce. Shanks rightly points to some of the advantages that human subjects provide for research in this area: the power of some of the models developed to account for experimental data obtained here is indeed impressive. But I predict that each area of research will derive even greater benefits from greater awareness of the strengths of the other. If I am right, the study of animal learning and cognition will soon be able to claim again an honored and important, even if not central, position in experimental psychology.

References

Darwin, C. (1871). *The descent of man and selection in relation to sex*. London: Murray.

Flaherty, C. F. (1985). *Animal learning and cognition*. New York: Knopf.

Hawkins, R. D. (1989). A biologically realistic neural network model for higher-order features of classical conditioning. In R. G. M. Morris (Ed.), *Parallel distributed processing: Implications for psychology and neurobiology* (pp. 214–247). Oxford: Oxford University Press.

Hebb, D. O. (1949). *Organization of behavior*. New York: Wiley.

Hodos, W., & Campbell, C. B. G. (1969). Scala Naturae: Why there is no theory in comparative psychology. *Psychological Review, 76*, 337–350.

Horn, G. (1985). *Memory, imprinting, and the brain: An inquiry into mechanisms*. Oxford: Clarendon Press.

Hull, C. L. (1952). *A behavior system*. New Haven, CT: Yale University Press.

Humphrey, N. K. (1976). The social function of intellect. In P. P. G. Bateson & R. A. Hinde (Eds.), *Growing points in ethology* (pp. 303–317). Cambridge: Cambridge University Press.

Jolly, A. (1966). Lemur social behavior and primate intelligence. *Science, 153,* 501–506.
Kohler, W. (1924). *The mentality of apes.* London: Routledge & Kegan Paul.
Konorski, J. (1948). *Conditioned reflexes and neuron organization.* Cambridge: Cambridge University Press.
Konorski, J. (1967). *Integrative activity of the brain: An interdisciplinary approach.* Chicago: University of Chicago Press.
Konorski, J. (1968). Foreword to 2nd edition of Konorski (1948).
Lovejoy, A. O. (1936). *The great chain of being.* Cambridge, MA: Harvard University Press.
Mackintosh, N. J. (1983). *Conditioning and associative learning.* Oxford: Oxford University Press.
Macphail, E. M. (1982). *Brain and intelligence in vertebrates.* Oxford: Clarendon Press.
McLaren, I. P. L., Leevers, H. J., & Mackintosh, N. J. (1994). Recognition, categorisation, and perceptual learning (or how learning to classify things together helps one to tell them apart). In C. Umilta & M. Moscovitch (Eds.), *Attention and performance XV: Conscious and nonconscious information processing.* Hillsdale NJ: Erlbaum.
Miller, S., & Konorski, J. (1928). Sur une forme particulière des reflexes conditionels. *Comptes Rendus des Seances de la Societe de Biologie et de ses Filiales, 99,* 1155–1157.
O'Keefe, J., & Nadel, L. (1978). *The hippocampus as a cognitive map.* Oxford: Clarendon Press.
Pearce, J. M. (1987). A model of stimulus generalization for Pavlovian conditioning. *Psychological Review, 94*(1), 61–73.
Rescorla, R. A., & Wagner, A. R. (1972). A theory of Pavlovian conditioning: Variations in the effectiveness of reinforcement and nonreinforcement. In A. H. Black & W. F. Prokasy (Eds.), *Classical conditioning II: Current research and theory* (pp. 64–99). New York: Appleton-Century-Crofts.
Roitblatt, H. L. (1985). *Introduction to comparative cognition.* New York: Freeman.
Roitblatt, H. L., Bever, T. G., & Terrace, H. S. (1985). *Animal cognition.* Hillsdale, NJ: Erlbaum.
Skinner, B. F. (1932). On the rate of formation of a conditioned reflex. *Journal of General Psychology, 7,* 274–286.
Small, W. S. (1900). An experimental study of the mental processes of the rat. *American Journal of Psychology, 11,* 133–165.
Spence, K. W. (1956). *Behavior theory and conditioning.* New Haven, CT: Yale University Press.
Sutton, R. S., & Barto, A. G. (1981). Toward a modern theory of adaptive networks: Expectation and prediction. *Psychological Review, 88,* 135–170.
Thomas, K. (1983). *Man and the natural world: Changing attitudes in England, 1500–1800.* London: Allen Lane.
Thompson R. F. (1986). The neurobiology of learning and memory. *Science, 233,* 941–947.
Thorndike, E. L. (1898). Animal intelligence: An experimental study of the associative processes in animals. *Psychological Monographs, 2.*
Tolman, E. C. (1932). *Purpose behavior in animals and men.* New York: Appleton-Century-Crofts.
Tolman, E. C. (1938). The determiners of behavior at a choice point. *Psychological Review, 45,* 1–41.
Wagner, A. R. (1981). SOP: A model of automatic memory processing in animal behavior. In N. E. Spear & R. R. Miller (Eds.), *Information processing in animals: Memory mechanisms* (pp. 95–128). Hillsdale, NJ: Erlbaum.
Wagner, A. R., & Brandon, S. E. (1989). Evolution of a structured connectionist model of Pavlovian conditioning (AESOP). In S. B. Klein & R. R. Mowrer (Eds.), *Contemporary*

learning theories: Pavlovian conditioning and the status of traditional learning theory (pp. 149–189). Hillsdale, NJ: Erlbaum.

Watson, J. B. (1914). *Behavior: An introduction to comparative psychology.* New York: Holt.

Watson, J. B. (1924). *Behaviorism.* Chicago: University of Chicago Press.

Yerkes, R. M. (1916). The mental life of monkeys and apes. *Behavior Monographs, 3.*

CHAPTER 2

Pavlovian Conditioning
Laws of Association

Geoffrey Hall

I. PRELIMINARIES

As a mechanism of behavioral adaptation, Pavlovian conditioning is a phenomenon of interest and importance in its own right. The conditioned response (CR) that is evoked by a once-neutral stimulus (the CS, conditioned stimulus) that has been experienced along with some motivationally significant event (the US, unconditioned stimulus) is often one that will enhance the ability of the animal to cope successfully when similar events occur in the future. But for the bulk of this chapter, I will not address at all the immediate behavioral consequences of Pavlovian conditioning. That the conditioning procedure frequently produces some change in behavior will be regarded as a happy accident: one that allows the experimenter, by monitoring the vigor of some CR, to make inferences about the changes that must occur within the organism as a consequence of the training to which it is being subjected. Adopting this strategy needs no apology if it can be shown (something that this chapter will attempt) that an analysis of these changes and of the mechanisms that bring them about will reveal principles of cognitive functioning that are fundamental in nature.

Our starting point will be the assumption that the Pavlovian conditioning procedure establishes an association between the CS and the US (or between the central representations of these events). This assumption is

Animal Learning and Cognition

widely, although not universally, accepted. Although there is now a consensus, among those who accept the associative notion, that the entities that become linked are indeed the CS and the US (see, e.g., Mackintosh, 1983, for a discussion of the relevant evidence), there has long been an important minority of theorists who reject the associative interpretation altogether. Traditionally the alternative was seen as being some form of "gestalt" hypothesis (see Greeno, James, DaPolito, & Polson, 1978, for a relatively recent discussion of this view), which holds that presenting the CS and US together will allow the subject to perceive these events as a unified whole; the animal's problem, given this starting point, might well be thought to be that of distinguishing the stimuli one from another rather than that of forming a link between them. The discovery that Pavlovian conditioning is sometimes possible when a substantial temporal interval intervenes between the CS and US gives grounds for the immediate rejection of the gestalt hypothesis; it also forms the basis of a more modern and influential non-associative account. The so-called "contingency" theory of conditioning assumes that animals are sensitive to temporal dependencies among events and that conditioned responding is a consequence of their ability to compute the change in probability of occurrence of the US that occurs in the presence of the CS. The traditional concept of an association as something created at a given moment in time by the conjoint occurrence of two events or entities finds no place here. (See the Chapters by Shanks and Gallistel, 12 and 8, respectively, for a fuller discussion of these matters.)

Contingency theory has been criticized on a priori grounds: "We provide the animal with individual events, not correlations or information, and an adequate theory must detail how these events individually affect the animal" (Rescorla, 1972, p. 10). But its proponents have been able to reply in similar terms, objecting to associative theory as trying to deal with how the animal may "adapt its behavior to temporal dependencies in its environment . . . without anything forming inside the organism isomorphic to the dependencies to which the animal's behavior becomes adapted" (Gallistel, 1990, p. 385). It is not our purpose in this chapter to attempt to address this dispute directly; we have begun by assuming the essential validity of the associative view (but see Durlach, 1989, for a discussion of the evidence and arguments that support this position). It will be necessary, nonetheless, to consider some of the empirical findings that have motivated the contingency theory, and when these challenge the customary assumptions of associative theory, to detail what modifications of these assumptions may be required. It is time now to outline what the central features of the associative account should be taken to be.

Associative theory had a long history in philosophy before falling into the hands of experimental psychologists (see Warren, 1921). And although no one philosopher can be taken as representing the views of all his peers, J. S. Mill's statement of his position perhaps constitutes the culmination of the phase of nonexperimental analysis. In his *System of Logic* of 1843 he wrote: "The subject . . . of Psychology, is the uniformities of succession, the laws . . . according to which one mental state succeeds another, is caused by, or at least is caused to follow, another" (1843/1974, p. 852). At the most general level, two principles are identified.

> First: Whenever any state of consciousness has once been excited in us . . . an inferior degree of the same state . . . resembling the former but inferior in intensity is capable of being reproduced within us, without the presence of any such cause as excited it at first . . .
> Secondly: These . . . secondary mental states are excited according to certain laws which are called Laws of Association. Of these laws the first is, that similar ideas tend to excite one another. The second is, that when two impressions have been frequently experienced . . . either simultaneously or in immediate succession, then whenever one of these impressions, or the idea of it, recurs, it tends to excite the idea of the other. The third law is, that greater intensity in either or both of the impressions, is equivalent, in rendering them excitable by one another, to a greater frequency of conjunction.
>
> *(J. S. Mill, 1843/1974, p. 852)*

Mill has little more to say on the topic, referring the reader to the works of others, in particular, and with proper filial piety, to those of James Mill "where the principal laws of association . . . are copiously exemplified, and with a masterly hand" (pp. 852–853). But in fact, Mill senior differs from his son on a number of points. In addition to contiguity, he allows (in the *Analysis* of 1829/1967) only frequency and "vividness" as associative principles, refusing a role for "resemblance" (similarity), a notion that his son had reclaimed from the works of Hume. Resemblance, says James Mill, is reducible to contiguity and frequency, since similar things are likely to have been experienced together often.[1] He also rejects the other main principle that Hume had espoused—that of causality. According to the *Treatise* of 1739, "there is no relation, which produces a stronger connexion . . . and makes one idea more readily recall another, than the relation of cause and effect between the objects" (Hume, 1739/1888, p. 11). But James Mill will have none of this, asserting that causality is just another name for successive contiguity (a possibility that, elsewhere in his writings, Hume himself subscribes to).

In spite of these disagreements over details, J. S. Mill's predecessors

---

[1] "Contrast," another of Hume's suggested principles, is dealt with equally dismissively. This can be reduced to resemblance, as contrasted objects will have aspects in common: a dwarf and a giant resemble each other in that both depart from the norm for size.

would probably have accepted his two general principles as constituting an adequate statement of the core doctrine of associationism. At any rate, we shall take them to be just that. Our task becomes, therefore, one of assessing these principles in the light of the large body of experimental evidence supplied by studies of Pavlovian conditioning. In particular we shall want to determine, first, whether the data confirm the validity of that feature of the associative account on which there appears to be general agreement—that is, the central role given to contiguity (and with it, to frequency and intensity). And second, what do the experimental data have to say on those "laws" (similarity and causality) about which there has been more dispute? Do these principles need to be included among the laws of association, and if they do, what are the implications for the laws of contiguity, frequency, and intensity?

II. CONTIGUITY, FREQUENCY, AND INTENSITY

A. Acquisition of a Conditioned Response

Figure 1 shows the development of a CR in rats trained in the conditioned suppression procedure. The rats experienced a series of presentations of a noise CS and a shock US, the offset of the noise being contiguous with the onset of the shock. The results show the sensitivity of classical conditioning to frequency—with repetition of the noise–shock conjunction, the CR (of response suppression in the presence of the noise) grows until an asymptote is reached. The figure also shows the effect of intensity. In Figure 1A, one group of subjects was trained with a shock of 0.49 mA, another with a

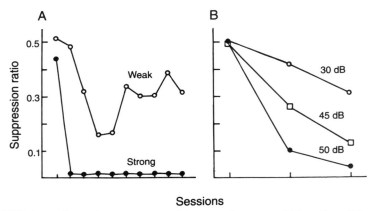

FIGURE 1   Effects of CS intensity and US intensity on acquisition of conditioned suppression. In (A), group strong was trained with an 0.85-mA shock, group weak with 0.49 mA (adapted from Annau & Kamin 1961.) In (B), the three groups were given differing intensities of a noise CS. From Kamin (1965).

shock of 0.85 mA. Suppression was acquired less readily in the former group, and the (somewhat unstable) asymptotic level was less profound. Figure 1B shows the effects of varying the intensity of the CS while keeping that of the US constant. Acquisition occurs more slowly with the lower intensity CSs. These results do not tell us what would have happened had training been continued, but there is reason to think that further training might have allowed the CR controlled by each of the CSs to reach the same final asymptote.

Figure 1 conveniently illustrates, for this training paradigm, the effects of frequency and intensity (both of CS and of US) on the conditioning produced by contiguous presentations of these stimuli. There is nothing special about this paradigm, and similar results could have been presented for a range of Pavlovian procedures. One minor complication to this simple picture should be mentioned. In an experiment analogous to that summarized in Figure 1B, Kamin (1965) investigated the effects of using a reduction in the intensity of a steady background noise as the CS. He found that acquisition of the CR proceeded more readily the greater the magnitude of the reduction: the more discriminable the change that constitutes the CS, the better. Intensity, then, in a law of association, should not be equated with intensity measured in units of physical energy. But it is safe to assume that the rat is equipped with some device that is sensitive to the magnitude of a change in stimulation (whether this be an increase or a reduction in the energy impinging on its receptors)—sensitive, in James Mill's terminology, to the "vividness" of events.

B. Associative Interpretation

1. A Simple Model and Some Complications

The essence of the associative interpretation of the data just described is exceedingly simple. Each of the events involved (CS and US) is assumed to induce activity in its own representation in the conceptual (and indeed in the real) nervous system—in what we shall refer to, for the conceptual system, as a *node*. The level of activity in a node is directly determined by the intensity (or vividness) of its stimulus. When two nodes are activated concurrently, the strength of a connection between them will be enhanced, the size of the increase in any moment of time being a direct function of the magnitude of the activity in the two nodes. The net change produced by a CS–US pairing will, of course, depend on the length of time over which concurrent activation is present. But for most purposes it has proved satisfactory and convenient to treat a pairing as if it were a punctate event and to summarize its effects thus:

$$\Delta V = \beta \alpha \lambda \tag{1}$$

where the change in strength of the connection (ΔV) is some function (determined by a learning-rate parameter β) of the levels of activation in the CS and US nodes (α and λ, respectively). It is easy to imagine a neuronal system (a Hebbian synapse) that might operate according to this equation.

The consequence of there being a connection between the two nodes is that presentation of one stimulus becomes able to evoke activity in the node to which it is linked. In studies of Pavlovian conditioning, the important thing is that the CS will be able to evoke activity in the US node, this induced activity being ultimately responsible for the observed CR. As they stand, these assumptions imply also the formation of a parallel connection that would allow presentation of the US to evoke activity in the CS node, but, for a CS lacking affective significance, no overt behavior is likely to result.

Not surprisingly, reasons for needing to complicate this simple model arise immediately. First, the model is concerned with the effects of concurrent activity in CS and US nodes, but in many studies of conditioning (including most done by Pavlov himself, as well as those discussed in Section II. A) it has been customary to arrange for the two events to succeed one another, the US being presented only after the CS has gone off. To accommodate the fact that this arrangement is effective in producing conditioning, we need to add the assumption that activity in a node might persist for some time after the eliciting stimulus has been terminated, thus ensuring some overlap in the times at which the two nodes are active.

Second, the simple model fails to account for the fact that conditioning proceeds to an asymptote (see Figure 1). Although in some cases this may simply reflect an inability of the animal to perform a more vigorous CR, in other cases this is clearly not so—rather, the strength of the CS–US link seems unable to grow beyond a certain point. Equation (1) provides no mechanism for this—repeated pairings of CS and US will continue to produce increments in associative strength, without apparent limit. But the empirical finding is that the effects of a pairing diminish as training proceeds, producing a negatively accelerated curve. From the work of Hull (1943) onward, mathematical models of learning have accommodated this fact by adopting some form of the following equation:

$$\Delta V = \beta\alpha(\lambda - V) \tag{2}$$

The increase in V produced by a CS–US pairing is thus limited by the growth of V and will be zero when V reaches λ.

2. An Elaboration

Equation (2) constitutes a primitive version of an "error-correcting" (or "delta") rule, in which the associative change is governed by the discrepancy (or error) between two values. The magnitude of the US sets a target value for associative strength, and increments in strength depend on the discrepancy between actual strength and that target.

The need for a rule of this sort is agreed; but how it should be incorporated into our model for conditioning has been the subject of some debate (e.g., Donegan, Gluck, & Thompson, 1989). At a minimum, it seems to require a relaxation of what has been called (McLaren & Dickinson, 1990) the "local activity constraint," a principle that has governed our thinking so far. This constraint gives expression to the most basic version of the contiguity principle; it holds first, that the strengthening of a connection between nodes will be determined solely by activity in those nodes; and second, that such activity will be directly determined in some simple manner by the properties of the stimuli that the nodes represent. To add to our model a system that computes a discrepancy would constitute a major departure from the first clause of the constraint. It proves possible, however, to accommodate the facts at the cost of committing only a minor transgression against the second clause of the constraint.

This transgression is a central feature of Wagner's SOP model of conditioning (Mazur & Wagner, 1982; Wagner, 1978; Wagner & Brandon, 1989). The model distinguishes two types of activity that can occur in a stimulus node. Presentation of the stimulus itself is held to evoke a primary state of activation (A1) that will decay after a time into a secondary state (A2), before the node returns to a state of inactivity. The activation excited in a node by means of an associative connection is assumed to be of type A2. Transition from A2 to A1 is not permitted and thus, to the extent that a node (or some elements of it) is in the A2 state, application of the relevant stimulus will not be fully effective in generating the primary state of activation. Adding the assumption that increments in associative strength require the concurrent presence of A1 activity in the two nodes means that the growth of strength will be self-limiting—the CS will become ever more potent at generating A2 in the US node, thus reducing the ability of the US itself to evoke A1.

The suggestion just described constitutes the basis of a possible conceptual nervous system embodying the principle of Equation (2). It would be misleading to stop at this point, however, as the distinction between A1 and A2 states has further implications, directly relevant to an understanding of how asymptote is achieved. The ability of one node to excite activity in another increases when both are in A1. It is legitimate to ask what the consequences may be of other patterns of activity. Wagner's theory postulates that having both nodes in A2 has no associative consequences. But when one is in A1 and the other is in A2, an *inhibitory* connection is strengthened, a connection that allows activation of the first of these nodes to oppose the effects that would otherwise be induced in the second by *excitatory* connections of the sort we have been considering so far. Since the standard conditioning procedure will necessarily result in some periods of overlap between A1 in the CS node and A2 in the US node (see Figure 2), it follows that the strengthening of an inhibitory connection will go on in parallel with the excitatory conditioning that occurs when both nodes are in A1. The final asymptote will reflect a balance between these two opposing influences.

3. Implications

I have referred to Wagner's distinction between primary and secondary states of activation as constituting only a minor modification to the simple model. It may already be evident, however, that the implications of this modification could turn out to be far reaching—a possibility that later sections of this chapter will certainly confirm. And it should be acknowledged immediately that, however minor the change, it amounts to an abandoning of the first of J. S. Mill's general principles. According to that principle, the state induced by an associative link differs from the real thing only in being "inferior in intensity." Clearly, this cannot be the case for the state evoked

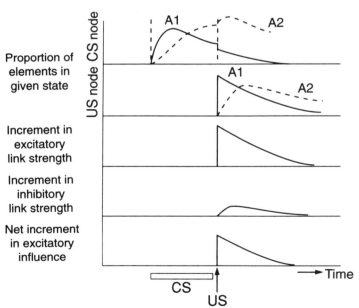

FIGURE 2   Wagner's (1978) SOP model applied to a conditioning trial in which a brief US is presented on termination of the CS. A1 and A2 represent the primary and secondary states of activation that can be induced in a node.

by the operation of an inhibitory link; but even the excitatory link breaches the principle when it produces a state (A2) that differs not only in quantity but also in quality from that (A1) evoked by direct activation of a node.

What remains is a strict adherence to the contiguity principle provided it is noted that the contiguity that this model requires is not between events in the world but (for excitatory conditioning) between two central states of primary activation. The magnitude of these states is not directly given by the intensity of their eliciting stimuli (it will depend, rather, on the "vividness" of the stimulus and, for a US node, on such associative influences as may be acting on the node) and to this extent the law of intensity must be qualified. And it is also possible that the critical feature of contiguity, temporal overlap between A1 states, might arise even when the eliciting environmental stimuli are not themselves contiguous. This possibility is discussed next.

C. Temporal Relationships

1. Phenomena

Temporal contiguity between CS and US is neither necessary nor sufficient for Pavlovian conditioning. Figure 3 shows the results of an experiment in which the interstimulus interval (ISI)—the interval between the onset of the CS and the onset of the US—was varied. Different groups of rats received a single training trial with a 4-s tone as the CS and a 4-s shock as the US. The figure shows the results of a test assessing for the various groups the extent to which the tone tended to suppress drinking. It is evident that learning

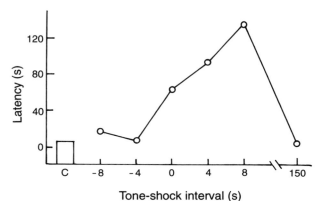

FIGURE 3   Suppression of licking (shown by long latencies) in the presence of a tone, after a single tone-shock pairing. The control subjects had received the shock but not the tone. (After Mahoney & Ayres, 1976.)

was best in those subjects for whom the onset of the tone preceded that of the shock by 8 s and that increasing the ISI to 150 s eliminated any evidence of learning. But conditioning was also poor when the ISI was reduced below 8 s—when the ISI was zero (the CS and US came on together) or negative (with the onset of the US preceding that of the CS). The absence of conditioning in these latter cases demonstrates the insufficiency of contiguity.

The effects of varying ISI have been determined for many conditioning procedures. Figure 4 shows the outcome of an experiment in which the CS was a flavor and the US a nausea-inducing procedure. The CR (rejection of the flavored substance on a subsequent test trial) grew less obvious as CS–US interval was lengthened, but an aversion was clearly present even at an interval of 12 hours. These results demonstrate that contiguity is not necessary for conditioning to occur.

2. Analysis

The inadequacy of the contiguity principle at the empirical level need not mean that we must abandon the theoretical principle that forms the basis for the model of conditioning outline in Section II.B.2. According to this model, a connection between nodes will be strengthened when they are activated concurrently, and the relevant form of activity can persist for some time after the termination of the stimulus. It is to be expected then, that some conditioning might be possible even when there is an interval between offset of the CS and onset of the US; also that increasing this interval, by

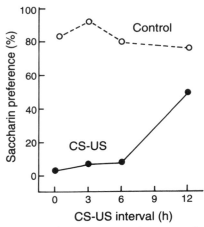

FIGURE 4 Preference for saccharin over water in rats previously exposed to saccharin followed by irradiation-induced illness. Control subjects experienced the saccharin without the illness. (After J. C. Smith & Roll, 1967.)

allowing time for activity in the CS node to wane, will reduce the size of any effect (as was the case even for the flavor-aversion results of Figure 4).

Although this interpretation may seem uncomfortably glib,[2] it is reasonable to ask whether any more convincing alternative has been offered. Discussions of delay of reinforcement have concentrated, for the most part, on explaining why conditioning should be so poor with a long delay and have emphasized the role played by events occurring in the interval between the experimenter's CS and the US. The occurrence of such events might, it is suggested (e.g., Dickinson, 1980), be what is responsible for the poor conditioning shown to the nominal CS. But to accept this suggestion is not to reject the contiguity principle. One interpretation of the suggestion assumes that the loss of activity in the CS node that occurs over time is a consequence of interference as well as of decay—that the magnitude of A1 activity in the CS node might be reduced when intervening events evoke A1 in their own nodes. A limited capacity system of this sort is entirely plausible and, indeed, was assumed in Wagner's version of the associative model described above (see Figure 2).[3]

What remains as a problem for the principle of contiguity is that any conditioning should occur at all when, as in the study shown in Figure 4, the CS–US interval is as long as 12 hours. It is easy enough to say that in this case, decay of the A1 state must have occurred slowly (or intervening events must have interfered rather little, or both). But to say this begs the question of *why* this flavor-aversion procedure should be so much less sensitive to the effects of the ISI than are other classical conditioning procedures. This issue will be taken up later.

The explanation offered by the contiguity principle for successful conditioning with a long ISI suffers from a lack of plausibility. But any failure of conditioning to occur when CS and US are presented simultaneously, or contiguously but in the reverse order, seems, at first sight, to constitute a rather more fundamental challenge. Here, however, appearances are deceptive. The first point to make is that there is now perfectly good evidence that excitatory conditioning can occur with the backward (US–CS) arrangement, given an appropriate choice of parameters (e.g., Ayres, Haddad, & Albert, 1987; Heth, 1976). When the parameters are right, the A1 state induced by the US may be presumed to persist for some time and thus

---

[2] It is certainly oversimplified: see Gormezano and Kehoe (1981) for a full account of the application of the contiguity principle to the results of experiments in which the ISI is manipulated.

[3] Another possibility—that intervening events interfere with conditioning to the nominal CS by virtue of acquiring associative strength themselves—follows from the theory to be described in Section III.B. This suggestion, too, depends on a contiguity principle, since intervening events can acquire strength preferentially only because they are more nearly contiguous with the US than is the nominal CS.

overlap at least the first part of the CS. Next, the contiguity principle is not embarrassed by cases of backward conditioning in which excitatory learning fails to occur. Having the US precede the CS increases the likelihood that A2 activity will be present in the US node when the CS is in the A1 state. Such conditions, it will be recalled, have been thought to foster the growth of an inhibitory CS–US connection, a proposal for which Wagner and Larew (1985) have presented plentiful evidence. When the conditioning parameters are such as to produce inhibitory learning, no CR can be expected, in spite of frequent contiguous occurrences of US and CS. The stimulus trained with a negative ISI in Figure 4 could well have acquired undetected inhibitory properties.

Elementary textbooks are often found to assert that conditioning does not occur when the CS and the US are presented simultaneously. The results for the zero ISI condition in Figure 3 demonstrate the falsity of this assertion, and similar results showing the effectiveness of simultaneous presentation are now available for a range of conditioning procedures (see, e.g., Rescorla, 1981). It remains to be explained, however, why simultaneous pairings should sometimes be ineffective (see, e.g., Smith, Coleman, & Gormezano, 1969); also, given that simultaneous presentation should surely be the optimum arrangement for ensuring overlap of activity in CS and US nodes, why the simultaneous procedure, when it is effective, should be less so than one in which the CS precedes the US (see Figure 3).

One possible answer to these questions has been developed by Rescorla (1981) who suggested that simultaneous presentation of CS and US might establish a perfectly strong association between the stimuli but that the existence of the association might prove rather difficult to detect. The presence of a US along with the CS is likely, he argued, to modify the way in which the latter is perceived. The consequence will be that the CS, when presented alone on a subsequent test trial, will be effectively a different stimulus from the one that has undergone training and thus will be poor at evoking the CR.

Evidence to confirm the validity of this suggestion comes from an experiment by Rescorla (1980). Rats received initial training in which a light CS (L) was presented along with an auditory cue (X), the LX compound being followed immediately by a different auditory cue, Y. In order to assess the associations formed in these conditions, it was necessary to give a further stage of training in which, for one group of subjects, cue X was presented along with a shock US, whereas, for a second group, cue Y was paired with shock. The ability of the light to activate a representation of the shock-related auditory cue can then be detected by assessing the extent to which the light, when presented alone, tends to evoke the CR established to that cue. Any decrement consequent on the fact that L presented alone might be perceived differently from L presented in compound with X will be suffered by both groups. In these conditions, Rescorla (1980) found, subjects for whom cue X had been paired with shock showed a substantially larger CR to the light than did subjects for whom Y had been paired with shock—that is, the simultaneous LX arrangement proved to be superior to the successive L–Y arrangement.

III. CAUSALITY

A. Phenomena

Hume's *Treatise* lists a series of "rules by which to judge of causes and effects." The first of these is that "cause and effect must be contiguous in space and time," and this is quickly followed by the assertion that "the cause must be prior to the effect" (Hume, 1739/1888, p. 173). The observation (see Figure 3) that conditioning proceeds most readily when CS precedes US has led to the suggestion (see Dickinson, 1980; Hall, 1990) that contiguity is important in association formation, not for its own sake, but because it serves as an important indicator of a possible causal relation between CS and US. Our demonstration (Section II.C.2) that the simultaneous training procedure can be effective in producing conditioning, seriously undermines any attempt to subsume the contiguity principle under one of causality. But it remains possible that a causality principle operates alongside that of contiguity and there is some evidence from conditioning studies to encourage this view.

The next of Hume's rules is that "the same cause always produces the same effect, and the same effect never arises but from the same cause . . ." (p. 173). An experiment by Rescorla (1968) demonstrates the application of this rule in Pavlovian conditioning. Four groups of rats received presentations of a tone during which the probability of receiving a shock US was 0.4. One group received no other shocks. For the remaining groups, shocks occurred in the absence of the tone with probabilities of 0.1, 0.2, or 0.4. As Figure 5 shows, subjects that received shocks only in the presence of the tone acquired a strong CR. But the presentation of shocks in the intertrial intervals led to a decrease in the size of the CR. The group that experienced shocks with equal probability in the presence and absence of the tone showed no evidence of conditioning, its performance being the same as that recorded for a control group that experienced the tone with no shocks at all. Evidently contiguous occurrences of CS and US are not enough to establish conditioning (group 0.4 in the figure experienced as many CS–US pairings as did group 0)—it seems rather that the CS must be a plausible cause of the US and that conditioning does not occur when the supposed effect is likely to occur in the absence of the presumed cause.

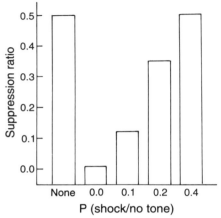

FIGURE 5   Conditioned suppression to a tone in rats that had experienced shocks with an 0.4 probability in the presence of the tone but different probabilities of shock in the absence of the tone. Control subjects received no shocks. (From data reported by Rescorla, 1968.)

Hume's next rule can also be shown to govern the behavior of animals in conditioning experiments. He writes, "where several objects produce the same effect, it must be means of some quality [that is] . . . common amongst them . . ." (1739/1888, p. 174). Table 1 presents an outline of two of the conditions studied in an experiment by Wagner, Logan, Haberlandt, and Price (1968). Both groups experienced training in which a compound CS consisting of a tone and a light was followed by the US on 50% of trials. Two different tones were used, and in the uncorrelated condition these occurred equally often on reinforced and nonreinforced trials. Figure 6 shows that the light acquired the ability to evoke a CR but that neither of the

TABLE 1   Design of Experiment

Group	Trial types[a]
Uncorrelated	T1L → US T1L → no US T2L → US T2L → no US
Correlated	T1L → US T2L → no US

[a] L represents a light; T1 and T2 are tones differing in frequency. The auditory cue and the light were presented as a simultaneous compound. From Wagner et al. (1968).

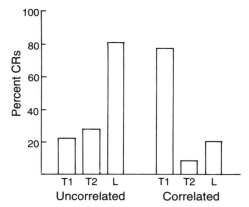

FIGURE 6 Conditioned responding in rabbits trained according to the procedures outlined in Table 1 and tested with tones (T1 and T2) and light (L) presented separately. (From data reported by Wagner, Logan, Haberlandt, & Price, 1968.)

tones did so. Although clearly not a sufficient cause of the US (the light sometimes occurs when no US follows), it is a quality common to those "objects" that precede the US and, in selectively associating the light with the US, the subjects appear to be following Hume's rule. In the correlated condition, on the other hand, the US occurred only on trials in which the light was accompanied by T1, and for the subjects given this treatment, T1 came to evoke the CR and the light (and T2) did not. These subjects appear to be following the complementary rule that "the difference in the effects of two resembling objects must proceed from that particular in which they differ" (Hume, 1739/1888, p. 174)—in this case, the difference between the two tones.

These results demonstrate that a pairing of a CS and US that is otherwise sufficient to establish an association will fail to do so unless the CS is also a good predictor (and thus a potential cause) of the occurrence of the US. The phenomenon of blocking (Kamin, 1968) provides a further, and much studied, instance of this general effect. In a blocking experiment, subjects are first trained to asymptote with a given CS (event A) and US. There follows a second stage in which a compound of A plus another CS (B) is paired with the US. A final test phase reveals that, in spite of having been paired with the US in stage two, CS B fails to elicit the CR when presented alone. Control subjects that receive just the stage of training with the AB compound show a CR to B—pretraining with A is said to have "blocked" learning about B. For animals given both stages of training, stimulus A is the only quality common to both, and thus A rather than B should be regarded as the cause of the US.

B. The Rescorla–Wagner Model

The empirical findings just outlined make it clear that our developing model for conditioning, which so far consists of no more than a slightly elaborated contiguity principle, must be modified to accommodate the fact that contiguity will not ensure association formation when the CS in question is an unreliable predictor of the US. The modification proposed by Rescorla and Wagner (1972) achieves this, without introducing some radically new principle of association. Their model is summarized in the following equation:

$$\Delta V = \beta\alpha(\lambda - \Sigma V) \qquad (3)$$

which differs from Equation (2) only in that it takes account of ΣV, the total associative strength of all CSs that happen to be present on a given training trial. According to Equation (2), the effectiveness of a US presentation in strengthening the link between the US representation and that of a given CS is reduced as the CS itself gains associative strength. Equation (3) (which now, it may be noted, constitutes a fully developed error-correcting rule, formally equivalent to the Widrow–Hoff delta rule—see Gluck & Bower, 1988) means that US effectiveness will be reduced by the presence of other CSs having associative strength. In terms of the SOP model, several CS nodes will have links with a common US node, and A2 activity in the US node evoked by any or all of these CSs will limit the ability of the US to evoke the A1 state.

The application of this model to phenomena demonstrating the importance of the predictive power of the CS in conditioning is most easily illustrated with respect to blocking. In this procedure, initial training allows the associative strength of CS A to reach asymptote. When B is added in the second stage, the effectiveness of the US on these compound trials, as given by the term $(\lambda - \Sigma V)$ of Equation (3), will be zero as the contribution of A to ΣV will itself amount to λ. Accordingly, CS B, redundant as a predictor of the US, will fail to gain associative strength. Given certain assumptions (see Rescorla & Wagner, 1972), the training procedures outlined in Table 1 are susceptible to a similar analysis that generates associative strengths to match the pattern of conditioned responding shown in Figure 6. Application to Rescorla's (1968) findings (see Figure 5) requires the assumption that the context in which training occurs can function as a CS. Once this is allowed, it follows that adding US presentations in the intertrial intervals will permit the context to acquire strength and thus, according to Equation (3), limit the ability of the experimenter's CS to do so.

The ease with which it is possible to accommodate these phenomena within an only slightly modified contiguity theory may seem almost suspiciously surprising. Lest it be thought that the Rescorla–Wagner model is, in some way, merely a disguised redescription of a causality principle, it is worth pointing out that there are circumstances in which the model and the principle of causality, as it is usually understood, make differing predictions. In particular, it is possible to conduct the blocking experiment with the order of the stages reversed, that is, with conditioning to the AB compound occurring before conditioning with A alone. An organism that operated by Hume's rules would not accept B as a potential cause of the US, and B should exhibit little associative strength. The fact that there is currently no properly established instance of such a "backward blocking" effect is enough to make us want to question the relevance of a causality principle to conditioning.[4] In addition, a failure to find backward blocking is just what would be expected on the basis of the Rescorla–Wagner model—pretraining with the AB compound will allow B to acquire a measure of associative strength that will be quite uninfluenced by subsequent training given to A alone.

C. Alternative Formulations

1. Attentional Processes and Blocking

The successes of the Rescorla–Wagner model have not suppressed the growth of theoretical rivals. In particular, it has often been argued that selective learning phenomena reflect the operation of an attentional process such as finds no place in the Rescorla–Wagner account. Independent evidence for the existence of such a process has been sought in studies of the phenomenon of latent inhibition (Lubow, 1973)—the well-established finding that prior nonreinforced exposure to the event to be used as a CS will dramatically retard the course of conditioning when subsequently this CS is paired with a US. Possible explanations of latent inhibition abound (see Hall, 1991, for a review), but one important class of theory attributes the effect to the operation of a learning process by which the subject comes to ignore an event that supplies no important information. Attempts to formalize this loosely stated notion are outlined below; but we may consider immediately the way in which it has been used in explaining selective association. As applied to blocking, the essence of the attentional account (e.g., Mackintosh, 1975; Pearce & Hall, 1980) is that the animal fails to learn about the added B stimulus because it quickly learns to ignore an event that supplies no information that A does not about the outcome of an AB trial. In the Rescorla–Wagner model, blocking occurs because of a loss of effec-

---

[4] At least in nonhuman subjects; Chapman and Robbins (1990) have recently presented some evidence of a backward effect in human subjects trained on a task in some respect analogous to those used in the study of Pavlovian conditioning in animals. (See also Shanks's Chapter 12, this volume.)

tiveness by the predicted US. The alternative attributes blocking to a change in CS effectiveness.

Evidence in favor of the attentional interpretation comes from investigation of a phenomenon known as "unblocking." It has been known for some time (Kamin, 1968) that pretraining with A will fail to block learning about B if the magnitude of the US is increased for the phase of training with AB. This unblocking effect follows readily enough from Equation (3); an increase in the US will give a positive value to $(\lambda - \Sigma V)$ by increasing the value of λ, thus allowing the CSs to gain strength. But Dickinson, Hall, and Mackintosh (1976) have shown that unblocking can be produced not only by increases in US magnitude but also by reducing the US for the compound stage. It will be evident that Equation (3) cannot predict excitatory conditioning to B in these circumstances. Now the basis of the attentional account of the standard blocking effect is that the effectiveness of the added stimulus (B) will decline substantially on the first compound conditioning trial as the occurrence of the US is fully predicted by the pretrained stimulus, A. Stimulus B will then be unable to gain associative strength on further compound trials. But if the outcome of the first compound trial is not what was predicted by stimulus A, the effectiveness of B will be maintained and blocking will not occur.

These results encourage acceptance of the proposal that changes in CS effectiveness play a role in selective association formation. But there are other results, also from experiments on blocking, to show that this cannot be the whole story. A feature of the attentional account of blocking is that it takes time, a minimum of one compound conditioning trial, for the animal to learn that the added B stimulus is uninformative. What follows is that learning about B should occur normally on this first trial. But the experimental results on this matter now seem clear—blocking will occur even with just a single compound trial (e.g., Balaz, Kasprow, & Miller, 1982; Dickinson, Nicholas, & Mackintosh, 1983). In order to accommodate all the results, therefore, it is necessary to allow that the training procedures used in blocking experiments induce changes both in CS and US effectiveness. The principles governing changes in US effectiveness have already been discussed; we turn now to those governing CS effectiveness.

2. Mechanisms

Perhaps the most elegant account of the mechanisms involved in changing CS effectiveness is that put forward by Wagner (first proposed by Wagner, 1978, but retained in some form in later versions of his theory). The effectiveness of the CS is held to be determined by processes that exactly parallel those determining US effectiveness. Just as the formation of an association between the US and the CS will lead to a reduction in the ability of the US to generate the primary state of activation in the US node, so the ability of a CS to activate its own node will decline if that CS becomes associated with some antecedent. Prior exposure to a CS, for example, might allow an association to form between that stimulus and the context in which it is presented; the reduced effectiveness of the CS would mean that subsequent conditioning in that context would then proceed only with difficulty—that is, the latent inhibition effect should be observed. The advantage of this theoretical proposal is that no new and special principles are required to explain either the learning mechanism by which CS effectiveness is changed or the process by which this change influences further association formation. Standard associative principles govern the formation of links between the CS node and its antecedent (the context in the example given above); and activity in these links then modulates the sensitivity of the CS node, making it less easily activated by presentation of the stimulus to which it corresponds. As before, a simple contiguity principle can then apply, with formation of the CS–US association depending on the level of concurrent activity in the two nodes. The only modification is that for the CS, just as for the US, activation of the relevant node will depend not only on the physical intensity of the stimulus but also on the extent to which the node is activated associatively by some antecedent event.

It is unfortunate, given the appealing simplicity of Wagner's account, that we must now turn to a consideration of observations that reveal some inadequacies and suggest new complexities. First, according to this interpretation, a CS will lose effectiveness when it is associatively linked to an antecedent—that is, effectiveness depends on how well the CS is predicted. But there is now a good deal of evidence to show that CS effectiveness is critically determined, rather, by the extent to which the CS itself has previously been a good predictor of its consequences. Hall (1991) reviews the relevant evidence, but it may be noted that the attentional account of blocking given above relies on this notion—the blocking effect may itself be taken as evidence that the effectiveness of a CS (the added B element in this case) will be reduced when the outcome of that stimulus is fully predicted. It is necessary to suppose, therefore, that some measure of the extent to which the US is predicted (for possible formalizations of this notion, see Mackintosh, 1975; McLaren, Kaye, & Mackintosh, 1989; Pearce & Hall, 1980) can be fed back to modulate what has been called the "associability" of the CS— a substantially more complex learning process than that postulated in Wagner's account.

A further complication comes from observations suggesting that changes in CS effectiveness are not to be explained solely in terms of variations in the sensitivity of the CS node itself. There is evidence suggesting that a stimulus that is low in associability will be quite capable of activating its central representation. Hall and Pearce (1979) found that a CS trained to asymptote with a given US suffers a loss of associability, forming further associations only with difficulty; but such a CS is perfectly capable of evoking the original CR. The modulatory signal that reflects the history of the CS cannot be acting directly on the CS node itself since such an arrangement would reduce the ability of the CS to control performance. This problem prompts consideration of the alternative assumption (see McLaren & Dickinson, 1990) that the modulatory signal acts at the point at which the CS and US make contact (i.e., at the synapse, for those who like to think in neuronal terms). Indeed, such an assumption seems to be required if we accept the arguments of Mackintosh (1975) who holds that changes in CS associability are specific to a given US (or class of USs). Support for this view comes from a further study of unblocking by Dickinson and Mackintosh (1979). This experiment demonstrated that unblocking could be produced by changing the outcome of the compound trials only when those changes were made to a US of the same type as that employed in the first stage of training—a change in the shock US on the compound trials would produce unblocking in subjects trained with shock in stage one, but the introduction of a food US during compound trials would not. These procedures appear to modulate not the general level of effectiveness of the CS but the extent to which it is associable with a particular type of US.

3. Further Implications

If we were to accept the principle of US-specific associability it would significantly complicate our account of the laws of association. Association formation now depends not just on the levels of concurrent activation in the CS and US nodes but also on the effects of a modulatory process, determined by the past history of the CS as a predictor of USs of that type, that acts on the link between them. There is, however, a compensating advantage in that this analysis helps to deal with an issue left unresolved in section II.C.2. The question raised there was why it should be that flavor aversions can be acquired with ease, even when CS and US are widely separated in time. The issue essentially concerns selectivity in association formation— why should it be that the US-induced illness becomes associated with a flavor consumed hours earlier rather than with the other events (such as those involved in administering the US) that occur much closer in time to the US? The results presented in Figure 7 make this point most forcefully. They come from an experiment by Garcia and Koelling (1966; see also Domjan, 1982, for confirmation and extension of the results) in which rats were allowed to drink a saccharin-flavored solution while a light and a noise were being presented. Half the subjects then received a nausea-inducing injection as the US, the others experienced a shock US. In a test phase, animals in the former group showed an aversion to saccharin but were quite

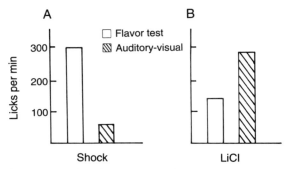

FIGURE 7   Suppression of drinking on test with flavored water or water associated with auditory and visual cues in rats given shock or a nausea-inducing injection in training. (After Garcia & Koelling, 1966.)

ready to drink plain water even when doing so was accompanied by the light and noise (Figure 7A). Animals in the second group (Figure 7B) showed the reverse pattern of results, drinking saccharin readily but refusing the "bright noisy water."

The finding that gastric distress is especially readily associated with taste cues, whereas an event that impinges on the body surface appears to have a special affinity for exteroceptive cues, has been interpreted by some (e.g., Dickinson, 1980) as a further indication of the role of causal relevance as a general principle in associative learning. Others have been prompted by the same finding to question the worth of attempting to discern general laws at all (see Shettleworth's Chapter 7, this volume). According to Garcia, Brett, and Rusiniak (1989), for example, at least two sets of laws will be needed— one to deal with the phenomena that have made up the bulk of this chapter and that come from studying the way in which animals learn about events like noises and shocks; a second to describe the operation of a specialized "gut-defense" system that is revealed in the phenomenon of flavor-aversion learning, but which is of fundamental importance in the adaptation of any animal to its environment. There are reasons, however, for wanting to question this distinction. Foremost is the fact that almost all of the phenomena that have been detected in standard Pavlovian conditioning procedures (second-order conditioning, blocking, sensory preconditioning, and so on) have also been demonstrated in the flavor-aversion paradigm (e.g., Revusky, 1977)—such results make it difficult to resist the implication that the same basic laws apply in both cases.

We need to ask, therefore, whether the principles that have been applied to cases of selective association in orthodox Pavlovian conditioning can provide any explanation for the phenomena of flavor-aversion learning. The notion of US-specific stimulus associability has the potential to do so. The everyday life of even the laboratory rat is likely to give it every opportunity to experience positive correlations between gastric events and what has recently been consumed, also, to experience the lack of correlation between gastric events and a range of other cues. Mackintosh's (1975) version of attentional theory predicts changes in stimulus associability in these circumstances that could generate selective learning of the sort shown in Figure 7. Whether such a mechanism is indeed responsible for the effects observed is another matter. There is certainly reason to think that it cannot be the whole story—changes in associability will depend on experience but the selective association of taste with illness rather than with shock is now known to occur in very young animals (Gemberling & Domjan, 1982, used rats only 24 hours old). At the very least, we need to allow that some further mechanism, in addition to learned changes in associability, can operate to produce selective association. The next section discusses a possible candidate.

IV. SIMILARITY

When Hume and J. S. Mill after him (see Section I) gave prominence to similarity or "resemblance" as an associative principle, they appear to have had in mind what we should call, in the terminology of learning theory, a factor governing performance rather than one governing acquisition. That is, they were not explicitly concerned with the possibility that similarity might play a role in promoting the *formation* of associations; rather, their principle was simply that one idea will tend to call up another when the second has something in common with the first.

It is possible to demonstrate the operation of such a principle in classical conditioning. Honey and Hall (1991) gave rats exposure, on quite separate occasions, to two different flavors, A and B. The two flavors were rendered similar, however, by the addition of a third flavor (X) to each of them. The rats then received aversion training with flavor A as the CS. A final test phase showed that the untrained flavor B was also capable of evoking the CR. Honey and Hall (1991) interpreted this result as showing that stimulus B was able to call up the representation of the CS, stimulus A; and they offered an explanation for this in terms of standard associative mechanisms. They pointed out that the first stage of training, although it could not establish associations between A and B, would allow the formation of associations between the elements of each compound, that is, A–X and B–X associations. Such associations could generate the test result as follows. Presentation of B on a test would activate the associatively linked representation of X, which in turn would be able to activate the A representation. The direct link between A and the US would then produce a CR. In general, the formation of associations between the unique features of a stimulus and features that it holds in common with another similar stimulus will allow the common features to mediate between the two so that presentation of one will activate the representation of the other.

The experiment just described demonstrates a role for similarity in determining performance; altogether more contentious is the suggestion that the acquisition of associations might be in part determined by the similarity of the events being associated, with events that share common features being more readily associated than those that do not. And it is this idea that has been put forward as a possible explanation for the special readiness with which flavor aversions are learned. As was pointed out by Testa and Ternes (1977), in the Garcia and Koelling (1966) experiment, the potential CSs (auditory-visual vs. flavor) differed not only in modality but also in spatial location—the flavor was an intrinsic attribute of the substance being ingested; the other cues came from elsewhere. Flavor may be learned about more readily, they suggested, because the interoceptive consequences of this CS occur in the same place as those of the LiCl injection. The association may be formed easily because the events being associated are similar, sharing a common feature, in this case a common spatial location.

Without independent empirical support, the proposal advanced above can be regarded as no more than a possibility. Rescorla and Cunningham (1979) report one of the few studies (but see also Testa, 1975) that has addressed the issue experimentally. They made use of a second-order conditioning procedure in autoshaping with pigeons. In a first stage of training all subjects learned that a given stimulus, X, was followed by food, whether X was presented on the left or the right key of a pigeon Skinner box. The properties acquired by X then allowed it to function as an effective US in a second stage of training. In this stage, presentation of a different stimulus (G) on the key was followed immediately by X, a second-order conditioning procedure that endows G, as a consequence of the formation of a G–X association, with the ability to evoke the CR previously conditioned to X. For some birds, pairings of G and X always occurred on the same key, that is, in the same spatial location. For other birds, when G occurred on the left key it was followed by X on the right, and vice versa. It was found that responding to G was acquired more readily in subjects trained with G and X on the same key.

These results constitute a *prima facie* case in favor of a role for spatial contiguity. But, as Rescorla and Cunningham (1979) themselves acknowledge, the case is not proved. Although their experimental design overcomes many of the problems found with previous investigations of the issue, it does not fully solve that of separating spatial from temporal contiguity. In particular, the procedure of presenting the critical stimuli on the same response key means that these stimuli are more likely to be experienced in close temporal contiguity than will be the case when the stimuli are presented on different keys. The difference between these two conditions could just as well be a consequence of temporal as of spatial factors.

The confound between spatial and temporal contiguity can be avoided, however, if some feature other than spatial location is used to establish similarity between the events that are to be associated. This was the strategy adopted by Rescorla and Furrow (1977). Table 2 shows the design of one of their experiments. All their subjects (pigeons) were given initial training in which presentations of two stimuli, a blue and a horizontal line, were reliably followed by food; two other stimuli, green and vertical lines, were not. Next, the birds were divided into two groups for a phase of second-order conditioning in which the stimuli from the first phase of training were presented as two pairs. The groups differed only in that for one, the stimuli that may be presumed to be similar were paired; the other received pairings of the dissimilar stimuli (see Table 2). Figure 8 shows the development of second-order conditioned responding to the first member of each stimulus pair in each group. It is evident that this responding, which may be assumed to depend on the formation of an association between the two members of the pair, develops more rapidly in subjects given pairings of similar stimuli. Rescorla and Furrow (1977) concluded that associations are formed more readily when the events to be associated are similar to each other, and, noting the parallel with the results of studies on selective association in flavor-aversion learning, suggested that adopting the similarity principle will provide a way of integrating such studies into our general theories of conditioning.

What remains to be determined is how the principle of similarity should itself be integrated into conditioning theory. From this theoretical perspective our starting point is the proposition that conditioning will occur more readily from a pairing of similar events (call them AX and BX, with X representing the common feature that establishes the similarity) than from a pairing of otherwise comparable dissimilar events (AX and BY). Putting

TABLE 2   Design of Experiment

Group	Stage 1	Stage 2
Similar	B+/G−[a]	G → B
	H+/V−	V → H
Dissimilar	B+/G−	G → V
	H+/V−	V → B

[a] B and G, blue and green; H and V, horizontal and vertical; +, stimulus followed by food; −, no food. From Rescorla and Furrow (1977).

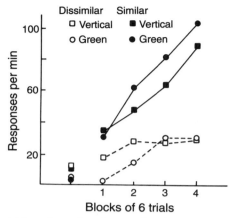

FIGURE 8 Acquisition of autoshaped responding by two groups of pigeons trained according to the procedures outlined in Table 2. The points at the left show responding at the end of stage 1. (After Rescorla & Furrow, 1977.)

the matter in this way makes one possible source of the effect immediately apparent. Saying that the stimuli are similar means that some aspect (X) of the US (BX) is also present in the CS (AX). The CS will, therefore, already, before the start of training, possess some tendency to evoke the same response as does the US; and to the extent that the CR resembles the UR, the development of conditioned responding will occur more rapidly than when the stimuli are dissimilar. This "head start" notion can thus provide a satisfactory account of at least some similarity effects.

It should be acknowledged immediately, however, that not all the effects observed experimentally (including those of Rescorla & Furrow, 1977) are susceptible to explanation in head start terms; see also LoLordo and Jacobs (1983). The theoretical and empirical analysis of similarity effects has only just begun and it may yet prove possible to deal with these results without any need to suppose that similarity fosters association per se. Clearly, however, we should be aware of the possibility that the similarity principle might need to be accorded *sui generis* status (like that granted to temporal contiguity), being allowed its own independent place among the laws of association.

V. CONCLUSION

A listing of the factors that have been shown empirically to determine the course of the acquisition of classically conditioned responding would be quite extensive. A place would have to be found for at least the following: the degree to which the CS and US are temporally contiguous; the number of pairings of these events; the intensity of the stimuli; the predictive validity of the CS; the past history of the CS; the nature of the events being used as CS and US; and so on. It is difficult to provide an abstract statement of what properties a "law of association" ought to possess; but clearly, to bestow the status of a law on each of the factors listed above would be to use the term too liberally. In this chapter I have endeavored to reduce these factors to the operation of a smaller number of more fundamental principles, and it is the resulting, more theoretical, account that I would offer as a statement of the laws of association. What emerges as central is a law of contiguity—the concurrent activation of the central representations of a pair of events will establish an associative link between them. The other factors play their part by determining the likelihood that application of the external stimulus will generate activity of an appropriate strength, in the appropriate place, at the appropriate time. The only apparent exceptions to this are found in the possibility that, first, previous experience of the relationship between CS and US, and second, the extent to which the two events are similar to one another, might act directly to determine the ease with which the link can form. These exceptions would repay a more extensive experimental and theoretical investigation.

If we accept J. S. Mill's view that the subject of psychology is the laws by which one mental state generates another, then the laws of association are the foundations of our discipline. The principles just outlined in the preceding paragraph may seem a flimsy basis on which to build, but might not the same once have been said of Mendel's laws?

Acknowledgments

Discussions with L. Aguado, R. C. Honey, E. M. Macphail, and especially C. Bonardi have contributed substantially to the writing of this chapter.

References

Annau, Z., & Kamin, L. J. (1961). The conditioned emotional response as a function of intensity of the US. *Journal of Comparative and Physiological Psychology, 54*, 428–432.

Ayres, J. J. B., Haddad, C., & Albert, M. (1987). One-trial excitatory backward conditioning as assessed by conditioned suppression of licking in rats: Concurrent observations of lick suppression and defensive behaviors. *Animal Learning and Behavior, 15*, 212–217.

Balaz, M. A., Kasprow, W. J., & Miller, R. R. (1982). Blocking with a single compound trial. *Animal Learning and Behavior, 10*, 271–276.

Chapman, G. B., & Robbins, S. J. (1990). Cue interaction in human contingency judgment. *Memory & Cognition, 18*, 537–544.

Dickinson, A. (1980). *Contemporary animal learning theory*. Cambridge: Cambridge University Press.

Dickinson, A., Hall, G., & Mackintosh, N. J. (1976). Surprise and the attenuation of blocking. *Journal of Experimental Psychology: Animal Behavior Processes, 2*, 313–322.

Dickinson, A., & Mackintosh, N. J. (1979). Reinforcer specificity in the enhancement of conditioning by posttrial surprise. *Journal of Experimental Psychology: Animal Behavior Processes, 5,* 587–612.

Dickinson, A., Nicholas, D. J., & Mackintosh, N. J. (1983). A re-examination of one-trial blocking in conditioned suppression. *Quarterly Journal of Experimental Psychology, 35B,* 67–79.

Domjan, M. (1982). Selective association in aversion learning. In M. L. Commons, R. J. Herrnstein, & A. R. Wagner (Eds.), *Quantitative analyses of behavior* (vol. 3, pp. 257–272). Cambridge, MA: Ballinger.

Donegan, N. H., Gluck, M. A., & Thompson, R. F. (1989). Integrating behavioral and biological models of classical conditioning. In R. D. Hawkins & G. H. Bower (Eds.), *Computational models of learning in simple neural systems* (pp. 109–156). San Diego: Academic Press.

Durlach, P. J. (1989). Learning and performance in Pavlovian conditioning: Are failures of contiguity failures of learning or performance? In S. B. Klein & R. R. Mowrer (Eds.), *Contemporary learning theories: Pavlovian conditioning and the status of traditional learning theory* (pp. 19–59). Hillsdale, NJ: Erlbaum.

Gallistel, C. R. (1990). *The organization of learning.* Cambridge, MA: MIT Press.

Garcia, J., Brett, L. P., & Rusiniak, K. W. (1989). Limits of Darwinian conditioning. In S. B. Klein & R. R. Mowrer (Eds.), *Contemporary learning theories: Instrumental conditioning theory and the impact of biological constraints on learning* (pp. 181–203). Hillsdale, NJ: Erlbaum.

Garcia, J., & Koelling, R. A. (1966). The relation of cue to consequence in avoidance learning. *Psychonomic Science, 5,* 123–124.

Gemberling, G. A., & Domjan, M. (1982). Selective association in one-day old rats: Taste-toxicosis and texture-toxicosis aversion learning. *Journal of Comparative and Physiological Psychology, 96,* 105–113.

Gluck, M. A., & Bower, G. H. (1988). From conditioning to category learning: An adaptive network model. *Journal of Experimental Psychology: General, 117,* 227–247.

Gormezano, I., & Kehoe, E. J. (1981). Classical conditioning and the law of contiguity. In P. Harzem & M. D. Zeiler (Eds.), *Advances in analysis of behaviour: Vol. 2. Predictability, correlation, and contiguity* (pp. 1–45). Chichester: Wiley.

Greeno, J. G., James, C. T., DaPolito, F., & Polson, P. G. (1978). *Associative learning: A cognitive analysis.* New York: Appleton-Century-Crofts.

Hall, G. (1990). Reasoning and associative learning. In D. E. Blackman & H. Lejeune (Eds.), *Behaviour analysis in theory and practice* (pp. 159–180). Hillsdale, NJ: Erlbaum.

Hall, G. (1991). *Perceptual and associative learning.* Oxford: Clarendon Press.

Hall, G., & Pearce, J. M. (1979). Latent inhibition of a CS during CS-US pairings. *Journal of Experimental Psychology: Animal Behavior Processes, 5,* 31–42.

Heth, C. D. (1976). Simultaneous and backward fear conditioning as a function of number of CS-US pairings. *Journal of Experimental Psychology: Animal Behavior Processes, 2,* 117–129.

Honey, R. C., & Hall, G. (1991). Acquired equivalence and distinctiveness of cues using a sensory-preconditioning procedure. *Quarterly Journal of Experimental Psychology, 43B,* 121–135.

Hull, C. L. (1943). *Principles of behavior.* New York: Appleton-Century.

Hume, D. (1888). *A treatise of human nature* (L. A. Selby-Bigge, Ed.). Oxford: Clarendon Press. (Original work published 1739)

Kamin, L. J. (1965). Temporal and intensity characteristics of the conditioned stimulus. In W. F. Prokasy (Ed.), *Classical conditioning: A symposium* (pp. 118–147). New York: Appleton-Century-Crofts.

Kamin, L. J. (1968). "Attention-like" processes in classical conditioning. In M. R. Jones (Ed.),

Miami symposium on the prediction of behavior: Aversive stimulation (pp. 9–33). Coral Gables, FL: University of Miami Press.

LoLordo, V. M., & Jacobs, W. J. (1983). Constraints on aversive conditioning in the rat: Some theoretical accounts. In M. D. Zeiler & P. Harzem (Eds.), *Advances in analysis of behaviour: Vol. 3. Biological factors in learning* (pp. 325–350). Chichester: Wiley.

Lubow, R. E. (1973). Latent inhibition. *Psychological Bulletin, 79,* 398–407.

Mackintosh, N. J. (1975). A theory of attention: Variation in the associability of stimuli with reinforcement. *Psychological Review, 82,* 276–298.

Mackintosh, N. J. (1983). *Conditioning and associative learning.* Oxford: Clarendon Press.

Mahoney, W. J., & Ayres, J. J. B. (1976). One-trial simultaneous and backward fear conditioning as reflected in conditioned suppression of licking in rats. *Animal Learning and Behavior, 4,* 357–362.

Mazur, J. E., & Wagner, A. R. (1982). An episodic model of associative learning. In M. L. Commons, R. J. Herrnstein, & A. R. Wagner (Eds.), *Quantitative analyses of behavior* (Vol. 3, pp. 3–39). Cambridge MA: Ballinger.

McLaren, I. P. L., & Dickinson, A. (1990). The conditioning connection. *Philosophical Transactions of the Royal Society of London, 329,* 179–186.

McLaren, I. P. L., Kaye, H., & Mackintosh, N. J. (1989). An associative theory of the representation of stimuli: Applications to perceptual learning and latent inhibition. In R. G. M. Morris (Ed.), *Parallel distributed processing: Implications for psychology and neurobiology* (pp. 102–130). Oxford: Oxford University Press.

Mill, J. (1967). *Analysis of the phenomena of the human mind.* New York: Augustus M. Kelley. (Original work published 1829)

Mill, J. S. (1974). *System of logic.* Toronto: University of Toronto Press. (Original work published 1843)

Pearce, J. M., & Hall, G. (1980). A model for Pavlovian learning: Variations in the effectiveness of conditioned but not of unconditioned stimuli. *Psychological Review, 87,* 532–552.

Rescorla, R. A. (1968). Probability of shock in the presence and absence of CS in fear conditioning. *Journal of Comparative and Physiological Psychology, 66,* 1–5.

Rescorla, R. A. (1972). Informational variables in Pavlovian conditioning. In G. H. Bower (Ed.), *The psychology of learning and motivation* (Vol. 6, pp. 1–46). New York: Academic Press.

Rescorla, R. A. (1980). Simultaneous and successive associations in sensory preconditioning. *Journal of Experimental Psychology: Animal Behavior Processes, 6,* 207–216.

Rescorla, R. A. (1981). Simultaneous associations. In P. Harzem & M. D. Zeiler (Eds.), *Advances in analysis of behaviour: Vol. 2. Predictability, correlation, and contiguity* (pp. 47–80). Chichester: Wiley.

Rescorla, R. A., & Cunningham, C. L. (1979). Spatial contiguity facilitates Pavlovian second-order conditioning. *Journal of Experimental Psychology: Animal Behavior Processes, 5,* 152–161.

Rescorla, R. A., & Furrow, D. R. (1977). Stimulus similarity as a determinant of Pavlovian conditioning. *Journal of Experimental Psychology: Animal Behavior Processes, 3,* 203–215.

Rescorla, R. A., & Wagner, A. R. (1972). A theory of Pavlovian conditioning: Variations in the effectiveness of reinforcement and nonreinforcement. In A. H. Black and W. F. Prokasy (Eds.), *Classical conditioning II: Current research and theory* (pp. 64–99). New York: Appleton–Century–Crofts.

Revusky, S. (1977). Learning as a general process with an emphasis on data from feeding experiments. In N. W. Milgram, L. Krames, & T. M. Alloway (Eds.), *Food aversion learning* (pp. 1–51). New York: Plenum.

Smith, J. C., & Roll, D. L. (1967). Trace conditioning with X-rays as the aversive stimulus. *Psychonomic Science, 9,* 11–12.

Smith, M. C., Coleman, S. R., & Gormezano, I. (1969). Classical conditioning of the rabbit's nictitating membrane response at backward, simultaneous, and forward CS-US intervals. *Journal of Comparative and Physiological Psychology, 69,* 226–231.

Testa, T. J. (1975). Effects of location and temporal intensity pattern of conditioned and unconditioned stimuli on the acquisition of conditioned suppression in rats. *Journal of Experimental Psychology: Animal Behavior Processes, 1,* 114–121.

Testa, T. J., & Ternes, J. W. (1977). Specificity of conditioning mechanisms in the modification of food preferences. In L. M. Barker, M. R. Best, & M. Domjan (Eds.), *Learning mechanisms in food selection* (pp. 229–253). Waco, TX: Baylor University Press.

Wagner, A. R. (1978). Expectancies and the priming of STM. In S. H. Hulse, H. Fowler, & W. K. Honig (Eds.), *Cognitive processes in animal behavior* (pp. 177–209). Hillsdale, NJ: Erlbaum.

Wagner, A. R., & Brandon, S. E. (1989). Evolution of a structured connectionist model of Pavlovian conditioning (AESOP). In S. B. Klein & R. R. Mowrer (Eds.), *Contemporary learning theories: Pavlovian conditioning and the status of traditional learning theory* (pp. 149–189). Hillsdale, NJ: Erlbaum.

Wagner, A. R., & Larew, M. B. (1985). Opponent processes and Pavlovian inhibition. In R. R. Miller & N. E. Spear (Eds.), *Information processing in animals: Conditioned inhibition* (pp. 233–265). Hillsdale, NJ: Erlbaum.

Wagner, A. R., Logan, F. A., Haberlandt, K., & Price, T. (1968). Stimulus selection in animal discrimination learning. *Journal of Experimental Psychology, 76,* 171–180.

Warren, H. C. (1921). *A history of the association psychology.* London: Constable.

CHAPTER 3

Instrumental Conditioning

Anthony Dickinson

I. INTRODUCTION

Instrumental behavior refers to those actions whose acquisition and maintenance depend on their consequences for the animal or, in others words, on the fact that the action is *instrumental* in causing some outcome. The functional significance of the capacity for instrumental action is so obvious as to require little comment; it is this capacity that allows us and other animals to learn to control our environment in the service of our needs and desires. Consider a purely Pavlovian animal equipped only with the capacity to detect and learn about the predictive relations between signals and important events in the world, but not about the contingencies between its actions and their consequences. Such an animal must rely on evolutionary processes to ensure that the responses elicited by a signal (*conditioned stimulus*) are appropriate for coping with the predicted event, and it will therefore be at the mercy of an unstable environment in which the consequences of its behavior may vary.

This point can be illustrated by considering simple approach to a food source. A hungry chick, for example, will, not surprisingly, rapidly learn to approach a food bowl as soon as it is presented. An instrumental analysis would argue that this simple form of conditioning arises from the chick's sensitivity to the relation between its approach behavior and access to food.

According to a Pavlovian account, by contrast, it is the predictive relationship between the stimulus of the bowl and food that is crucial; once established as signal for food, the sight of the bowl elicits approach irrespective of the actual consequences of this action. While these two accounts cannot be separated in a normal, stable environment, the extent to which the chick can adapt to changes in the causal structure of the environment depends on which relation controls the behavior.

Consider an environment in which the normal relationship between locomotion and spatial translation is reversed so that in order to gain access to the food bowl the chick has to run away from it. Hershberger (1986) arranged such a "looking glass" world by employing an unusual runway in which the food bowl receded twice as fast as the chick ran toward it and drew near at twice the speed that the chick ran away from it. For a perfect instrumental agent such an environment should present no great problem; sensitivity to the instrumental consequences of attempted approach and withdrawal will ensure that the animal rapidly learns to run away from the bowl. By contrast, a purely Pavlovian animal, being insensitive to the consequences of its actions, would never be able to adapt to the "looking glass" world. For as long as the food bowl remained a signal for food, the animal should continue to perform the response elicited by such signals, namely attempted approach. And this was in fact the behavior pattern observed by Hershberger—the chicks showed little evidence of learning to run away from the food bowl across 100 minutes of training. Thus, this simple approach response would not appear to be controlled by its instrumental relation to outcome, rendering the chick unable to adapt to the novel environment.

Although both instrumental (Thorndike, 1911) and Pavlovian conditioning (Pavlov, 1927) were studied at the turn of the century, it took students of learning some time to appreciate the critical difference between the two forms. Miller and Konorski (1969) are usually credited with being the first to make the distinction in 1928. They passively flexed a dog's leg in the presence of a stimulus and paired this compound event with the presentation of food. After a number of such pairings, the dog started to lift its leg spontaneously when the stimulus was presented, a conditioned response that, they argued, is at variance with Pavlov's principle of stimulus substitution. According to this principle, exposure to stimulus–outcome pairings endows the stimulus with the capacity to act as a substitute or surrogate for the outcome and thereby to elicit the same responses; but, as Miller and Konorski noted, the ability of the stimulus to control leg flexion could not be explained in terms of it becoming a substitute for food. For this reason they argued for a second form of conditioning which they termed Type II. Four years later Skinner (1932) also argued for a Type II conditioning for much the same reason, specifically that Pavlov's principle could not explain why hungry rats learn to press a freely available lever for food.

Although these studies clearly provided a challenge to stimulus substitution as a universal principle of conditioning, what they did not demonstrate was the *instrumental* character of Type II conditioning, namely that it is controlled by the relation between the conditioned action and outcome. That behavior can be controlled by this relation was first demonstrated by Grindley in a paper also published in 1932. Grindley trained restrained guinea pigs to turn their heads either to the left or to the right and then back again to the center when a buzzer sounded in order to gain the opportunity to nibble at a carrot. What established that this behavior was under the control of the action–outcome relation was the fact that the animals would reverse the direction of the head turn when the instrumental contingency was reversed. In other words, when the stimulus–outcome relation between the buzzer and the carrot was kept constant,[1] the behavior was controlled by its relation to the outcome and, other things being equal, such *bidirectional* conditioning can be taken as the critical assay of instrumental control.

Traditionally, behavioral psychologists have identified types of conditioning in terms of the experimental contingencies rather than in terms of the relations that actually control performance, and for this reason spatial behavior in runways and mazes has typically been classified as instrumental. As we have seen, however, there are good reasons to doubt whether such behavior is in fact controlled by its instrumental relation to the outcome. Similar concerns have also been expressed about the instrumental status of other widely studied behaviors, such as the pigeon's key peck (e.g., Moore, 1973). Consequently, the present discussion will focus primarily on the *free-operant* procedure introduced by Skinner in 1932 in which rats learn to press a freely available lever that yields an attractive outcome, such as a food pellet. There is good evidence that this action is under the control of the instrumental relation. In contrast to the approach behavior of Hershberger's (1986) chicks, free-operant lever pressing by rats is sensitive to a reversal of the action–outcome relation. Having trained their rats to lever press for food, Davis and Bitterman (1971) changed the contingency so that each lever press now postponed a food delivery that would otherwise have occurred. This *omission* contingency reduced responding more rapidly than a simple noncontingent schedule under which the food was delivered independently of pressing (but see, e.g., Uhl, 1974). But perhaps the best evidence that the manipulation of a lever by rats on a free-operant schedule is under instrumental control comes from a study of *punishment* by Bolles, Holtz, Dunn, and Hill (1980). They trained rats both to press down and to push up a lever for food pellets. The schedule was such that sometimes a press was required for the next pellet and sometimes a push in a manner that was unpredictable to the animal, and consequently the rats learned to intersperse presses and pushes. Bolles et al. then attempted to punish one category of this bidirectional behavior by following either pushes or presses with a shock. Although the introduction of the punishment contingency suppressed both actions to a certain extent, the category upon which the shock was contingent was performed at a significantly lower level. By implementing this bidirectional assay, Bolles et al. (1980) were able to demonstrate that these actions are sensitive to their instrumental relation to the outcome.

This experiment also illustrates the fact that actions can enter into various instrumental relations to outcomes. Skinner's (1932) original study arranged a positive contingency between the action and outcome in that pressing the lever increased the likelihood of access to a food pellet. Because this action was strengthened or reinforced through a positive contingency with the outcome, this form of conditioning is referred to as an example of *positive reinforcement* and the outcome is identified as a *reward* or positive *reinforcer*. By contrast, the Bolles et al. (1980) punishment procedure was a case in which a positive contingency between action and outcome led to a reduction in performance. Corresponding to these two cases are ones in which a negative relation is arranged between an action and outcome so that performing the action causes the omission of an event that would otherwise have occurred (avoidance) or the cessation of a stimulus (escape). When such a relation serves to increase the probability of the action, then we have an example of *negative reinforcement* with the omitted (or terminated) event being referred to as a negative reinforcer. Avoidance has received considerable experimental attention (e.g., Sidman, 1966) not only because of its obvious functional importance but also because of the special theoretical problems it raises (e.g., Herrnstein, 1969; Seligman & Johnston, 1973). These issues must lie beyond the scope of this chapter, however, which will address exclusively what appears to be the least problematic case, namely that of reward conditioning or positive reinforcement.

II. ACTIONS AND HABITS

When Grindley (1932) challenged the ubiquity of Pavlov's principle of stimulus substitution, he attributed the development of instrumental conditioning to the formation of a stimulus–response connection, as had Thorndike (1911) previously in the formulation of his "Law of Effect." The presentation of an effective outcome following an action, Thorndike argued, reinforces a connection between the stimuli present when the action is performed and the action itself so that subsequent presentations of these stimuli elicit the instrumental action as a response. The most perverse feature of such stimulus–response (S–R) theories has always been the claim that knowledge about the instrumental contingency between action and outcome plays no role in the performance of the action. According to this S–R/reinforcement theory, an instrumental action is simply a habitual response triggered by the training stimuli.

While acknowledging a role for habits, our folk psychology gives a quite different account of goal-directed, instrumental actions. In everyday discourse we explain such actions in terms of cognitive or intentional processes, specifically the interaction of an instrumental belief about the causal relationship between an action and its consequent outcome with a desire for that outcome. Thus, for example, the act of operating a light switch in a dark room is to be explained by the agent's belief that performance of this action will produce illumination and his or her desire for light. It is the interaction of these two mental states that produces the immediate mental antecedent to behavior, namely an intention to operate the switch. Throughout the heyday of S–R theory (e.g., Guthrie, 1952; Hull, 1943; Spence, 1956), Tolman (1932, 1959) defended a position akin to that of folk psychology. An instrumental belief about the consequences of action he identified as a "means–end readiness," which, when activated in the form of an "expectancy," could interact with the "value" that the agent assigns to the outcome to determine instrumental performance. Although there may be some doubt about the intended ontological status of Tolman's concepts (Amundson, 1986), when interpreted from the standpoint of a mental realist, his expectancy-value psychology maps onto the cognitive belief-desire explanation of folk psychology.

Traditionally, the conflict between cognitive and S–R theories has been fought on the battlefield of latent learning. In a typical latent learning study, the animal is trained initially to perform some instrumental action for an outcome. The value of that outcome is then changed in some way without allowing the animal to re-experience the action–outcome relation before, finally, the propensity to perform the action is assessed. If the initial training simply established an S–R habit reinforced by the outcome, a subsequent change in the value of the outcome should have no impact on performance in the absence of any further experience with the action–outcome contingency. As the relation between action and outcome is not encoded or represented within an S–R habit, once the outcome has served the function of reinforcing the habit during the initial instrumental training, any subsequent change in its properties should have no effect on performance. By contrast, an intentional or cognitive account would anticipate an immediate impact of outcome revaluation on performance; instrumental action, being mediated by the animal's current desire for the outcome, should directly reflect any change in the value of the outcome.

Although latent learning was intensively studied many years ago, almost without exception these classic studies employed spatial learning tasks (see MacCorquodale & Meehl, 1954; Thistlethwaite, 1951, for reviews), the instrumental status of which we have seen is ambiguous. Adams and Dickinson (1981) were the first to distinguish successfully between the S–R and cognitive accounts of the rat's lever pressing using an outcome devaluation procedure. The basic rationale of this procedure was similar to that underlying latent learning studies. The animals were initially trained to press a lever using two types of pellets, standard mixed composition food pellets and sugar pellets; one type of pellet, the reinforcer, was delivered contingent upon lever pressing, whereas the other type was delivered noncontingently.[2] According to the cognitive theory, this training should have established a belief that lever pressing causes the delivery of the reinforcer (but not of the noncontingent pellet), and consequently any subsequent change in the value of the reinforcer should be immediately expressed in instrumental performance. S–R habit theory, by contrast, would simply regard the lever presses as responses elicited by the contextual stimuli of the training situation and thus impervious to a revaluation of the reinforcer.

To test these contrasting predictions, one type of pellet was then devalued. The devaluation procedure capitalized on the fact that if consumption of a flavored food is followed by gastric illness, induced in this case by an injection of lithium chloride (LiCl), animals develop an aversion to the food in that it will no longer function as a reward. Thus, immediately after instrumental training, an aversion was established to the reinforcer for some of the animals but not for the others. The lever was not present during this aversion conditioning and the two types of pellets were presented independently of any instrumental action on alternate days. Animals in the paired condition received LiCl injections following exposure to the reinforcer, whereas the injections occurred following exposure to the noncontingent food in the unpaired group. Thus, this design yielded four groups: P–S, P–F, U–S, and U–F, where the first term refers to whether the LiCl injection was paired (P) or unpaired (U) with exposure to the reinforcer during aversion training and the second term to whether the reinforcer was the sucrose (S) or standard food pellets (F).

As Figure 1A shows, when the impact of this treatment on instrumental performance was assessed by giving access to the lever once again, the animals for whom the reinforcer had been devalued, Groups P–S and P–F, pressed less than the respective control groups, Groups U–S and U–F, for whom the aversion was conditioned to the noncontingent food rather than to the reinforcer. Two features of this devaluation procedure are worthy of note. The first is that this test was conducted in the absence of any outcomes (i.e., in extinction). If Adams and Dickinson had actually presented the food pellets contingent on lever pressing during the test, this behavior could have been punished by the now-aversive outcome. By testing in extinction, however, Adams and Dickinson ensured that the differential performance of the two groups must have reflected the interaction of knowledge acquired during instrumental training with the relative values of the two outcomes following devaluation.

The second important feature of the design is that the only difference

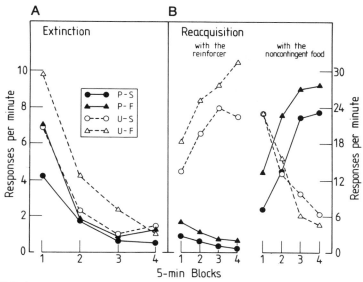

FIGURE 1   The mean rate of lever pressing during the extinction test (A) and during reacquisition with the reinforcer and the noncontingent food (B) in Experiment 2 reported by Adams and Dickinson (1981). The reinforcer was a sugar pellet for two groups of rats (P–S and U–S) and a mixed composition food pellet for the remaining animals (P–F and U–F). The aversion was conditioned to the reinforcer in the paired groups (P–S and P–F) and to the noncontingent food in the unpaired groups (U–S and U–F).

between the treatment received by the two groups was the relation between the devalued and nondevalued pellets and lever pressing during instrumental training. Thus, any differential effect of the devaluation treatment in the two groups must have been mediated by the experience with the contingency between lever pressing and the outcomes during the initial training. Colwill and Rescorla (1985) established the same point by training their rats to perform two different actions, lever pressing and chain pulling, for different outcomes before devaluing one of the outcomes. When the animals were subsequently given a choice between the performance of the two actions in extinction, they showed a preference for that associated with the nondevalued outcome during the original training.

Taken together, these two features of the devaluation effect clearly favor a role for cognitive processes in instrumental action. This does not mean, however, that S–R habits play no role in instrumental conditioning. It is notable that both Adams and Dickinson (1981) and Colwill and Rescorla (1985) observed residual performance of the action associated with the devalued food outcome even though the animals would not eat the food following aversion conditioning (note the performance of Groups P–S and

P–F in the first 5-min block of the extinction test in Figure 1). It is unlikely that this residual performance reflected a failure to produce a total devaluation of the outcomes. As Figure 1B shows, when Adams and Dickinson attempted to re-establish performance by presenting, contingent upon lever pressing, first the original reinforcer and then the noncontingent food, the devalued outcomes failed to act as effective reinforcers. Thus, the residual responding, observed during the extinction test in Figure 1A, suggests that the instrumental training established lever pressing partly as a goal–directed action, mediated by knowledge of the instrumental relation, and partly as an S–R habit impervious to outcome devaluation. And, as we shall see, there are reasons for believing that the nature of the action–outcome contingency may well determine the character of instrumental performance in this respect.

III. INSTRUMENTAL KNOWLEDGE

Whatever the contribution of a habit mechanism, the outcome devaluation effect establishes that the rat's lever press must at least in part be controlled by the type of instrumental expectancy or belief envisaged by Tolman. The problem with Tolman's theory, however, is that it never specified the psychological *mechanism* by which expectancies or beliefs interact with desires or values to cause instrumental behavior, leaving it open to Guthrie's (1952, p. 143) famous jibe that Tolman's rats are left "buried in thought." Two such mechanisms have been proposed, both associative in nature[3]: the first is Pavlov's bidirectional theory, whereas the second, which I refer to as the associative–cybernetic model, was offered by Thorndike (1931) as an alternative to the Law of Effect.

A. Bidirectional Theory

Although only briefly addressed by Pavlov (1932), the bidirectional account of instrumental conditioning was more fully developed by his students, such as Asratyan (1974; see Gormezano & Tait, 1976, for a review). When translated into associative terms, the basic idea is that the pairing of two events establishes not only a *forward* connection from the representational unit(s) activated by the first event, E1, to the unit(s) activated by the second event, E2 (which mediates standard Pavlovian conditioning), but also a *backward* connection from E2 to E1. And it is this backward connection that mediates instrumental conditioning when E1 is the instrumental action and E2 the reinforcer. Thus, following instrumental training, placing the animal in the experimental context should excite a representational unit for the reinforcer through its Pavlovian association with the contextual stimuli, and this excitation in turn should activate the response unit through the backward association brought about by the instrumental contingency, thus producing performance of the response. When interpreted in cognitive terms, the excitation of the reinforcer unit represents the value assigned to the outcome, and the activation of the connection from reinforcer to the response unit the expectancy of the outcome given performance of the instrumental action.

This account subsumes instrumental learning under the general claim that pairing of two events establishes bidirectional excitatory connections between their representational units and, indeed, the fact that backward excitatory conditioning can be observed with Pavlovian procedures (see Chapter 2 by Hall, this volume) could be regarded as independent evidence for the backward connections that are critical for instrumental conditioning. It should be noted, however, that in such demonstrations E2 is a neutral conditioned stimulus in contrast to the reinforcing event employed as E2 in instrumental conditioning. In an attempt to provide independent evidence for backward associations when E2 is a reinforcer, Gormezano and Tait (1976) trained water-deprived rabbits with intra-oral water delivery as E2 and a corneal air puff, which elicits an eyelid response, as E1. With this procedure backward associations would be manifest by the development of conditioned eyelid responses to the water delivery. Gormezano and Tait (1976) observed no evidence for such backward conditioning, however; the percentage of eyelid responses evoked by the water presentations was no greater than that observed in a control group in which the two events were unpaired and, moreover, the level of responding actually declined across training.

Not only is the empirical evidence problematic for the bidirectional account, it is also far from clear how this theory would account for punishment. In this case, the backward association from the representational unit for E2, the punisher, to that for E1, the instrumental action, cannot be excitatory, for the effect of a punishment contingency is to suppress performance of the action. An obvious solution is to assume that when E2 is an aversive event the backward connection is inhibitory rather than excitatory so that activating the unit representing the punisher inhibits the response unit. However, in the absence of any empirical evidence for either excitatory or inhibitory backward connections under conditions analogous to an effective instrumental contingency, the bidirectional account is not a plausible candidate for instrumental conditioning.

B. Associative-Cybernetic Model

As an alternative to his Law of Effect, Thorndike considered what he called "the representative or ideational theory." In the context of maze learning, Thorndike argued that:

> . . . this theory . . . would explain the learning of a cat who came to avoid the exit S at which it received a mild shock and to favor the exit F which led it to

food, by the supposition that the tendency to approach and enter S calls to the cat's mind some image or idea or hallucination of the painful shock, whereas the tendency to approach and enter F calls to mind some representation of the food, and these representations respectively check and favor these tendencies. (Thorndike, 1931, pp. 47–48)

This account can be characterized both as associative, because it involves the formation of a connection between a representation of the instrumental action (approach and entering S or F) and that of the outcome (S or F), and as cybernetic, because the activation of these outcome representations feed back to modulate performance. Although Thorndike was prepared to entertain this account for "deliberative" human behavior, he dismissed it as a general explanation of instrumental performance; others (e.g., Mowrer, 1960a,b), however, have subsequently argued for such an account and, indeed, Sutton and Barto (1981) presented a simulation of the performance of an associative-cybernetic model in the context of a classic latent learning procedure.

Figure 2 presents a cartoon of a possible architecture of an associative-cybernetic system. Actions have their origin in what I have called a habit memory, which consists of an array of stimulus-detecting units linked to an

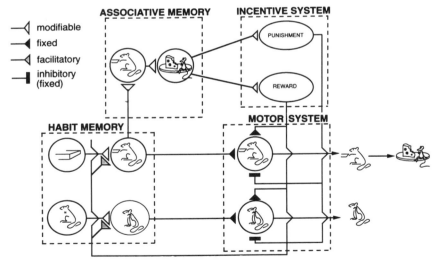

FIGURE 2   A cartoon of the associative-cybernetic model of instrumental action. The cartoon illustrates stimulus units in the habit memory that are excited by the presence of a lever and an "itch" sensation, which have connections via response units in the habit memory to motor units for lever pressing and scratching, respectively. Under an instrumental contingency, performance of a lever press gives access to food (a plate of cheese), which in turn activates corresponding representational units in the associative memory.

array of response units corresponding to the animal's untrained (or pre-trained) reactions to these stimuli. The activation of the response units, which is assumed to be mutually exclusive, is transmitted to corresponding units in the motor system whose activation causes output of the relevant actions. Studies of instrumental conditioning normally choose actions, such as lever pressing, that have a relatively low baseline level before training so that an increase in performance can be observed under the contingency. This fact is reflected in the model by ensuring that the input from a response unit in the habit memory is not normally sufficient to fire the corresponding motor unit reliably. Because of fluctuations in either or both the strength of the input and the threshold of the motor unit, however, the unit will be activated occasionally, thus producing, for example, a baseline level of lever pressing as a spontaneous untrained response.

If the animal can detect and represent the occurrence of the action, performance of the action has an important consequence. Within the model the representation of the performance of an action is instantiated by activation of a unit in the animal's associative memory. Consequently, performance of this action will result in contiguous and contingent excitation of the response unit in the habit memory and the corresponding action unit in associative memory. If we assume that this pattern of firing in the two units is sufficient to form and strengthen a connection between them, then, on subsequent occasions when the habit response unit is excited, the corresponding action unit in the associative memory will be activated even if no overt action occurs.

The presence of an action unit in associative memory is crucial for *instrumental learning* about the relation between action and outcome. In the presence of an effective instrumental relation, performance of the action and the consequent outcome will lead to the contiguous and contingent activation of the action and outcome units in the associative memory, which, it is assumed, is sufficient for the formation of a connection between them. It is this connection that represents the animal's knowledge of the instrumental relation, in that activation of the outcome unit by the action unit corresponds to the Tolmanian expectancy of the outcome.

The final component of the model, the incentive system, is designed to handle the cybernetic function that implements the role of desires or outcome values in instrumental action. Any unit in the associative memory activated by an event of motivational significance has connections to units in the incentive system. The connection is to a reward unit if the event is attractive, such as the presentation of food to a hungry rat, but to a punishment unit if the event is aversive or noxious so that these connections represent the animal's desire for or value assigned to such outcomes. I shall argue that these connections are also acquired through a process referred to as *incentive learning* (Dickinson & Balleine, 1994). Thus, it is experience with the food outcome while hungry that establishes a connection between the representational unit for this outcome in the associative memory and the reward unit. Correspondingly, aversion conditioning in a devaluation procedure establishes a connection from the outcome unit to the second, punishment unit in the incentive system.

The function of these incentive units is to exert a general and *indiscriminate* influence on all units in the motor system, an excitatory one in the case of the reward unit and an inhibitory one in the case of the punishment unit.[4] It must be noted, however, that activity in the reward unit is not sufficient by itself to fire the motor units. In order to activate a motor unit, the feedback influence from the reward unit must sum with a temporally contiguous input from the habit memory. Thus, even under the influence of the incentive system, the particular motor unit that is activated at any one time is determined by the input from this memory. For this reason it might be best to think of the incentive feedback as altering the thresholds of the motor units, lowering them in the case of a reward feedback and raising them in the case of punishment.

In summary, this associative–cybernetic model explains instrumental conditioning in term of the formation of three associative connections, which serve to open a positive feedback loop through the associative memory and incentive system that converges with the input from habit memory onto the motor unit for the instrumental action. The first is the connection from the response unit in this memory to the corresponding action unit in the associative memory formed by the animal's ability to detect and represent its own behavior. The second is that between the action and outcome units in associative memory brought about by experience of the instrumental relation between action and outcome. Finally, experience with the attractive outcome during training will establish the connection between the outcome unit in associative memory and the reward unit in the incentive system. Clearly, as these connections strengthen during training, so will the positive feedback on the motor units, thus enhancing the reliability with which the original eliciting stimulus triggers the instrumental action. Outcome devaluation counteracts this positive influence by opening a parallel negative feedback loop through the formation of a connection between the outcome unit in the associative memory and the punishment unit in the incentive system.

In the remainder of the chapter, I shall consider the evidence relating to the associative connections comprising the feedback loop. But, before doing so, it is worth noting that habit formation can be accommodated within this model by assuming that the output of the reward unit not only affects the motor system but also exerts a facilitatory influence on the connections between the stimulus and response units in the habit memory (see Figure 2).

The effect of this influence is to increase the strength of a connection when it occurs at the same time as activity in a stimulus unit. Once activity in the stimulus unit begins to elicit the instrumental action reliably via the feedback loop, this activity will be consistently paired with reward facilitation, thus strengthening the connection in the habit memory. And once this connection becomes sufficiently strong to allow the stimulus element to trigger the motor unit independently of the feedback influence, the instrumental action will have become a habit impervious to outcome devaluation. It is this habit mechanism that may account for the residual responding observed by Adams and Dickinson (1981) and Colwill and Rescorla (1985) after outcome devaluation.

IV. REPRESENTATION OF INSTRUMENTAL ACTIONS

Central to the feedback loops is the unit in the associative memory representing the instrumental action. It is this unit that interfaces the instigator of action, the stimulus input to the habit memory, with knowledge of the instrumental relation, namely the connection between action and outcome units in associative memory. Thus, according to the model, the susceptibility of a behavioral activity to instrumental conditioning should depend on whether the animal detects and represents the occurrence of that behavior. Shettleworth (1975) compared the sensitivity of a variety of behavior patterns to conditioning with a food reward in hamsters. Whereas certain actions, such as rearing, were responsive to the instrumental contingency, others, such as face washing and scratching, were more resistant to conditioning. Morgan and Nicholas (1979) subsequently confirmed this pattern in the rat and, furthermore, provided evidence that it may reflect a difference in the degree to which performance of these various actions is detected and represented. Two levers were inserted into the experimental chamber on occasions when the rats either reared or face washed, and pressing one lever was rewarded after a bout of rearing and the other after a bout of face washing. The fact that the rats learned to make this discrimination indicates that they could represent their performance of at least one of the activities in that their choice was controlled by the preceding behavior. By contrast, the animals found it much more difficult to learn a discrimination between scratching and face washing. This difference in the discriminability of rearing and scratching, when contrasted with face washing, clearly maps onto the relative sensitivity of these actions to an instrumental contingency. Thus, according to Morgan and Nicholas's (1979) data, rats do not readily represent their own scratching, a fact instantiated in the model by the absence of any unit in the associative memory that is excited by this activity. For this reason Figure 2 shows a stimulus unit activated by an "itch" and a connected response unit for scratching in the habit memory, which mediates the elicitation of this activity via the motor system. There is, however, no corresponding unit for this activity in associative memory.

To accommodate recent evidence on observation learning of instrumental relations, the model has to assume that representational units for actions can be activated not only by performance of the action itself but also by observation of a conspecific executing the same behavior. Heyes and her colleagues (Heyes & Dawson, 1990; Heyes, Dawson, & Nokes, 1992) allowed observer rats to watch previously trained demonstrators pushing a pole suspended from the ceiling of the chamber either to their left or their right for a food reward. When the observers were subsequently given access to the pole, they tended to push it in the same direction as their demonstrator even though they would have earned food by pushing it in either direction. Even more striking is the fact that this directional bias was maintained even though the pole was placed in a new position in the chamber between the observation and testing periods, suggesting that the observers were reproducing the specific action performed by the demonstrator. Given the bidirectional nature of the action, it is not clear how to explain this example of observational learning without assuming that the observer encodes the instrumental relation between action and outcome during exposure to the demonstrator. And to encompass this result with the model, it has to be assumed that observing an action is sufficient to excite the corresponding representational unit so that it can enter into association with the outcome unit.

V. INSTRUMENTAL LEARNING

In describing the associative-cybernetic model, the formation of the action–outcome connection in associative memory was attributed to experience with a contiguous and contingent or causal relation between action and outcome, simply because in a standard instrumental conditioning procedure performance of the action does, in fact, cause the immediate presentation of the outcome. There is good empirical evidence, however, that both these variables determine instrumental acquisition and performance.

A. Contiguity

From a purely functional perspective, the sensitivity of instrumental conditioning to action–outcome contiguity is surprising given that it is the causal rather than contiguous relation between action and outcome that is critical in allowing control over the environment. There is no doubt, however, that delaying the outcome after the action that caused it has a profound effect on instrumental acquisition. Dickinson, Watt, and Griffiths (1992), for example, studied the acquisition of lever pressing by rats when each outcome was delayed by intervals ranging up to 64 s after the press that caused it. With this procedure each lever press simply caused the delivery of a food pellet after the appropriate delay. Figure 3 illustrates the mean rate of lever pressing on the tenth 20-min session as a function of the mean delay between the delivery of an outcome and the immediately preceding lever press (which may not have been the action that actually caused the outcome). As can be seen, performance decreased systematically with the mean experienced delay.

Although these data certainly indicate that performance is degraded by delaying the reward, they also suggest that strict action–outcome contiguity is not necessary for conditioning in that acquisition was observed at intermediate delays. Given the procedure used by Dickinson et al., however, we cannot be certain that acquisition at these intermediate delays was not the result of fortuitous contiguities between a lever press and a reward caused by prior pressing. To control such fortuitous contiguities, Lattal and Gleeson (1990) employed a procedure in which each lever press that occurred during a delay period postponed the delivery of the next reward so that the actual interval between an outcome and the last lever press always matched the programmed delay. With this procedure Lattal and Gleeson reported acquisition with delays up to 30 s.

B. Contingency

Although the Lattal and Gleeson (1990) study provides convincing evidence that animals can detect an instrumental contingency across a delay, it still remains the case that acquisition is degraded by lengthening the action–outcome interval. The functional reason for this sensitivity to contiguity

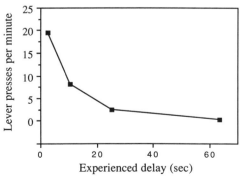

FIGURE 3 Mean rate of lever pressing as a function of the mean experienced delay between each press and the next outcome. (From Dickinson, Watt, & Griffiths, 1992.)

may well lie with the problem that animals face in discriminating a causal relation between action and outcome from a noncontingent schedule in which an outcome occurs relatively frequently but independently of their behavior. One way in which such a discrimination could be achieved is by contrasting in some way the likelihood of the outcome in periods when the target action is performed with that when the animal refrains from this behavior. To the extent that outcomes are more probable in periods when the action is performed, then the animal has causal control over the outcome. Such a contrast depends, however, on discriminating outcomes that occur during periods of action from those that occur in the absence of an action, a discrimination that can be based only on the temporal relation between action and outcome.

That animals make this discrimination is demonstrated in a classic study by Hammond (1980). Hammond developed a schedule in which he could manipulate independently the probability of an outcome given that an action had occurred in the preceding second [P(O/A)] and the probability of an outcome in the absence of an action in the preceding second [P(O/−A)]. Thus, P(O/A) refers to the probability of a contiguous outcome and P(O/−A) to the likelihood of a noncontiguous outcome. Figure 4 illustrates the terminal levels of performance observed by Hammond under various combinations of P(O/A) and P(O/−A) when lever pressing by thirsty rats was the action and water delivery the outcome.

The first point to note is that performance increases with the probability of a contiguous outcome in the absence of any noncontiguous outcomes (cf. Mazur, 1983). With P(O/−A) set to zero, the rats pressed at a higher rate

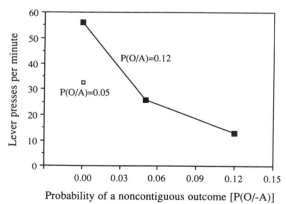

FIGURE 4   Mean rate of lever pressing on the last session of training as a function of the probability of a noncontiguous outcome [P(O/−A)] for different probabilities of a contiguous outcome [P(O/A)]—(the data points were estimated graphically from Figure 2 of Hammond, 1980, for the last session of training).

when $P(O/A)$ was 0.12 rather than 0.05. This finding, of course, accords with the predictions of a contiguity-based learning process; the higher this probability the greater will be the proportion of actions that are immediately followed by an outcome. More important for the present discussion, however, is the effect of varying the probability of noncontiguous outcomes. Unless the animals could discount the effect of these noncontiguous outcomes in some way, increasing their frequency by raising $P(O/-A)$, if anything, should have augmented instrumental performance by acting as a source of delayed reinforcement. In fact, exactly the opposite effect was observed; when $P(O/A)$ was 0.12, raising the probability of noncontiguous outcomes $[P(O/-A)]$ reduced performance until when $P(O/A)$ equaled $P(O/-A)$ the rats pressed relatively infrequently. This low level of performance is, of course, entirely appropriate from a causal perspective, for under such a noncontingent schedule pressing has no effect on the likelihood of the outcome—on average, the animals would gain exactly the same number of water presentations whether they pressed in every second or never pressed at all.

Although it is tempting to explain this contingency effect in terms of some form of direct sensitivity to the causal relation between action and outcome, an account is available in terms of a simple contiguity-based learning process. Of necessity, the presentation of the noncontiguous outcomes must have been paired with the action of approaching the water source and, as a result, may have served to enhance the performance of this behavior at the expense of lever pressing. In others words, the decrement in lever pressing under a noncontingent schedule may have been simply due to behavioral competition or interference from approaches to the water source reinforced by the noncontiguous outcomes.

In an attempt to address this behavioral competition account, Dickinson and Mulatero (1989) followed up an initial study of Colwill and Rescorla (1986) by training hungry rats to press two levers, one of which delivered food pellets and the other a sugar solution at the same source. The instrumental contingency was then degraded by scheduling noncontiguous presentations of one type of outcome, either the food pellets or the sugar solution, following any second in which the animal pressed neither lever. According to the competition account, there is no reason why any competing behavior established by the noncontiguous outcomes should have interfered more with one action rather than the other. And yet Dickinson and Mulatero (1989; see also Williams, 1989) found that these noncontiguous outcomes produced a greater reduction in pressing on the lever associated with the same outcome. Moreover, this difference was not due to a selective satiety for the noncontiguous outcome, for it persisted into a test session in which no outcomes were presented.

From a causal perspective this finding is not unexpected. It is true that the animals cannot affect the overall frequency of outcomes by pressing on either lever under this schedule. What they can alter, however, is the relative frequency of the two types of outcome. In the case in which the noncontiguous outcome is the food pellets, for example, if the animal presses neither lever it would receive only this type of outcome, a distribution that is not affected by pressing the pellet lever. By pressing the sucrose lever, however, the animal can substitute sugar deliveries for pellet presentations, and thus retain control over the relative frequency of the two outcomes. What the Dickinson and Mulatero (1989) result shows is that rats are sensitive to this contingency, a sensitivity that cannot be explained in terms of simple behavioral competition.

C. Surprise and Learning

Even this sensitivity to causality does not require us to abandon the claim that the crucial feature of an effective instrumental relation is action–outcome contiguity, because the general pattern of results observed across variations in temporal and contingency parameters is that anticipated by contemporary theories of associative learning in Pavlovian conditioning (see Hall, Chapter 2; Shanks, Chapter 12, this volume). At the heart of such theories is the claim that the efficacy of a contiguous outcome depends on the extent to which its occurrence is predicted, whether this be because of some error-correcting process (e.g., Rescorla & Wagner, 1972), or the modulation of learning either by attentional and associability processes (e.g., Mackintosh, 1975; Pearce & Hall, 1980) or by generalization (Pearce, 1987; Chapter 5, this volume). Predicted outcomes are less effective than unexpected or surprising ones in bringing about the formation of an associative connection with a preceding event.

An extension of this general principle to instrumental conditioning provides a ready explanation of the contingency effect observed by Hammond. The presentation of the noncontiguous outcomes on the noncontingent schedule results in conditioning to the contextual cues so that an animal learns to expect the outcome in this context. As a result, the occurrence of a contiguous outcome, the event responsible for instrumental learning, is predicted by the contextual cues and thus fails to engage the learning process that is necessary to acquire and maintain the associative connection(s) mediating instrumental performance. A comparable attenuation does not occur in the absence of noncontiguous outcomes, at least in part because the overall number of outcomes presented in the context is lower so that less conditioning accrues to the context.

There is no doubt that this account has much merit; not only does it preserve a contiguity-based learning process but it also predicts the conditions under which we should be able to induce an "illusion of control" on a noncontingent schedule. As the deleterious impact of noncontiguous outcomes is attributed to contextual conditioning, if we could in some way prevent or minimize this conditioning we should be able to maintain the effectiveness of the contiguous outcomes and thus also performance under a noncontingent schedule. One way to do so is to capitalize on the phenomenon of "overshadowing." Pavlov (1927) observed that the conditioning accruing to a stimulus was reduced or overshadowed when it is reinforced in compound with another stimulus. Thus, if each noncontiguous outcome is signaled by a discrete stimulus, say, a brief light, then the amount of conditioning to the contextual cues should be reduced. This in turn should have the effect of maintaining the surprising nature of the outcomes paired with the action and hence their ability to bring about instrumental learning. Both Hammond and Weinberg (1984) and Dickinson and Charnock (1985) have reported just such an effect; signaling each noncontiguous outcome elevated the rate of lever pressing by rats on a noncontingent schedule even though this action had no causal effect on the overall frequency of outcomes.

D. Interval Schedules

In summary, we have seen that a simple contiguity-based learning process can provide an account of the sensitivity of instrumental performance to variations in the causal effectiveness of an action. All we need to assume is that the occurrence of each unexpected, contiguous outcome strengthens the underlying associative connections, whereas performance of the action in the absence of an effective, contiguous outcome weakens these connections. As a consequence, the primary determinant of performance should be the probability of an outcome or, in other words, the probability of reinforcement. What is less clear, however, is whether such a process can also explain the effect of another major determinant of instrumental performance—outcome or reinforcement rate. The point at issue can be illustrated by considering two different sources of outcomes. So far we have discussed an inexhaustible and nondepleting source; each execution of the action has a fixed probability of causing an outcome. This type of source is usually referred to as a *ratio* schedule in that the contingency maintains, on average, a constant ratio between the number of actions performed and the number of outcomes generated. Ratio schedules are characterized by a linear function between the rate at which the action is performed and that at which the outcome occurs. The specific example illustrated in Figure 5 has a ratio value of 20 so that the probability of reinforcement is 0.05.

Although many important resources in the world conform to a ratio contingency, there is another class that depletes and then regenerates with time. Organic food sources are typically of this class. Such sources can be modeled by assuming that an outcome becomes available with some fixed probability in each time interval so that a source can be characterized by average temporal interval between successively available outcomes and, for this reason, the contingency is referred to as an *interval* schedule. The feedback function for the interval schedule represented in Figure 5 generates an available outcome on average once every 17.5 s so that sufficiently high action rates produce an outcome rate of about 3.4/min.

The feedback functions displayed in Figure 5 illustrate that ratio and interval contingencies arrange quite different causal relations between work rate and payoff rate. Whereas an agent can generate a higher rate of return by working faster on a ratio schedule, there is little or no payoff for performing faster under an interval contingency once the action rate is sufficiently high to collect all the outcomes that are available in a given time period. Consequently, if animals are sensitive to these different feedback functions, one might expect them to perform faster on a ratio schedule relative to a matched interval contingency, a prediction confirmed in numerous studies. Dawson and Dickinson (1990), for example, trained hungry rats to pull a chain for food on a ratio schedule under which the probability of reinforcement for each pull was 0.05. Their mean rate of pulling over the last five sessions of training when performance was stable is shown in Figure 5. Matched or yoked to each of these ratio animals was another rat trained on an interval schedule that yielded the same outcome or reinforcement rate. For theoretical reasons beyond the scope of this discussion, Daw-

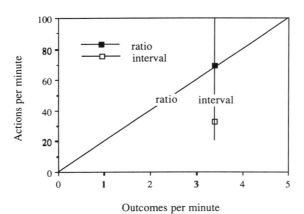

Outcomes per minute

FIGURE 5 Illustration of the feedback functions for a ratio schedule with a probability of an outcome of 0.05 and an interval schedule with an outcome rate of 3.4/min. The outcome rates have not been defined at low action rates for the interval schedule because the way in which the feedback function approaches the origin depends on the particular type of interval schedule in force. The data points represent the mean rates of chain pulling maintained by rats under these schedules during the last 5 sessions of training in Experiment 2 of Dawson and Dickinson (1990).

son and Dickinson implemented the interval schedule by keeping a running record of the animal's rate of pulling over the last 50 pulls and then setting the probability of reinforcement for the next pull at a value that ensured the appropriate outcome rate if the animal continued to perform at the same rate. As can be seen in Figure 5, the procedure was successful in matching the reinforcement rates on the ratio and interval schedules.

The important finding in this study, however, was that the animals performed about twice as fast on the ratio schedule as under the interval contingency (see Figure 5). This difference is entirely reasonable given the nature of the feedback functions for the two schedules—if the ratio rats had pulled the chain at the same rate as the interval animals they would have earned only about half as many outcomes, whereas the interval animals could not have increased their outcome rate by pulling the chain any faster. This result is problematic, however, for a simple contiguity-based learning process, which, as we have seen, predicts that the strength of the action should increase with the probability of reinforcement. At variance with this prediction is the fact that the interval schedule maintained a lower action rate even though the probability of reinforcement was about twice that on the ratio contingency.

Following Baum (1973), we could argue that animals are sensitive to the way in which the number of outcomes generated in a series of relatively short time windows (e.g., of 10 s or so) varies with changes in the number of actions performed in each time window, thus allowing them to assess the contingency or correlation between action and outcome *rates* across a series of time windows. The learning process envisaged by this account is sensitive not to the contiguity between individual actions and outcomes, but rather to the contingency or correlation between local rates of performance and outcomes. Interval schedules generate relatively low contingencies and thus low rates of performance, as do ratio schedules with noncontiguous outcomes. Moreover, this account also predicts the deleterious effect of delaying outcomes; the longer the delay the more likely it is that outcome will be delivered in a different and subsequent time window to that containing the action that caused it, with the effect of reducing the local contingency.

The problem with this correlational theory is to explain why interval schedules maintain as high a rate of performance as they do given the low correlation with outcome rate. If we assume that this correlation determines instrumental knowledge or, in terms of the associative-cybernetic model (see Figure 2), the strength of the connection between the action and outcome units in the associative memory, this connection should be weak under an interval contingency, leaving performance primarily determined by the habit processes, which is represented in the model by the strengthening of the connection between the stimulus and response units in habit memory. This means that unlike ratio performance, instrumental behavior on an interval schedule should be primarily an S–R habit rather than an action based on knowledge of the action–outcome relation. An obvious prediction of this analysis is that performance established by an interval schedule should be relatively insensitive to outcome devaluation. Dickinson, Nicholas, and Adams (1983) evaluated this prediction by comparing the effect of outcome devaluation following comparable training on ratio and interval schedules. Whereas the standard devaluation effect was observed following ratio training, conditioning an aversion to the outcome had no detectable effect on instrumental performance in subsequent extinction tests when the rats had been initially trained to press the lever on an interval schedule.

Although this result accords with the idea that interval performance can be primarily controlled by an S–R habit process, what this claim does not explain is why such performance is an orderly function of the rate at which the outcome occurs. Herrnstein (e.g., de Villiers & Herrnstein, 1976) established that the response rate on interval schedules is a negatively accelerated function of the reinforcement rate given by:

$$B = aR \ / \ (b + R)$$

where B is the response rate, R the reinforcement rate and a and b parameters. On the basis of Herrnstein's equation, it has been argued that reinforcement rate exerts a direct effect on instrumental performance. According to Killeen (1982), for example, outcomes not only reinforce immediately preceding actions but also enhance performance in general for a period after their presentation. Clearly, on this account, the higher the frequency of outcomes the greater will be the overall level of this motivational influence.

VI. PAVLOVIAN–INSTRUMENTAL INTERACTIONS

Killeen also suggested that this motivational influence can be conditioned to contextual cues through their Pavlovian pairings with the outcome. The potential role of such conditioning has already been discussed in the analysis of the effect of noncontiguous outcomes, which acted to attenuate the impact of contiguous outcomes on instrumental *learning* through contextual conditioning. In addition, contextual conditioning also exerts a potentiating influence on instrumental *performance,* an idea that in fact has a venerable history in the form of two-process learning theory (Rescorla & Solomon, 1967). Within one version of this theory, contiguous outcomes are assumed not only to reinforce immediately preceding actions but also to condition a motivational influence to accompanying stimuli so that such stimuli will augment behavioral output.

The canonical evidence for two-process theory comes from studies of

Pavlovian–instrumental transfer. Lovibond (1983), for example, initially trained hungry rabbits to lift a lever for a delivery of sugar solution directly into their mouths through an oral fistula. The lever was then removed for a second, Pavlovian stage in which a 10-s stimulus was paired with sugar deliveries. Finally, Lovibond assessed the effect of presenting the Pavlovian stimulus while the animals were engaged in the instrumental task. If the Pavlovian conditioning endows the stimulus with motivational properties, the rate of lever lifting should be elevated during the stimulus, and this is what Lovibond observed.

Whether or not this transfer effect really provides compelling evidence for the motivational version of two-process theory is, however, a matter of controversy. An alternative, favored, for instance, by Trapold and Overmier (1972), argues that the Pavlovian stimuli act by reinstating conditions that are more similar to those in which the instrumental action was trained. This idea is probably best illustrated by considering another Pavlovian-instrumental transfer study, this time conducted by Colwill and Rescorla (1988). In the initial, Pavlovian stage a stimulus was established as a signal for a particular outcome; for half the hungry rats the outcome was a food pellet, whereas the remainder received sugar solution as the outcome. Each rat then received instrumental training in which lever pressing caused one of the outcomes and chain pulling the other outcome in *separate sessions*. The design of that part of their study relevant to the present discussion is illustrated in the top half of Table 1 for the animals receiving the pellets for lever pressing and the sugar for chain pulling (the remaining animals received the opposite action–outcome assignments during training).[5]

As a result of this training the animals learned to press the lever at a time when the contextual cues were associated with pellets and pull the chain when these cues were paired with the sugar solution. In the subsequent test the animals were for the first time given the opportunity both to press the lever and to pull the chain in the presence of the stimulus but in the absence of any outcomes (extinction). For the animals tested with the stimulus paired with pellets, this condition should have been most similar to that in which lever pressing was trained and, consequently, if Pavlovian stimuli control instrumental performance by reinstating the training conditions, the rats should have pressed the lever more than they pulled the chain. Correspondingly, the reverse pattern should have been observed when the stimulus was paired with the sugar solution. This is the result reported by Colwill and Rescorla (1988, see Table 1).

Although there is little doubt that Pavlovian stimuli can control instrumental actions trained with a common outcome, this does not mean that we should abandon the motivation hypothesis. Dickinson and Dawson (1987) also conducted a Pavlovian-instrumental transfer experiment using food pellets and sugar solution as outcomes but in this case each was paired with

TABLE 1   Design of Pavlovian-Instrumental Transfer Studies

Group	Training	Test
Colwill and Rescorla (1988)		
Pel	S → Pel, Lp → Pel, Cp → Sug[a]	S: Lp > Cp
Sug	S → Sug, Lp → Pel, Cp → Sug	S: Lp < Cp
Dickinson and Dawson (1987)		
Hunger:	S1 → Pel, S2 → Sug, Lp → Pel	Hunger: S1: Lp > S2: Lp Thirst: S1: Lp < S2: Lp

[a] S, Pavlovian stimulus; Lp, lever press; Cp, chain pull; Pel, food pellet; Sug, sugar solution.

a different Pavlovian stimulus during training (see bottom half of Table 1). In addition, all the rats were also taught to press a lever in the absence of either stimulus but in this case for *food* pellets only. Although all the animals were trained hungry, only half of them were tested in this state. This test assessed the rate of lever pressing in the presence of the two stimuli, again in the absence of any outcomes. In accord with the pattern observed by Colwill and Rescorla (1988), the stimulus associated with the same outcome as the action—the pellets—controlled the higher rate of pressing when the animals were hungry during the test.

The critical group, however, was composed of animals tested while thirsty rather than hungry. The point of the shift from training under hunger to testing under thirst was to alter the motivational relevance of the two outcomes. Whereas both outcomes are relevant to the state of hunger, the sugar solution not surprisingly acts as a more effective outcome than food pellets when animals are thirsty. Consequently, we should expect a stimulus paired with the solution rather than the pellets to exert a greater motivating effect when the animals are thirsty. Thus, the test under thirst pitted this motivating effect of the Pavlovian stimulus against the reinstatement of the training condition. In this case the motivational influence won out since the thirsty animals pressed more in the stimulus paired with the sugar solution. Dickinson and Balleine (1990) have observed a similar effect following a shift from thirst to hunger.

It seems clear, therefore, that Pavlovian stimuli exert an influence on instrumental performance by two processes. The first is engaged when the stimulus and action share common outcomes and operates by reinstating the conditions in which the action was conditioned. By contrast, the second depends on the relevance of the Pavlovian outcome to the agent's motivational state and is manifest as general potentiating influence on instrumental behavior.[6]

VII. DISCRIMINATIVE CONTROL

The fact that the contextual stimuli in which an instrumental action is trained can exert control over the performance of that action is a specific example of the general phenomenon of *discriminative control*. If lever pressing delivers food in the presence of stimulus but not in its absence, not surprisingly, rats rapidly learn to press only when the stimulus is present. As the animals use the stimulus to discriminate periods when there is an instrumental contingency between the action and outcome from periods in which there is no contingency it is referred to as a *discriminative stimulus*.

Given that stimuli associated with an outcome can potentiate instrumental performance, discriminative control may represent nothing more than the operation of a direct Pavlovian association between the discriminative stimulus and the outcome. Alternatively, a stimulus may acquire control over an instrumental response through the habit mechanism envisaged by the Law of Effect. Thus, in terms of the associative-cybernetic model (see Figure 2), reinforcing the response in the presence of the stimulus would act to strengthen a connection between the input unit activated by the stimulus in the habit memory and the response unit. A recent outcome devaluation study by Colwill and Rescorla (1990), however, makes clear that discriminative control can involve processes that transcend both these mechanisms.

The design of their study is illustrated in Table 2. In the first, discrimination training stage, the rats were trained to perform two actions, lever pressing and chain pulling, concurrently to receive two outcomes, sugar solution and food pellets, in the presence of two stimuli, a noise and light. In one stimulus lever pressing produced the sugar solution and chain pulling the food pellets, whereas these action–outcome assignments were reversed in the second stimulus. In the second stage, one of the outcomes (O2 in Table 2) was devalued by pairing its consumption with the induction of gastric illness until the intake was suppressed. Finally, the animals were given a choice between the two actions during separate presentations of both stimuli in an extinction test. The Pavlovian account predicts that the devaluation treatment should have no effect on the relative performance of

TABLE 2   Design of Experiment

Discrimination training	Outcome devaluation	Test
S1: A1 → O1, A2 → O2 S2: A1 → O2, A2 → O1	O2 → illness	S1: A1 > A2 S2: A1 < A2

Source: From Colwill and Rescorla (1990).
[a] S, discriminative stimulus; A, instrumental action; O, outcome.

lever pressing and chain pulling in the two stimuli; as both stimuli were paired with the sugar solution and food pellets, devaluing one of these outcomes should have an equivalent effect on performance in the two stimuli. Moreover, as was pointed out in the initial discussion of the Law of Effect, devaluing an outcome should have no impact on subsequent performance of a habit.

At variance with these predictions, during testing, the animals were reluctant to perform the action trained with the devalued outcome—in the presence of S1 they performed A2 less than A1, whereas a reverse pattern was observed during testing in S2 (see Table 2). On the basis of this result, Colwill and Rescorla (1990) argued that discriminative control can reflect encoding of the three-term relationship between the discriminative stimulus, the instrumental action, and the outcome. Within associative theory, one way of representing such a relation is to appeal to a configural unit, which is optimally excited by a compound of two of the terms of the triad (see Pearce, Chapter 5, this volume). Thus, we could replace the action unit in the memory of the associative-cybernetic model (see Figure 2) by one that is optimally excited by a compound of the discriminative stimulus and an action input, thus ensuring that the action is performed more reliably in the presence of the stimulus.[7] According to this analysis, the discrimination training given by Colwill and Rescorla (1990) would have involved four configural units (see Table 2): S1A1 and S2A2 units each connected to the O1 unit and S1A2 and S2A1 units each connected to the O2 unit. As a consequence, devaluing O2 would mean that activation of either the S1A2 or the S2A1 units would indirectly excite the punishment unit thus reducing the performance of A2 in S1 and A1 in S2.

As an alternative to this configural account of discriminative control, Rescorla (1990) has pointed out that another way of analyzing the triadic relation between a discriminative stimulus, instrumental action, and outcome is to view the stimulus as signaling that the action–outcome contingency is in operation. On this analysis, discriminative control may reflect learning of the conditional relation between the stimulus and the binary action–outcome relation. In an attempt to provide evidence for this conditional account, Rescorla (1990) studied the effect of disrupting the instrumental contingency following discrimination training. Without going into the details of the full counterbalanced and controlled design (which is exceedingly complex), he used a training procedure similar to that illustrated in Table 2: an action produced one outcome in the presence of one stimulus (S1: A → O1) and a different outcome in the presence of another stimulus (S2: A → O2). In addition, the animals were also trained to perform this action for one of the outcomes, say, O1, in the presence of a third stimulus (S3: A → O1). The purpose of this third stimulus was to provide a context in which he could subsequently disrupt the contingency between the action and the first outcome, O1. So, following discrimination training, Rescorla allowed the rats to perform the action in the presence of the third stimulus, but in the absence of any outcome (S3: A-), until performance extinguished.

The question of primary interest is whether this disruptive treatment will have any effect on performance in the presence of the first (S1) and second (S2) discriminative stimuli. According to a simple configural account, it should not. While destroying the connection between an S3A configural unit and the O1 unit, those between S1A and S2A units and the outcome units should remain intact. In fact, Rescorla (1990) found that this extinction treatment produced a relative reduction in performance in the discriminative stimulus that signaled the same action–outcome contingency, specifically in S1. This finding follows reasonably directly from the conditional theory. During training the animals should have learned three conditional relations involving the action: S1 → (A → O1); S2 → (A → O2); and S3 → (A → O1). Performing the action in the absence of any outcome in S3 will, it is argued, lead the animal to believe that the action in general no longer causes the first outcome, O1, while leaving its representation of the action–O2 contingency intact. Consequently, when the animal is once again exposed to the discriminative stimuli, it should continue to perform the action in S2 but not in S1. Thus, this study suggests that discriminative control can be mediated by encoding the conditional relation between the discriminative stimulus and the action–outcome contingency.[8]

In relating the associative-cybernetic model (see Figure 2) to the cognitive or intentional account of instrumental action, I argued that the connection between action and outcome units in the associative memory corresponds to the belief that the action causes the outcome, and the discussion so far has concentrated primarily on the nature of the learning processes underlying the acquisition of this binary belief. What Rescorla's analysis of discriminative control suggests is that instrumental performance can be based on a triadic belief that represents the conditional relation between the action–outcome contingency and the discriminative stimulus. The general problem of how to represent such conditional knowledge within an associative structure remains a matter of current dispute (e.g., Holland, 1992).

VIII. INCENTIVE LEARNING

The second component of the cognitive account of instrumental action is the agent's desire for the outcome. Whatever the strength of an instrumental belief, an animal should not perform the relevant action in the absence of a desire for the outcome, which reflects the value that is currently assigned to the outcome. In elaborating the associative-cybernetic model, I suggested that this value could be represented by direct associative connections between the outcome unit in the associative memory and the reward and punishment units in the incentive system (see Figure 2). I also suggested that such connections were acquired through a process of incentive learning (Dickinson & Balleine, 1994).

Perhaps the most straightforward example of incentive learning is that seen in the outcome devaluation effect. Devaluing the outcome by aversion conditioning, I argued, sets up a direct connection from the outcome unit in the associative memory to the punishment unit in the incentive system. At first sight, this may seem a surprising suggestion given that the devaluation was brought about by pairing consumption of the food outcome with gastric malaise. If the associative structures in memory represent beliefs about the contingencies and relations between events in the world, why, one might ask, is not that between consumption of the food and consequent gastric illness also represented? According to Garcia (1989), however, aversions conditioned by gastric malaise are not based on the knowledge of the relation between the food and illness. Rather, he argued that conditioning brings about a latent change in the affective reaction elicited by the food, which is then manifest when the animal is re-exposed to the food. Thus, following aversion conditioning, the rat does not refrain from eating sugar pellets because it knows that they cause illness, but rather because they now taste noxious. According to this account, the change in value of an outcome does not occur at the time of aversion conditioning but rather when the rat is subsequently re-exposed to food and experiences that it is now distasteful. And it is this experience that leads to the formation of the direct connection between the outcome and punishment units without any representation of the relation between the food and illness. Following aversion conditioning, contact with the food produces conjoint activation of the outcome and punishment units, which is sufficient to establish the connection between them.[9]

The rats in the Adams and Dickinson (1981) devaluation study had the opportunity for such incentive learning because it required a number of pairings of the outcome and illness to suppress consumption. Thus, they experienced the progressively more noxious character of the outcome across the series of pairings. Balleine and Dickinson (1991) provided direct evidence for the role of incentive learning in the outcome devaluation effect by conditioning the aversion with a single pairing. They trained rats to press a lever for a sugar solution in a single session, which was followed immediately by a LiCl injection to induce gastric malaise. When these animals were subsequently tested (in extinction), they showed no evidence of a devaluation effect; the rats pressed just as much as control animals that had not received the toxin. This did not reflect a failure of the aversion procedure, for these animals would not work for the sugar solution when it was once again presented contingent upon lever pressing. The absence of a devaluation effect with this procedure is anticipated, however, by the incentive learning account. Without re-exposure to the sugar solution before testing, the animals did not have an opportunity to experience the change in their reactions to the solution with the result that no connection should have been formed between the outcome and punishment units (see Figure 2). Of course, if given such an experience, a connection would have been established and a devaluation effect evident. In accord with this prediction, Balleine and Dickinson (1991) found a robust devaluation effect in a second group of rats given the opportunity to drink the sugar solution between aversion conditioning and testing.

This demonstration of incentive learning does not imply that knowledge about the relation between outcomes and other significant events never plays a role in controlling outcome value; indeed there is good evidence that it does (e.g., Balleine & Dickinson, 1992; Rescorla, 1994). The reason for emphasizing incentive learning is that it appears to be the process by which primary motivational states, such as hunger and thirst, control outcome value. It is well established that the motivational state of an animal is a major determinant of instrumental performance; not surprisingly, hungry animals work more vigorously for food reward than do sated ones. And yet the motivational state of the animal does not feature in the determination of outcome value in the associative–cybernetic model presented in Figure 2. In this model the positive value of an outcome is simply determined by the strength of the connections between the outcome unit and the reward and punishment units.

There are, in fact, good empirical reasons for arguing that motivational states do not directly control outcome value. Balleine (1992) trained hungry and sated rats to lever press for a novel food reward before shifting their motivational state and testing their performance in the absence of any outcomes (extinction). Although the hungry animals pressed faster than the sated ones during training, their motivational state on test had no effect on their performance during the extinction test. The animals trained hungry continued to press at a high rate when tested sated and, correspondingly, those trained sated pressed at a low rate when tested hungry. As in the case of aversion conditioning, motivational state appears to bring about a change of outcome value only when the animals are actually exposed to the outcome. Presumably, exposure to a food outcome during training produces high level of activation of the reward unit in the incentive system when the animal is hungry, which in turn results in the formation of a strong connection between the outcome and reward units (see Figure 2). In the absence of further exposure to the outcome, a shift to the sated state does not affect the strength of this connection and hence the value of the outcome. Similarly, only a weak connection to the reward unit is established by training in the sated state, which then continues to determine performance when the animal is hungry in the absence of any further exposure to the outcome.

This does not mean, however, that motivational states cannot gain control over outcome value if the animals are given the opportunity to learn about this value in a particular state. Balleine (1992) gave his rats such an opportunity by pre-exposing them to the food reward in the sated state prior to instrumental training under hunger. When these animals were then sated and tested in extinction, there was an immediate reduction in performance. On the basis of their experience of the outcome in both the sated (during pre-exposure) and hungry states (during instrumental training), the animals appear to have learned to assign different and appropriate incentive values to the outcome depending on their current motivational state (see also Lopez, Balleine, & Dickinson, 1992). Balleine (1992) also reported an analogous incentive learning effect for an upshift in motivational state; following training in the sated state, hungry rats pressed faster than non-deprived animals in an extinction test only if they have been previously exposed to the food outcome when hungry. Thus, it would appear that animals can learn that outcome value is conditional on their motivational state just as they can learn that an instrumental contingency is conditional on the presence of a discriminative stimulus. And, of course, the representation of such conditional knowledge is equally problematic in the two cases.

IX. COGNITION AND INSTRUMENTAL ACTION

Within intentional psychology, goal-directed actions are the final common pathway for the behavioral expression of all cognition. Whatever the complexity of an inferential chain, if it does not yield a conclusion that rationalizes a certain course of action in relation to a goal, the cognitive processes underlying the inference must remain behaviorally silent. In this sense goal-directed action represents the most basic behavioral marker of cognition. And the evidence reviewed in this chapter suggests that the standard instrumental or operant procedures of the animal conditioning laboratory can establish goal-directed actions, mediated by the interaction of knowledge of the action–outcome relation with knowledge about status of the outcome as a goal. The former is acquired through experience with a causal relation between action and outcome, although it remains unclear whether the critical determinant of acquisition is the probability of a contiguous, unexpected outcome or the correlation between action and outcome rates. Knowledge of the value of the outcome as a goal of an instrumental action also appears to be acquired through a process of incentive learning.

The problem with a purely cognitive account of instrumental action, couched in terms of the interaction of intentional mental states, lies with the absence of any process by which such cognitive determinants of action can interact with mechanistic influences on behavior. The resistance of instrumental performance to outcome devaluation indicates a role for such influences as does the motivational effect of Pavlovian stimuli. An associative-cybernetic model was developed in an attempt to address this problem. Within this model, habit formation is explained in terms of traditional S–R/reinforcement theory. The output of this habit system, however, not only determines behavior directly but also provides an input to the cognitive system in the form of an associative memory in which is represented knowledge of the environmental contingencies. The output of this associative memory re-engages with noncognitive influences on behavior through a feedback loop onto the output of the habit system. This loop operates via an incentive system, and it is at this point that the consequences of cognitive processes can interact with other, mechanistic influences such as the motivational effect of Pavlovian stimuli.

Whatever the merits of this particular attempt to integrate cognitive and mechanistic processes, the important feature of instrumental action remains its intentional character, even in the humble rat. It is this feature that distinguishes goal-directed, instrumental action from habits, reflexes, fixed-action patterns, and the like. Unlike all these other classes of behavior, an instrumental action can stand in a unique, reciprocal relation to cognition; to the extent that they are goal directed, such actions can be "decomposed into a representation on the one side, and a piece of operant behavior on the other, with the former causing the latter and the latter satisfying the former" (Danto, 1976, p. 24). If by learning we change our representations to fit the world, then in pursuing our desires through instrumental action we change the world to fit our representations.

Endnotes

1. This analysis assumes that stimuli are defined allocentrically rather than egocentrically. If one is prepared to accept a behaviorally contingent definition of stimuli, it is possible to give an account of Grindley's (1932) observation in terms of stimulus–outcome relations. Thus, it could be argued that the instrumental contingency between head turning, say, to the left, and the outcome established a stimulus–outcome relation between the contextual stimuli on the left and the carrot, which would result in the animal orienting to that side. Bindra (1972, 1978), for example, has offered such an account of instrumental conditioning. However, as the definition of such stimuli is conditional upon an action, sensitivity to stimulus–outcome and action–outcome relations are indistinguishable at a behavioral level, and in the present discussion I shall assume that stimuli are defined independently of behavior.

2. In fact these noncontingent pellets were delivered in a negative relation to lever pressing to preclude fortuitous pairings (see Adams & Dickinson, 1981), but for the purposes of the present exposition they can be treated as occurring independently of lever pressing.

3. I shall ignore the myriad of problems arising from the assumption that associative structures can carry the representational or intentional content of mental states such as beliefs and desires. In fact, the operation of processes, such as excitation and inhibition, on associative connections constitutes about the only mechanistic psychology available.

4. This pattern of feedback from the incentive system is appropriate for appetitive and investi-

gatory responses directed toward the stimulus. However, the behavioral repertoire represented within the habit memory and motor system should include defensive responses for which the pattern of feedback is reversed. For such defensive responses, it is the punishment element that should exert the excitatory influence on the respective motor elements with the reward element having an inhibitory influence.

5. This is an abbreviated presentation of the design of Experiment 3 presented by Colwill and Rescorla (1988) and omits the fact that each animal was also trained and tested with a discriminative stimulus.

6. The reader may wonder why I have not attempted to explain Pavlovian-instrumental transfer effects in terms of the associative-cybernetic model. It could be assumed that the Pavlovian conditioning sets up a connection in associative memory from a unit excited by the Pavlovian stimulus to the outcome unit, which would certainly have the effect of enhancing the positive feedback on the motor units in the presence of the stimulus, and hence instrumental performance. The reason for resisting this suggestion lies in the fact that the effects on performance mediated by stimulus–outcome and action–outcome relations can be dissociated by motivational manipulations. Recall that Dickinson and Dawson (1987) showed that rats lever pressed more in the presence of a stimulus associated with a sugar solution than during one associated with food pellets when their motivational state was shifted from hunger to thirst (see Table 1). It is notable, however, that this Pavlovian effect occurred even though the animals had never previously experienced these two outcomes when thirsty. This direct effect of motivational state contrasts with that mediated by the action–outcome contingency which, as we shall see below, depends on incentive learning. Dickinson and Dawson (1988, 1989) demonstrated that incentive learning is necessary if a shift from hunger to thirst is to alter the relative incentive values of the sugar solution and food pellets as instrumental outcomes. For this reason, the learning mediating a motivationally based Pavlovian-instrumental transfer effect cannot be represented with the same memory system as that encoding the instrumental contingency.

7. It is interesting to note that this was essentially the account of instrumental conditioning offered by Miller and Konorski (1969) when they first distinguished this form of conditioning from the Pavlovian variety. They suggested that the animal performs the instrumental action in order to form a compound event with the discriminative stimulus that reliably predicts the outcome.

8. This is not to say that a more elaborate but plausible configural theory could not give a satisfactory account of Rescorla's (1990) result. If it is assumed that extinction of the action in S3 established an inhibitory association between the S3A configural stimulus and O1, one might well argue that this inhibition should generalize to the S1A and S2A configural stimuli on the basis of the common A element. This generalized inhibition should impact on performance only in S1, however, as it is only in this stimulus that performance is controlled by an excitatory association with O1.

9. Of course, some learning must take place at the time of aversion conditioning to mediate the altered affective reactions experienced during re-exposure to the outcome. It is assumed, however, that this learning does not take place at the cognitive level and therefore is not represented in terms of connections within the associative memory that encodes instrumental action–outcome relations.

References

Adams, C. D., & Dickinson, A. (1981). Instrumental responding following reinforcer devaluation. *Quarterly Journal of Experimental Psychology, 33B,* 109–122.

Amundson, R. (1986). The unknown epistemology of E. C. Tolman. *British Journal of Psychology, 77,* 525–531.

Asratyan, E. A. (1974). Conditioned reflex theory and motivated behavior. *Acta Neurobiologiae Experimentalis, 34,* 15–31.

Balleine, B. (1992). Instrumental performance following a shift in primary motivation depends upon incentive learning. *Journal of Experimental Psychology, 18,* 236–250.

Balleine, B., & Dickinson, A. (1991). Instrumental performance following reinforcer devaluation depends upon incentive learning. *Quarterly Journal of Experimental Psychology, 43B,* 279–296.

Balleine, B., & Dickinson, A. (1992). Signalling and incentive processes in instrumental reinforcer devaluation. *Quarterly Journal of Experimental Psychology, 45B,* 285–301.

Baum, W. M. (1973). The correlational-based law of effect. *Journal of the Experimental Analysis of Behavior, 20,* 137–143.

Bindra, D. (1972). A unified account of classical conditioning and operant training. In A. H. Black & W. F. Prokasy (Eds.), *Classical conditioning II: Current research and theory* (pp. 453–482). New York: Appleton-Century-Crofts.

Bindra, D. (1978). How adaptive behavior is produced: A perceptual motivational alternative to response-reinforcement. *Behavior and Brain Sciences, 1,* 41–52.

Bolles, R. C., Holtz, R., Dunn, T., & Hill, W. (1980). Comparison of stimulus learning and response learning in a punishment situation. *Learning and Motivation, 11,* 78–96.

Colwill, R. C., & Rescorla, R. A. (1985). Postconditioning devaluation of a reinforcer affects instrumental responding. *Journal of Experimental Psychology: Animal Behavior Processes, 11,* 120–132.

Colwill, R. C., & Rescorla, R. A. (1986). Associative structures in instrumental learning. In G. H. Bower (Ed.), *The psychology of learning and motivation* (Vol. 20, pp. 55–104). Orlando, FL: Academic Press.

Colwill, R. M., & Rescorla, R. A. (1988). Associations between the discriminative stimulus and the reinforcer in instrumental learning. *Journal of Experimental Psychology: Animal Behavior Processes, 14,* 155–164.

Colwill, R. M., & Rescorla, R. A. (1990). Evidence for the hierarchical structure of instrumental learning. *Animal Learning and Behavior, 18,* 71–82.

Danto, A. C. (1976). Action, knowledge and representation. In M. Brand & D. Walton (Eds.), *Action theory.* Dordrecht, The Netherlands: Reidel.

Davis, J., & Bitterman, M. E. (1971). Differential reinforcement of other behavior (DRO): A yoked-control comparison. *Journal of the Experimental Analysis of Behavior, 15,* 237–241.

Dawson, G. R., & Dickinson, A. (1990). Performance on ratio and interval schedules with matched reinforcement rates. *Quarterly Journal of Experimental Psychology, 42B,* 225–239.

de Villiers, P. A., & Herrnstein, R. J. (1976). Towards a law of response strength. *Psychological Review, 83,* 1131–1153.

Dickinson, A., & Balleine, B. (1990). Motivational control of instrumental performance following a shift from thirst to hunger. *Quarterly Journal of Experimental Psychology, 42B,* 413–431.

Dickinson, A., & Balleine, B. (1994). Motivational control of goal-directed action. *Animal Learning and Behavior, 22,* 1–18.

Dickinson, A., & Charnock, D. J. (1985). Contingency effects with maintained instrumental reinforcement. *Quarterly Journal of Experimental Psychology, 37B,* 397–416.

Dickinson, A., & Dawson, G. R. (1987). Pavlovian processes in the motivational control of instrumental performance. *Quarterly Journal of Experimental Psychology, 39B,* 201–213.

Dickinson, A., & Dawson, G. R. (1988). Motivational control of instrumental performance: The role of prior experience of the reinforcer. *Quarterly Journal of Experimental Psychology, 40B,* 113–134.

Dickinson, A., & Dawson, G. R. (1989). Incentive learning and the motivational control of instrumental performance. *Quarterly Journal of Experimental Psychology, 41B,* 99–112.

Dickinson, A., & Mulatero, C. W. (1989). Reinforcer specificity of the suppression of instrumental performance on a non-contingent schedule. *Behavioral Processes, 19*, 167–180.

Dickinson, A., Nicholas, D. J., & Adams, C. D. (1983). The effect of the instrumental training contingency on susceptibility to reinforcer devaluation. *Quarterly Journal of Experimental Psychology, 35B*, 35–51.

Dickinson, A., Watt, A., & Griffiths, W. J. H. (1992). Free-operant acquisition with delayed reinforcement. *Quarterly Journal of Experimental Psychology, 45B*, 241–258.

Garcia, J. (1989). Food for Tolman: Cognition and cathexis in concert. In T. Archer & L.-G. Nilsson (Eds.), *Aversion, avoidance and anxiety* (pp. 45–85). Hillsdale, NJ: Erlbaum.

Gormezano, I., & Tait, R. W. (1976). The Pavlovian analysis of instrumental conditioning. *Pavlovian Journal of Biological Sciences, 11*, 37–55.

Grindley, G. C. (1932). The formation of a simple habit in guinea pigs. *British Journal of Psychology, 23*, 127–147.

Guthrie, E. R. (1952). *The psychology of learning.* New York: Harper.

Hammond, L. J. (1980). The effects of contingencies upon appetitive conditioning of free-operant behavior. *Journal of the Experimental Analysis of Behavior, 34*, 297–304.

Hammond, L. J., & Weinberg, M. (1984). Signaling unearned reinforcers removes suppression produced by a zero correlation in an operant paradigm. *Animal Learning and Behavior, 12*, 371–374.

Herrnstein, R. J. (1969). Method and theory in the study of avoidance. *Psychological Review, 76*, 49–69.

Hershberger, W. A. (1986). An approach through the looking glass. *Animal Learning and Behavior, 14*, 443–451.

Heyes, C. M., & Dawson, G. R. (1990). A demonstration of observational learning in rats using a bidirectional control. *Quarterly Journal of Experimental Psychology, 42B*, 59–71.

Heyes, C. M., Dawson, G. R., & Nokes, T. (1992). Imitation in rats: Initial responding and transfer evidence. *Quarterly Journal of Experimental Psychology, 45B*, 229–240.

Holland, P. C. (1992). Occasion setting in Pavlovian conditioning. In D. L. Medin (Ed.), *The psychology of learning and motivation* (Vol. 28, pp. 69–125). San Diego: Academic Press.

Hull, C. L. (1943). *Principles of behavior.* New York: Appleton-Century-Crofts.

Killeen, P. R. (1982). Incentive theory. In D. J. Bernstein (Ed.), *Nebraska symposium on motivation: Vol. 29. Response structure and organization* (pp. 169–216). Lincoln: University of Nebraska Press.

Lattal, K. A., & Gleeson, S. (1990). Response acquisition with delayed reinforcement. *Journal of Experimental Psychology: Animal Behavior Processes, 16*, 27–39.

Lopez, M., Balleine, B., & Dickinson, A. (1992). Incentive learning and the motivational control of instrumental performance by thirst. *Animal Learning and Behavior, 20*, 322–328.

Lovibond, P. F. (1983). Facilitation of instrumental behavior by a Pavlovian appetitive conditioned stimulus. *Journal of Experimental Psychology: Animal Behavior Processes, 9*, 225–247.

MacCorquodale, K., & Meehl, P. E. (1954). Edward C. Tolman. In W. K. Estes et al. (Eds.), *Modern learning theory* (pp. 177–266). New York: Appleton-Century-Crofts.

Mackintosh, N. J. (1975). A theory of attention: Variations in the associability of stimuli with reinforcement. *Psychological Review, 82*, 276–298.

Mazur, J. E. (1983). Steady-state performance on fixed-, mixed- and random-ratio schedules. *Journal of the Experimental Analysis of Behavior, 39*, 293–307.

Miller, S., & Konorski, J. (1969). On a particular form of conditioned reflex. *Journal of the Experimental Analysis of Behavior, 12*, 187–189. (Translation by B. F. Skinner, Sur une forme particulière des reflex conditionnels. *Comptes Rendus des Seances de la Société Polonaise de Biologie, 1928, 49*, 1155–1157)

Moore, B. R. (1973). The role of directed Pavlovian reactions in simple instrumental learning in the pigeon. In R. A. Hinde & J. Stevenson-Hinde (Eds.), *Constraints on learning* (pp. 159–188). London: Academic Press.

Morgan, M. J., & Nicholas, D. J. (1979). Discrimination between reinforced action patterns in the rat. *Learning and Motivation, 10,* 1–22.

Mowrer, O. H. (1960a). *Learning theory and behavior.* New York: Wiley.

Mowrer, O. H. (1960b). *Learning theory and the symbolic processes.* New York: Wiley.

Pavlov, I. P. (1927). *Conditioned reflexes.* Oxford: Oxford University Press.

Pavlov, I. P. (1932). The reply of a physiologist to a psychologist. *Psychological Review, 39,* 91–127.

Pearce, J. M. (1987). A model of stimulus generalization for Pavlovian conditioning. *Psychological Review, 94,* 61–73.

Pearce, J. M., & Hall, G. (1980). A model for Pavlovian learning: Variations in the effectiveness of conditioned but not of unconditioned stimuli. *Psychological Review, 87,* 532–552.

Rescorla, R. A. (1990). Evidence for an association between the discriminative stimulus and the response-outcome association in instrumental learning. *Journal of Experimental Psychology: Animal Behavior Processes, 16,* 326–334.

Rescorla, R. A. (1994). A note on depression of instrumental responding after one trial of outcome devaluation. *Quarterly Journal of Experimental Psychology, 47B,* 27–37.

Rescorla, R. A., & Solomon, R. L. (1967). Two-process learning theory: Relationship between Pavlovian conditioning and instrumental learning. *Psychological Review, 74,* 151–182.

Rescorla, R. A., & Wagner, A. R. (1972). A theory of Pavlovian conditioning: Variations in the effectiveness of reinforcement and non-reinforcement. In A. H. Black & W. F. Prokasy (Eds.), *Classical conditioning II: Current research and theory* (pp. 64–99). New York: Appleton-Century-Crofts.

Seligman, M. E. P., & Johnston, J. C. (1973). A cognitive account of avoidance learning. In F. J. McGuigan & D. B. Lumsden (Eds.), *Contemporary approaches to conditioning and learning* (pp. 69–110). Washington, DC: Winston.

Shettleworth, S. J. (1975). Reinforcement and the organization of behavior in golden hamsters: Hunger, environment, and food reinforcement. *Journal of Experimental Psychology; Animal Behavior Processes, 1,* 56–87.

Sidman, M. (1966). Avoidance behavior. In W. K. Honig (Ed.), *Operant behavior: Areas of research and application* (pp. 448–498). New York: Appleton-Century-Crofts.

Skinner, B. F. (1932). On the rate of formation of a conditioned reflex. *Journal of General Psychology, 7,* 274–285.

Spence, K. W. (1956). *Behavior theory and conditioning.* New Haven, CT: Yale University Press.

Sutton, R. S., & Barto, A. G. (1981). An adaptive network that constructs and uses an internal model of its world. *Cognition and Brain Theory, 4,* 217–246.

Thistlewaite, D. (1951). A critical review of latent learning and related experiments. *Psychological Review, 48,* 97–129.

Thorndike, E. L. (1911). *Animal intelligence: Experimental studies.* New York: Macmillan.

Thorndike, E. L. (1931). *Human learning.* New York: Century.

Tolman, E. C. (1932). *Purposive behavior in animals and man.* New York: Century.

Tolman, E. C. (1959). Principles of purposive behavior. In S. Koch (Ed.), *Psychology: A study of a science* (Vol. 2, pp. 92–157). New York: McGraw-Hill.

Trapold, M. A., & Overmier, J. B. (1972). The second learning process in instrumental learning. In A. H. Black & W. F. Prokasy (Eds.), *Classical conditioning II: Current research and theory* (pp. 427–452). New York: Appleton-Century-Crofts.

Uhl, C. N. (1974). Response elimination in rats with schedules of omission training, including yoked and response-independent reinforcement comparisons. *Learning and Motivation, 5,* 511–531.

Williams, B. A. (1989). The effect of response contingency and reinforcement identity on response suppression by alternative reinforcement. *Learning and Motivation, 20,* 204–224.

CHAPTER 4

Reinforcement and Choice

Ben A. Williams

I. INTRODUCTION

The goal of an explanation of behavior in terms of the Law of Effect is the specification of the functional relationships between the strength of behavior and the various attributes of reinforcement. Such specification would seem to be a straightforward, and perhaps even pedestrian, task. In fact, however, these relationships have often proven to be extremely complex.

The difficulties of specifying how the most common measure of response strength, response rate, is related to the various parameters of reinforcement can be illustrated by an examination of the effects of magnitude of reinforcement on response rate maintained by two common schedules of reinforcement, variable interval (VI) and variable ratio (VR). With VI schedules reinforcers become available for the first response after a temporal interval has elapsed, and this temporal interval varies over some specified range (determined by the mean value of the schedule) between successive reinforcers. Because responses that occur before the prevailing interval has elapsed have no effect, reinforcement rate is independent of response rate as long as the animal responds occasionally to obtain any reinforcers that may have been scheduled. With VR schedules, reinforcers depend on a required number of responses, which also varies across successive reinforcers around some mean value. Here, reinforcement rate is directly determined by the

Animal Learning and Cognition

rate of responding. Fixed value versions of both schedules may also be used (FI and FR, respectively), but the advantage of the VI and VR is that they produce substantially more regular rates of responding.

The traditional assumption for both interval and ratio schedules is that response rate should be monotonically related to both the frequency and the amount of reinforcement. Contrary to this assumption, Reed (1991) found that response rate maintained by a VI schedule (with rats as subjects) decreased with larger numbers of food pellets as the reward, while rate maintained by a VR schedule increased with larger rewards. According to Reed's interpretation, the reason the different schedules produced different functional relations between response rate and reinforcement magnitude was that the larger reward facilitated the discrimination of the different molecular contingencies of reinforcement entailed by the different schedules. With VI schedules, longer interresponse times (IRTs) have a higher probability of reinforcement than shorter IRTs, because the longer the animal has waited since its last response, the more likely the interval scheduling the next reinforcer has elapsed. Discrimination of this temporal contingency should thus produce more long-IRT responses, and a lower response rate because of the inverse relation between IRT and rate. With VR schedules, there is no differential IRT contingency because reinforcement is dependent solely on the number of responses. Moreover, the reinforcement contingency for number of responses entails that the probability of reinforcement is higher for bursts of responding than for isolated responses. Greater discrimination of this contingency then causes more bursts to occur, which results in higher overall response rates. The different effects of reward magnitude on the behavior maintained by the different schedules was thus determined by the different schedules producing different response units. Incommensurate units of measurement vitiate differences in response rate as indicators that different schedules produce differences in response strength. Reed's analysis raises the additional concern that the units of measurement may also vary with the value of a single parameter of reinforcement, so that differences in response rates as a function of high versus low parameter values (e.g., in amount of food) may correspond poorly to the underlying differences in the strength of behavior.

The problem of incommensurate response units is only one of the obstacles confronting the specification of simple functional relations between the parameters of reinforcement and behavior. Much recent attention has been given to how reinforcement parameters interact with "economic principles," the most basic of which is how the reinforcement contingent on behavior is related to the total amount of reward consumption that is available to the subject. For example, in "open" economies, the total amount of food consumption is not dependent on the animal's behavior during an experimental session, in that the subject is fed additional amounts after the experimental session to maintain it at some fixed level of deprivation (e.g.,

80% of its free-feeding weight). In "closed" economies, all of the food is obtained during the experimental session. Considerable evidence has developed showing that the functional relations between response rate and reinforcement parameters can be radically different under these different regimens. For example, response rates in closed economies are often inversely related to the magnitude of the reward, while in open economies they are more often directly related (e.g., Collier, Johnson, Hill, & Kaufman, 1986). Such effects have been interpreted as demonstrating that response rate does not map onto the strength of behavior in a generally consistent fashion; instead rate measures are subordinate to how different patterns of behavior allow the subject to optimize its overall ecological strategy. However, this interpretation has also been disputed: other investigators have interpreted such effects as the result of differences in food deprivation, in that closed economy experiments typically have allowed the motivational level of their subjects to vary within experiment sessions, because the subject either lives continuously in the experimental chamber or is fed to satiation within experimental sessions (Timberlake & Peden, 1987; but also see Collier, Johnson, & Morgan, 1992, and Hall & Lattal, 1990).

While many of these issues remain unsettled, it is already evident that variation in the parameters of reinforcement may produce complex effects not easily understood from traditional perspectives that entail some simple mapping of response rate onto an underlying dimension of response strength. For some investigators, this complexity has inspired a search for alternative measures of response strength. An important example of such efforts is the development of the concept of "resistance to change" by Nevin and his colleagues (Nevin, 1979, 1992; Nevin, Mandell, & Atak, 1983) based on the analogy with physical concept of momentum. Behavior that is difficult to change, either in terms of increasing or decreasing its rate, is thus assumed to have a greater response strength than behavior that is easy to change. Unlike response rate itself, resistance-to-change measures generally correspond in an ordinal fashion to variation in the various parameters of reinforcement, although not without other complexities (e.g., the analysis of the partial reinforcement extinction effect: Nevin, 1988).

The present treatment will focus on an alternative approach, by examining how variations in reinforcement affect choice between alternative behaviors that are simultaneously available. As will be seen, effects on choice, like resistance to change but unlike simple response rate, do correspond in intuitively plausible ways to variations in reinforcement value.

II. CHOICE AS A MEASURE OF RESPONSE STRENGTH

The use of choice as a measure of the effects of reinforcement, rather than simple response rate, may seem paradoxical given the added degree of complexity that choice measures seem to entail. At least from a historical perspective, choice has been considered a derivative topic. Given knowledge of the associative strengths of the alternatives being chosen among, the outcome of the choice has been assumed to result from the comparison of those relative values, with the more fundamental issue being the determinants of the strength of each separate alternative (for a recent rendition of this perspective, see Skinner, 1986). Accordingly, an adequate choice theory consists of a set of rules by which individual responses are incremented/decremented in value, and then a rule for distributing the behavior given the set of response strengths for the available alternatives. Various modern treatments continue to be based on these general assumptions (Couvillon & Bitterman, 1985; Daly & Daly, 1982; Mazur, 1992).

It may be that choice can ultimately best be explained in terms of an analysis of the separate response strengths of the constituents of the choice situation. It should be evident from the preceding discussion, however, that "response strength" is an intervening variable, which is not currently connected in any clear way to simple measures of behavior. Thus, the outcome of choice procedures provides an alternative index of the underlying dimension of response strength, which turns out empirically to behave in a more reasonable fashion than measures of responding of individual responses considered in isolation. There is also an additional reason for believing that choice measures provide a better measure of the effects of reinforcement. As argued by Hernstein (1970), all behavior is choice, in the sense that there are always alternatives other than the response measured by the experimenter. Thus, the animal is always "deciding" which response to perform, so that supposed measures of "absolute strength" of a given response studied in isolation (e.g., speed in a straight alley) are in fact the result of choice dynamics. The implication is that any understanding of response rates of individual responses depends critically on the mechanisms of choice that may not be easily evident from the analysis of simple conditioning, because the alternatives competing with the response measured by the experimenter are not under direct experimental control.

Research on choice in the past three decades has been dominated by free-operant procedures involving concurrent schedules of reinforcement. With this procedure animals choose between two alternatives that are continuously available, with each alternative correlated with its own independent reinforcement schedule. The impetus for the large amount of research with this procedure was the discovery of the "matching law" (Hernstein, 1961). Pigeons were presented two simultaneously available response keys, each reinforced according to independent VI schedules. The sum of the reinforcement rates for the two schedules was held constant at 40/hour, while the number allocated to one key or the other was systematically varied across experimental conditions. The results revealed a remarkably orderly relation between the choice measure and the reinforcement rates: the proportion of the total behavior to either key was approximately equal to the proportion of the total reinforcement allocated to that key. The matching relation is captured algebraically by Equation (1), in which B_1 and B_2 represent the response rates for the two behaviors, and R_1 and R_2 represent the corresponding reinforcement rates.

$$B_1/(B_1 + B_2) = R_1/(R_1 + R_2) \tag{1}$$

It is important to recognize that the matching relation is not generally constrained by the reinforcement schedules, per se. Because VI schedules entail a weak correlation between the rate of responding and the obtained rate of reinforcement, many different response patterns would have produced similar distributions of obtained reinforcement. The fact that only one particular distribution of behavior occurred out of all of those possible constitutes a significant discovery.

III. RELATIVE VALUE AS THE BASIS OF THE MATCHING LAW

Although the matching relation was first established with food-deprived pigeons as subjects, its generality has been confirmed with many other species (e.g., humans, fish, monkeys, rats) and types of reinforcers (e.g., money, brain stimulation, cocaine, verbal approval, water). These data have been reviewed elsewhere in more extensive treatments (Davison & McCarthy, 1988; deVilliers, 1977; Williams, 1988). More important theoretically is the extension of the matching law to parameters of reinforcement other than rate of reinforcement. The most important claim that has been advanced is that matching of behavioral allocation occurs with respect to the relative "value" of each choice alternative, with different reinforcement parameters all converging in their effects on the value dimension (Rachlin, 1971). Given this assumption, all of the major parameters of reinforcement, when appropriately scaled, should produce matching when studied in isolation. Equations (2)–(4) depict these relationships in the simplest possible way for the separate parameters of rate (R), amount (A), and delay (D) of reinforcement (or more precisely, the inverse of delay, immediacy of reinforcement):

$$B_1/B_2 = R_1/R_2 \tag{2}$$

$$B_1/B_2 = A_1/A_2 \tag{3}$$

$$B_1/B_2 = (1/D_1)/(1/D_2) \tag{4}$$

The most general version of the matching law combines these different parameters into a single expression that corresponds to the "relative value" of the two alternative behaviors, as given by Equation (5).

$$B_1/B_2 = R_1/R_2 \times A_1/A_2 \times (1/D_1)/(1/D_2) = V_1/V_2 \tag{5}$$

The reason for giving a formal expression of the most general form of the matching law is that it makes evident several substantive assumptions that warrant separate evaluation. Not only does Equation (5) assume that matching applies with respect to each individual parameter of reinforcement, it also assumes that all reinforcement parameters are fundamentally similar in the manner in which they exert their effects. This assumption is in contrast to the common intuition that some parameters, such as delay, depend on associative factors, while others, such as amount, depend on motivational dynamics. The notion of an underlying dimension of value entails that the effects of different parameters of reinforcement may complement or compete against each other in specified ways, an implication that has sponsored several important theoretical extrapolations to topics outside the animal laboratory, such as self-control (see Rachlin & Green, 1972), and various "irrational" features of human decision making (see Rachlin, Logue, Gibbon, & Frankel, 1986).

Not only does Equation (5) require that the different parameters of reinforcement combine their effects multiplicatively, it further assumes that the effects of a given parameter on choice depend on its relative value with respect to each choice alternative, independent of the absolute value of the parameter and independent of the absolute value of other parameters with which it is combined. Some assessment is needed of the validity of these assumptions, especially given that some learning theories have assumed that differences in the values of response strength, not their ratios, determine choice allocation (e.g., Daly & Daly, 1982; Spence, 1936).

Whether matching applies to each separate parameter of reinforcement when studied in isolation is the issue most extensively studied. Reviews of this literature have been provided elsewhere (Davison & McCarthy, 1988; de Villiers, 1977; Williams, 1988), so we will not consider the issue in detail here. A large number of studies have been reported showing deviations from matching with all of these parameters, but many of these can clearly be explained away by nonoptimal procedural characteristics. For example, "undermatching" with respect to relative rate of reinforcement (which means that the choice proportions are less extreme than the reinforcement proportions) has often been reported. To capture such deviations, the "generalized matching law" has been employed (see Baum, 1974), which adds the additional parameters of bias (b) and sensitivity (s) to the basic matching expression, as shown in Equation (6). Undermatching implies that the value of s is less than 1.0, most often in the range 0.8–0.9.

$$B_1/B_2 = b(R_1/R_2)^s \qquad (6)$$

We now know that the likelihood of obtaining undermatching is affected by the method of constructing the distribution of interreinforcement intervals constituting the different VI schedules. Two types of distributions have been used: arithmetic distributions in which the short and long intervals are symmetrically distributed around the mean schedule value, and exponential distributions, such that the distribution of intervals is skewed toward longer durations. When the literature is divided according to which type of schedule was used (Taylor & Davison, 1983), the result is that close approximations to perfect matching have occurred with exponential schedules, while substantial undermatching has occurred with arithmetic schedules.

While this finding may at first seem only to reveal the abstruse nature of research on schedules of reinforcement, it in fact demonstrates a fundamental constraint on interpreting molar choice proportions. The important difference between arithmetic and exponential distributions is that the latter is random with respect to time, so that the probability of a reinforcer being scheduled is independent of the time of occurrence of the preceding reinforcer. With arithmetic schedules, in contrast, different time periods after the preceding reinforcer are associated with different probabilities of reinforcement. To the extent the subject discriminates between these temporal intervals, a VI schedule composed of an arithmetic distribution is in effect a multiple schedule in which different time periods of the schedule are correlated with different rates of reinforcement. The molar choice proportion is then the result of summing over the different components of the schedule. The result is that perfect matching could apply to each component of the schedule independently, but then produce undermatching with respect to the total behavior in the situation simply because of the error inherent in the mechanics of averaging. A detailed analysis of why this is true has been presented by Baum (1979) and Williams (1988). The implication is that perfect matching is necessarily an idealization, applying only when the conditions of reinforcement are homogeneous over the entire time period over which the choice proportions are calculated.

Extensive analysis of the various procedural factors that cause deviations from matching can be found in the previous reviews cited above. For present purposes, we will note only the major conclusions. The matching law clearly does apply to rate of reinforcement after making allowance for various procedural artifacts. It also describes the results of many studies of amount of reinforcement, although deviations from matching have been more frequent with amount than with rate as the independent variable. This may be due in part to procedural difficulties in presenting different amounts unconfounded with differences in delay (i.e., the last portion of a large amount of food is necessarily more delayed in time from the response than the first portion of the same amount). But there also appear to be other factors involved, notably that the effects of relative amount seem more sensitive to the buildup of proactive interference effects from presenting several different conditions in succession to the same subject (Keller & Gollub, 1977).

Unlike the effects of rate and amount of reinforcement, there are major difficulties for the description of delay of reinforcement, at least in terms of the form specified by Equation (4). Because the response-reinforcer temporal relation has historically been considered the critical feature of conditioning, it is of interest to consider the effects of relative delay value in some detail.

The matching relation with respect to the relative immediacy of the two alternatives was first advanced by Chung and Herrnstein (1967). Pigeons were presented a concurrent schedule in which the two responses were rewarded according to equal VI schedules but with the reinforcers delayed for varying times. One alternative had a constant delay (for most conditions 8 s), while the other had both shorter and longer delays in successive experimental conditions. The same stimulus conditions occurred during all delays, as all lights in the chamber were off. Although the analysis by Chung and Herrnstein (1967) suggested that Equation (4) provided a good description of their results, re-analysis of their data by Williams and Fantino (1978), plus a replication of their procedure, revealed that choice deviated systematically away from the matching relation as a function of the absolute size of the delay that varied across conditions. In general, greater absolute delay values produced deviations in the direction of "overmatching," meaning that choice proportions were more extreme than the simple ratio of delay values (in terms of Equation (6), producing values of s substantially greater than 1.0).

It is difficult to know at this point how fundamental is the just-noted violation of Equation (4). The reason is that other evidence in favor of matching of relative immediacy has been obtained by Mazur (1984) using a discrete-trial rather than free-operant procedure. On each trial, a choice was defined by a single response. Forced-choice trials with each delay value were interspersed with free-choice trials with both alternatives in order to ensure experience with the delay values operative on any given trial for each alternative. Most important, a titration procedure was used, such that one alternative was associated with a constant standard delay while the other alternative had its delay value increased or decreased across trials as a function of the last series of choices. When the standard delay condition was a set of different delays (e.g, 2s and 18s with equal probability), the titrating delay was then adjusted to produce an indifference level of 50% in the choice allocation. By using a series of different delay intervals as the standard in different experimental conditions, Mazur was able to determine the form of the function that must apply to predict the obtained indifference levels. That function is given by Equation (7), in which V refers to the value of a choice alternative, A to the value of the reinforcer at a zero delay, and p to the probability that a given delay will occur out of the set of delays contingent on a response.

$$V = \Sigma \, p_j[A/(1 + KD_j)] \tag{7}$$

The only free parameter is K, which defines the sensitivity of the subject to delay, with high values of K corresponding to rapid drops in value as delay increases, and low values of K with slower changes in value. In most cases, an adequate fit by Equation (7) to choice data (where the values of K would be equal for the two alternatives) was possible with the value of K set to 1.0. The important observation for present purposes is that Equation (7) assumes that the value of a response alternative is an inverse function of the delay of reinforcement contingent on that behavior, a conclusion inconsistent with the free-operant results of Williams and Fantino (1978) noted above. To date, there is no clear resolution of the basis of the disparity in outcomes of the different procedures.

A perhaps related difficulty regarding the effects of delay of reinforcement involves its interaction with the effects of amount of reinforcement. As specified by Equation (5), their respective effects have been assumed to be independent and multiplicative, and, as noted above, this assumption has served as the basis for important theoretical extrapolations to the problem of "self control." Unfortunately, considerable evidence, from the concurrent chains procedure, indicates that the effects of delay and amount are not independent. With this procedure, the subject chooses between the initial links of two independent chain schedules, with the consequence being access to the terminal link of whichever chain is chosen. Different parameters of reinforcement can then be varied in the terminal links of the two schedules, which allows a determination of how the different parameters combine to determine the relative value of the two terminal links.

Ito (1985) presented rats a concurrent chains procedure in which the initial links were each VI 30 s. The corresponding terminal links were equal-valued FI schedules, which were 5 s in one experiment and FI 20 s in a second experiment. In both studies, one choice alternative led to a single food pellet at the end of the FI schedule, while the other choice response led to a larger number of reinforcers, which varied form one to five pellets across experimental conditions. The issue was whether the choice proportions during the initial link would match the relative number of pellets contingent on the two responses. Contrary to Equation (5), the effect of the differing amounts varied strongly with the absolute value of the FI schedules. With the FI 5-s schedule, substantial undermatching occurred; with the FI 20-s schedule, substantial overmatching occurred. Similar results with pigeon subjects have been reported by White and Pipe (1987).

In contrast to the interactions between reinforcement parameters noted in the above studies, Mazur (1987) has made a strong case for the simple multiplicative rule given in Equation (5). Using the delay titration procedure described above, his standard alternative was associated with a fixed amount of reinforcement with a fixed delay. The titrating alternative was associated with a different fixed amount of reinforcement and its delay was determined by whatever value was needed to produce indifference between the two combinations of delay and amount. For example, a typical pigeon was indifferent between 2 s of food delivered after a 6-s delay and 6 s of food delivered after a 17-s delay. By then using different delay values for the fixed-value alternative while keeping constant the different amounts for the two alternatives, Mazur determined how much the large–reinforcer delay had to be increased to maintain indifference. The resulting plots of large-reinforcer delays as a function of equally preferred small-reinforcer delays was linear in all cases, indicating that the interaction between amount and delay was multiplicative. It should be noted that Mazur used only one pair of differential amounts, so that it is uncertain how the slopes of the functions would have been changed by variations in the absolute size of the amount variable.

The previous discussion makes clear that the validity of the assumptions underlying the strongest version of the matching law, Equation (5), is still uncertain. With the standard free-operant choice procedure, the absolute values of the reinforcement parameters appear to play an important role in determining the approximation to matching, both when studied in isolation and when in combination with other parameters. Some of these deviations from matching may reflect procedural artifacts, such as how the absolute parameters change the degree of discrimination of the local contingencies. Alternatively, the deviations may reflect scaling problems, in that there is no necessary reason that the psychological scale for different delays or amounts corresponds to the physical scale, and it is the psychological scale to which matching occurs. But it is also possible that the problem is more fundamental, as the combination rule for different reinforcement parameters may be something other than multiplicative. Future research will be needed to decide between these alternatives.

IV. THEORIES OF MATCHING

Quite apart from its importance for the understanding of the relationships between the different parameters of reinforcement, the matching law has served as a rich empirical framework for the development of different theories of choice. Two interrelated issues have been contested: the choice rule by which behavior is allocated between alternatives, and the measure of value on which the choice rule is applied. In principle these issues are separable, but in practice they have been totally intertwined.

These theoretical issues arise because the matching law is defined in terms of molar measures of behavioral allocation, typically the total response output from an entire session. The matching law says nothing directly about the individual responses constituting the molar measures. Given that choice obeys the matching law in the aggregate, what is controlling the individual pecks or lever presses?

The most common approach has been to assume that the best estimates of the local response probabilities are their molar probabilities. Matching is then the automatic result of each alternative response occurring in proportion to its own strength without further specification (e.g., Catania, 1973; Herrnstein, 1970; Myerson & Miezen, 1980). This notion is conceptually consistent with Skinner's concept of "emitted behavior" in that the different response alternatives are analogous to different sources of radioactivity, where the rate of emission of particles from the different sources is a function only of the current "strengths" of the alternative sources.

A second approach assumes that each individual response is determined by the contingencies operating at that moment, so that matching in the aggregate is the result of adjustments to changing contingencies. It is then the nature of these adjustments, rather than matching itself, that reveals the fundamental principles of behavior (e.g., Herrnstein & Vaughan, 1980; Hinson & Staddon, 1983; Shimp, 1966).

Still a third approach is derived from optimality theory, which assumes that matching is the behavioral distribution that maximizes the total reward in the situation (e.g., Baum, 1981; Rachlin, Battalio, Kagel, & Green, 1981; Staddon & Motheral, 1978). This approach makes no attempt to deal with the molecular processes generating the molar maximizing, but instead provides a normative description of the optimal pattern of behavior, given some set of values for the various commodities/activities available in the environment. For some investigators, this assumption is axiomatic, so that the task of the experimenter is to discover the value structure of the subject that corresponds to optimization, given the particular distributions of responding that are observed.

A review of the previous literature pertaining to these different theoretical approaches has been provided by Williams (1988), so we will concentrate our attention on recent findings that add substantively to that previous review.

A. Optimality Theory

Perhaps the most frequently used procedure for testing the predictions of optimality theory has been the concurrent VI VR schedule. Because the reinforcers from the VI alternative are held until obtained by the next response, while the reinforcers from the VR schedule are directly proportional to the number of responses to that alternative, the optimal strategy with such a schedule is for the subject to spend most of its effort on the VR alternative and occasionally sample the VI alternative to obtain any reinforcers that may have been scheduled while the subject was working on the VR. In other words, time allocation to the VR alternative not only earns VR reinforcers directly, it also makes available the VI reinforcers as well, which then can be obtained with a single response to the VI alternative. VI responses, by comparison, make available and earn only the VI reinforcers. Thus, optimality theory predicts a significant bias toward the VR schedules given equal rates of reinforcement for the the VR and VI alternatives.

Previous results with the concurrent VI VR schedule generally have not agreed with the predictions of optimality theory. In general, excellent approximations to matching have been obtained, in that fits of the results by Equation (6) have produced values of the sensitivity parameter, s, very near 1.0. Most important, the bias toward the VR alternative required by optimal performance has not been found, as the fits by Equation (7) have produced values of b also near 1.0 (DeCarlo, 1985; Herrnstein & Heyman, 1979; Heyman & Herrnstein, 1986; Vyse & Belke, 1992; Williams, 1985). One exception to this pattern of results was reported by Green, Rachlin and Hanson (1983), who did obtain a significant bias toward the VR schedule. The critical difference between their study and the others appears to be that Green et al. did not use a change-over-delay (COD), which prevents alternation between the schedules from being immediately reinforced and is generally regarded as necessary for matching to be obtained. On the basis of the apparently critical role of the COD, Rachlin, Green, and Tormey (1988) have argued that optimality theory can be salvaged if it is assumed that the reinforcements from the VI alternative that accumulate while the subject is working on the VR are devalued because they are temporally separated from the VR behavior that actually makes them available. Thus, while subjects are not optimizing with respect to the maximum number of reinforcers that are delivered in the course of a session, they may be optimizing in terms of the rescaled units of value that are produced when the temporal discounting is included.

The role of temporal separation between choices was explicitly investigated by Williams (1992), who presented rats a discrete-trial version of a concurrent VI VR schedule. For the VR alternative the probability of reinforcement was constant across trials; for the VI alternative, a different probability schedule was used, which held any scheduled reinforcers until they were obtained on the subsequent trials. This latter procedure mimics the contingencies of a free-operant VI schedule, in which the probability of reinforcement for a VI choice increases the greater the time (or number of trials) since the last choice of the VI alternative. Different pairs of probabilities were then used across different conditions of the experiment.

The major variable of interest was the time between successive trials. In one condition, a short (5 s) intertrial interval (ITI) was used; for the second condition a longer ITI (30 s) was employed. The results were that a substantial bias toward the VR alternative occurred with the short ITI, while a much smaller (and nonsignificant) bias occurred with the long ITI. The results were thus consistent with the analysis provided by Rachlin et al. (1988), in that a significant bias toward the VR occurred only when the VI reinforcers were sufficiently close in time to the preceding VR choices. In addition, the approximation to matching was significantly better under the long ITI condition.

The best explanation of the results of Williams (1992) appears to be in terms of the effects of delayed reinforcement. Given that VI responses were more likely to be reinforced after a preceding choice of the VR alternative than after a choice of the VI alternative, while the occurrence of VR reinforcers was equally probable after the two types of choices, the delayed effects of the VI reinforcers should differentially strengthen the VR choice on the preceding trial over and above the strengthening effects of the reinforcers immediately contingent on each response. This additional delayed reinforcement would then provide the basis for the bias toward the VR response with the short ITI, but be absent with the longer ITI when the delay exceeded the limits of the delay-of-reinforcement gradient. Accordingly, deviations from matching were greater with the short ITI, because the delayed reinforcement effects that were influencing choice were not included in the calculation of the relative reinforcement rate determining the response allocation.

Support for this analysis comes from the related results of Mazur and Vaughan (1987), using a concurrent fixed-ratio progressive ratio schedule, which, like the concurrent VI VR schedule, frequently has been used as a method for testing the predictions of optimality theory (e.g., Lea, 1981). For the FR alternative, a fixed number of responses is required to produce reinforcement on a given trial. For the progressive-ratio alternative (PR) the number of responses is initially small, and then increases by some constant increment after each successive choice of the PR. For example, the initial value might be 1 response, and then increase by increments of 10 thereafter. The critical feature of the procedure is that the value of PR schedule resets to its initial level after each choice of the FR alternative. The issue has been whether the subject drives the PR value up to a level equal to the FR value before switching, or whether it switches to the FR before the PR reaches the FR value. Optimal performance requires the subject to switch to the FR substantially before the PR value reaches the FR level. For example, with a FR value of 41, an initial PR value of 1, and increments in the PR value of 10, the subject should switch to the FR schedule after completing the PR 11, rather than continuing on the PR schedule to PR 41. Thus, optimal performance requires that the animal switch to the FR alternative on a trial in which the next reinforcer will occur more quickly on the PR rather than the FR alternative.

Previous studies with primates had shown that near-optimal performance does in fact occur with this procedure (Hineline & Sodetz, 1987; Hodos & Trumbule, 1967). To understand the dynamics of this behavior, Mazur and Vaughan (1987) presented pigeons a choice between an FR 81 schedule and a PR schedule that began with a value of 1 and then increased by 10 after each successive PR choice. In one set of conditions successive choices were allowed to occur without an intervening ITI; in a second set of conditions an ITI of 25 or 50 s was used. The dependent variable was the mean value of the last PR requirement completed before the switch to the FR alternative. Optimal performance required a reset ratio value of 21. The obtained reset ratio in the absence of the ITI was 46.2, indicating that while optimal performance was not obtained, the pigeons nevertheless switched their preference to the FR substantially before the PR value reached the FR value. When the 50-sec ITI was used, the mean value was between 75 and 80, close to the FR value. Mazur and Vaughan further demonstrated that their entire pattern of results could be described by Equation (7), listed above, when it was assumed that the variable controlling the strength of the FR choice was not only the reinforcer immediately contingent on the FR choice, but also the next PR reinforcer that followed it. When no ITI was used, the next PR reinforcer occurred after a single response, and hence was close in time to the FR choice on the preceding trial. When the ITI was employed, this delay was sufficiently large to eliminate any meaningful impact of the PR reinforcer on choice of the FR alternative, which resulted in the abolition of any approximation to optimal performance. Combined with the results from the concurrent VI VR schedules described above, the results of Mazur and Vaughan make a strong case that putative optimality effects can be reduced to much simpler reinforcement mechanisms. Thus, rather than optimality theory being an explanatory framework for understanding choice, approximations to optimal performance seem better explained as the result of the operation of contingencies of delayed reinforcement.

This does not mean that optimality theories of behavior are incorrect, since, as noted above, optimality can be considered axiomatic, with the task of the experimenter being to determine the units of value that produced the obtained distributions of behavior. For example, Staddon (1992) has argued that much of the data taken to be contrary to optimality theory can be encompassed if it is assumed that the quantity being maximized is the average delay to reward, rather than the total quantity of reward itself. Given that the dimensions of value determining optimal performance map directly onto long-established principles of reinforcement that determine response strength, such as delay of reinforcement, any clear distinction between the two approaches ceases to exist.

B. Momentary Maximizing

This theoretical approach assumes that molar measures of choice are aggregates of responses that individually are determined by the subject choosing the alternative with the highest momentary reinforcement probability (Hinson & Staddon, 1983; Shimp, 1984; Silberberg, Hamilton, Ziriax, & Casey, 1978). Despite this all–or–none decision rule, matching occurs because of the nature of the changes in reinforcement probability as a function of time since the last response for concurrent VI VI schedules. Because the probability of reinforcement for a given alternative continuously increases the longer the time since the last response to that alternative, at some point the local probability of reinforcement for the schedule with the lower average value will equal or exceed the local probability of reinforcement for the schedule with the higher average value. It can be shown analytically that a momentary maximizing strategy implies a particular response sequence. For example, if the schedule for the right key is VI 1 min, while that for the left key is VI 3 min, the sequence that follows the maximizing strategy is RRRLRRRL, assuming constant and equal interresponse times. The result is that the molar distribution of responding matches the distribution of reinforcement, but only because the subject's behavior tracks the dynamic changes in reinforcement probability.

A critical feature of momentary maximizing theory is that performance depends on the subject's memory of the preceding response(s). The implication is that variables that facilitate memory should produce greater adherence to the optimal response strategy, and thus a closer approximation to matching. The most obvious variable that should affect the degree of memory is the time between trials. When explicit response strategies such as "win–shift, lose–stay" are differentially reinforced, adherence to those strategies does in fact decrease with longer ITIs (Shimp, 1976; Williams, 1991, Experiment 1). The question, therefore, is how changes in the efficacy of the memory of the preceding response correspond to changes in the approximation to the matching law.

Williams (1991, Experiment 2) presented to rats a discrete-trial analogue to concurrent VI VI schedules by arranging differential reinforcement probabilities for repeating a response versus switching to the alternative, with both types of probabilities different for the two alternatives. For example, in one condition, the probability of reinforcement for a switch from the high-valued alternative to the low-valued alternative was 0.25, while the probability of reinforcement for a stay on the high-valued alternative was 0.2; the probability of reinforcement for a switch to the high-valued alternative from the low-valued alternative was 0.6 while the probability of reinforcement for a stay on the low-valued alternative was 0.1. It can readily be seen that such contingencies imply an optimal strategy of alternating between responses on each trial. Unlike the concurrent VI VI schedule, however, this strategy would not produce matching because the resulting response ratio would be 1:1, while the reinforcement ratio would be 7:3. Thus, the local contingencies of reinforcement, which momentary maximizing theory assumes to be fundamental, were pitted against the molar reinforcement contingencies. At issue was which type of contingency would dominate the behavior. The further issue was how variation in the ITI would affect this trade-off.

Figure 1 shows the degree of control by the local contingencies from the condition just described, by depicting the probability of a choice of the lever with the highest overall reward probability (referred to as the majority lever in the legend) as a function of the different types of preceding trial. Perfect adherence to a momentary maximizing strategy would have produced a value of zero after either a reinforced or nonreinforced choice of the majority lever, and a value of 1.0 after either a reinforced or nonreinforced choice of the minority lever, as the highest probability of reward was always for an alternation away from the lever chosen on the last trial. Although far from complete, considerable adherence to this strategy did occur with the 5-s ITI: choice of the majority lever was substantially reduced when a choice of that lever had occurred on the previous trial, although this pattern was complicated by the tendency to perseverate to that lever when the preceding choice had been reinforced. With the 30-s ITI, in contrast, little control by the nature of the preceding trial is evident, as the majority lever was chosen

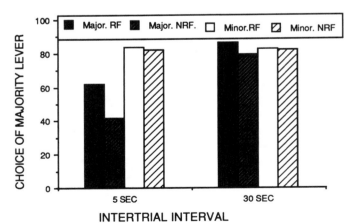

FIGURE 1   Data from Williams (1991) showing the percentage of choices of the lever associated with the higher molar reinforcement probability (referred to as the "majority" lever), subdivided according to the location and outcome of the preceding trial. Shown on the left are the means of eight subjects trained with a 5-s ITI; on the right are the mean results when the ITI was 30 s. RF, Reinforced; NRF, nonreinforced.

approximately equally across all types of trials. Thus, longer ITIs did abolish control by the memory of the preceding trials. The critical issue then becomes how the inability to conform to the momentary maximizing strategy affected conformity to the matching law. This was assessed by fitting Equation (6) to the results of a series of conditions with different local probabilities that produced different overall reinforcement rates for the different levers. The outcome was that adherence to matching was substantially greater with the 30-s ITI, as the fits of Equation (6) produced values of the sensitivity parameter, s, of 0.56 with the 5-s ITI (which would have been zero if complete control by the local contingencies had occurred) but 0.84 with the 30-s ITI. The results thus demonstrate a direct competition between the local and molar contingencies: when a strong memory of the preceding trial is available, local contingencies may play an important role and may decrease or increase the adherence to the matching of the molar distribution of reinforcement, depending on whether the local and molar contingencies are congruent or incongruent. The critical test is what occurs when no such memory is available, since in its absence momentary maximizing theory has no basis for predicting that matching to the molar reinforcement rates should occur. As shown by Williams (1991), however, the longer ITI, which independently had been shown to greatly decrease the memory of the preceding trials, did result in an approximate adherence to matching.

Further evidence against momentary maximizing theory was provided by the discrete-trial analogue of concurrent VI VR of Williams (1992) that was described above. The local contingencies of a concurrent VI VR schedule are somewhat simpler than those of a concurrent VI VI, because in the former, the reinforcement probability of only one of the choice alternatives is varying across trials, so that the pattern of behavior that follows a momentary-maximizing strategy should be learned more easily. Consider, for example, the case for the discrete-trial schedule when the VR probability of reinforcement is 0.20, whereas the VI probability is 0.10. Assuming the choice on the preceding trial was to the VI alternative, the higher probability of reinforcement would be for the VR alternative (0.20 vs. 0.10). If the last VI choice were two trials earlier, which means there were two opportunities rather than one opportunity for the VI reinforcer to be scheduled, the probability of reinforcement is only slightly in favor of the VR (0.20 vs. 0.19). With three trials since the last VI response, the probability of reinforcement favors the VI alternative (0.20 vs. 0.27), with the difference favoring the VI becoming progressively larger with each successive trial intervening since the last VI response (0.34, 0.41, 0.47 for 4, 5, and 6 trials, respectively). Thus, the strategy predicted by momentary maximizing would be choice of the VR alternative on the first trial after the last VI response, approximate indifference on the second trial, and clear preference for the VI alternative thereafter. The actual results obtained were that some control by these local contingencies did occur, in that the probability of a VI response did increase with the time since the last VI response in the majority of conditions with the 5-s ITI, although the degree of this sensitivity was far below that predicted by the theory. Most important, little evidence of control by the local probabilities occurred with the 30-s ITI, and it was the 30-s ITI that produced the best approximation to matching (s from Equation (6) was 0.84). Once again, therefore, the results indicated that the local contingencies may play some role when the subject remembers the preceding set of trials, but nevertheless cannot be regarded as the basis of matching. Matching itself seems independent of such local contingencies because it is better approximated whenever the memory of the preceding trials, and hence the role of the local contingencies, is abolished.

C. Melioration

A different type of molecular theory of matching is "melioration" (Herrnstein, 1982; Herrnstein & Vaughan, 1980; Vaughan, 1985; also see Rachlin, 1973), which, like molecular maximizing, assumes that choice responses are determined by local reinforcement contingencies that change continuously throughout a session. Unlike momentary maximizing, the local probabilities are due not to fluctuating stimulus states defined by the set of responses during the preceding moments, but rather to how the local rate of payoff for a given alternative changes with the proportion of time devoted to that alternative. To see how the melioration process is assumed to work, consider a concurrent VI 1-min VI 2-min schedule with behavior equally distributed to each alternative at the start of training. Assuming that all scheduled reinforcers are obtained (which would occur as long as the animal occasionally samples both alternatives), the local rate of reinforcement is defined by the scheduled rate of reinforcement divided by the proportion of the total time allotted to that schedule. For the VI 1-min schedule, the scheduled rate is 60 reinforcers/hour. If the animal allots 1/2 of its time to that schedule at the start of training, initial local rate of reinforcement would be 120 reinforcers/hour (60 reinforcers/0.5 hour). For the VI 2-min alternative, the scheduled rate is 30 reinforcers/hour, so an allocation of 1/2 of the time to that schedule would produce a local rate of 60 reinforcers/hour (30 reinforcers/0.5 hour). Behavior would thus shift to the VI 1-min alternative. Now consider the same schedule but with 90% of the time allocated to the VI 1-min alternative. The local rate of reinforcement for the VI 1-min alternative would be 66.7 reinforcers/hour (60/0.9 hour) while that for the VI 2-min alternative would be 300 reinforcers/hour (30 reinforcers/0.1 hour). Behavior would then shift toward the VI 2-min alternative. Equilibrium between the two types of shifts occurs only when the local rates of reinforcement are equal, which is uniquely defined by the point at which matching occurs (in this case when 2/3 of the time is allocated to the VI 1-min alternative, when obtained local reinforcement rates would be 90 reinforcers/hour for both alternatives). The critical feature of this account is that choice is controlled by the obtained local reinforcement rates, not the overall reinforcement distribution, per se.

Williams and Royalty (1989) tested melioration theory by determining whether choice was controlled by the scheduled reinforcement rates (which is equivalent to the number of reinforcers delivered over an entire session) or by the local reinforcement rates. Pigeons were trained on a multiple schedule in which different concurrent schedules occurred in each component. In component A, choice was between VI 20-s and VI 120-s schedules; in component B, choice was between VI 60-s and VI 80-s schedules. Training continued until matching of relative response rate to relative reinforcement rate occurred in both separate components. As in the example presented above, matching approximately equalized the local reinforcement rates for the two alternatives in each concurrent schedule. Thus, both the VI 20-s and VI 120-s alternatives were associated with local rates of reinforcement of approximately 210 reinforcers/hour, while both the VI 60-s and VI 80-s alternatives were associated with local rates of reinforcement of approximately 105 reinforcers/hour. Then, during probe trials, the subjects chose between the stimuli correlated with the VI 60-s and VI 120-s schedules. Whereas the scheduled rate of reinforcement (and the absolute number of reinforcers delivered for each alternative) favored the VI 60-s alternative, the obtained local reinforcement rate favored the VI 120-s alternative. The scheduled rate of reinforcement was the controlling variable: preference during the probe trials was in favor of the VI 60-s alternative. The same general pattern of results, using both free-operant and discrete-trial procedures, was replicated by Williams (1993) and Belke (1992). Apparently, choice is not controlled directly by the local rate of reinforcement.

It should be recognized that the failure of local rate of reinforcement as the controlling variable has ramifications that extend beyond the evaluation of melioration theory. Local rate of reinforcement is equivalent to probability of reinforcement when the local rates of responding are equal for the different choice alternatives, as they typically are in concurrent VI VI schedules when choice has reached stability (see Williams, 1988, for a discussion), and is equivalent to obtained probability of reinforcement in discrete-trial procedures in which a single choice is allowed on each trial, such as those used by Williams and Royalty (1989, Experiments 2 and 3) and Williams (1993). This equivalence is noteworthy because it implies that melioration theory is nothing more than the hypothesis that the subject chooses the response alternative with the highest obtained probability of reinforcement. Accordingly, matching does not reflect some higher order principle but is the result of the simplest possible conception of how behavior is controlled by reinforcement.

A major puzzle posed by the results of Williams and Royalty (1989) is how the failure of obtained probability (or local rate) of reinforcement to control behavior in their choice procedures is to be reconciled with other evidence showing it to be the controlling variable when stimuli are presented in isolation. For example, Graf, Bullock, and Bitterman (1964) presented pigeons with a discrete-trial, single-key procedure in which two stimuli were individually presented while response latency to each was measured. Across several manipulations, probability of reinforcement (i.e., responses/reinforcer) was the major determinant of response latency despite large differences in the absolute number of reinforcers. Similar results were obtained by Vom Saal (1972), who used choice between the different stimuli after their separate training, rather than response latency, as the dependent variable.

To better understand the differences between the results after single-stimulus pretraining and the findings of Williams and Royalty (1989) described above, Williams (1993) compared a choice procedure similar to that used by Williams and Royalty (1989) with a second procedure in which only a single response alternative was present on a given trial, but with the frequency of the different stimuli, and the probability of reinforcement associated with them, determined by those obtained by the choice subjects. Thus, the choice subjects chose between two responses with reinforcement probabilities of 0.20 versus 0.05, with reinforcers from both schedules held until obtained, as with concurrent VI VI procedures. The second component of the schedule presented a single response alternative with a reinforcement probability of 0.10. During the choice component, the mean choice proportion of the 0.20 alternative was 0.75, slightly below the 0.80 predicted by perfect matching. As a result of this choice pattern, the obtained reinforcement probabilities for the three different schedules were 0.22, 0.18, and 0.11, for the 0.20, 0.05, and 0.10 schedules, respectively. For the second group of subjects, the "0.20" stimulus was then presented on 75% of the trials, and the "0.05" stimulus was presented on 25% of the trials of the "choice" component, while the 0.10 stimulus continued to be presented on every trial of the alternating component. The reinforcement probabilities associated with each stimulus were those obtained by the choice subjects, noted above. Both sets of subjects were then given probe tests involving a choice between the 0.20 versus 0.10 stimulus and a choice between the 0.10 versus 0.05 stimulus. The choice subjects performed like those of Williams and Royalty (1989), in that the 0.20 stimulus was preferred to the 0.10 stimulus, but the 0.10 stimulus was preferred to the 0.05 stimulus, despite the obtained reinforcement probabilities for the 0.20 and 0.05 stimuli being similar. The single-stimulus subjects, in contrast, preferred both the "0.20" and "0.05" stimuli over the 0.10 stimulus, indicating that they were controlled by the probability of reinforcement obtained with each stimulus. The results thus demonstrate that the availability of the 0.20 alternative for the choice subjects somehow devalued the 0.05 stimulus from that which would be expected simply on the basis of the obtained pattern of reinforcement/nonreinforcement associated with that stimulus in isolation.

The failure of the obtained probability of reinforcement with respect to an individual stimulus to predict the preference for that stimulus has major implications for the analysis of discrimination learning, quite apart from the theory of the matching law. For example, models of the acquisition of preference between different reinforcement schedules have been developed by Mazur (1992) and by Couvillon and Bitterman (1985), which have described the pattern of acquisition for both pigeons and honeybees in surprising detail. Both models (like most others) rely on the assumption that the value of an individual choice alternative is determined solely by its own history of reinforcement/nonreinforcement independent of the contingencies associated with the other choice alternative. The results of Williams (1993) show clearly that this critical assumption is incorrect. Just how the influence of the presence of the other choice alternative exerts its effect is a vital theoretical issue that remains to be clarified.

D. Matching as Fundamental?

None of the three most prominent theories that have attempted to explain the matching law in terms of more fundamental principles has withstood the scrutiny of empirical testing. It is now time to consider the possibility that matching is itself fundamental, in the sense that it represents the choice rule by which behavior is allocated to response alternatives of different strengths. This assumption is in fact implicit in many different theoretical formulations, as the theoretical tactic has been first to derive the basis for the strengths of the individual responses, and then assume that behavioral output simply reflects their proportional values (e.g., Catania, 1973; Herrnstein, 1970; Killeen, 1982). The sources of disagreement among such approaches have been the nature of how individual response strengths are determined, not the matching principle itself.

If matching is the result of the separate strengths of each response, an important implication is that responses that are first stabilized separately should conform to the matching law the first time they occur together. Herrnstein and Loveland (1976) were the first to test this implication, by training pigeons with four different VI schedules, each associated with a separate response key. Initial training consisted of separate presentations of four of the six possible pairs for 8-min periods, with each of the four keys involved in two of the training pairs. The remaining two pairs were then presented during 30-s test periods in which no reinforcement was available. Given that each member of the new pairs was familiar, did responding to the new pairs obey matching? Matching was not obtained; instead, there was a strong tendency for the response key associated with the higher valued schedule to be preferred exclusively. A similar tendency toward overmatching was obtained by Edmon, Lucki, and Gresham (1980), who tested pigeons with new concurrent schedules after each component had first been trained in isolation as members of a multiple schedule.

A similar combining of separately trained response alternatives was studied by Young (1981) using differential amounts of reinforcement rather than differential frequencies. One response alternative was always associated with a probability of 0.5 of receiving either 10 food pellets or 0 pellets. The second alternative was associated with a probability of 1.0 of receiving a constant amount of food, ranging from 1 to 10 pellets across different experimental conditions. The critical feature of the procedure was that each alternative was presented individually on a long series of forced-choice trials (either 39 or 15 depending on the condition) before being paired together. The results of those choice trials, averaged over subjects, are shown in Figure 2, in terms of preferences for the probabilistic alternative. There was a substantial bias in favor of the probabilistic alternative even when the same average number of pellets was available for both responses, as shown by the preference value of 66% when five pellets were the reward for the alternative with the certain outcome. Quite apart from that bias, preference for the 0.5 probability was inversely related to the amount of the certain reinforcer. Of critical importance for understanding the choice rule is the nature of that relationship. If a maximizing rule were used, a step function should occur,

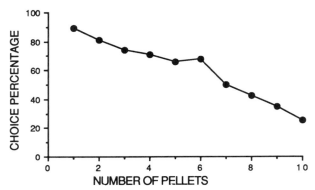

FIGURE 2   Data from Young (1981), showing the average preference for the choice alternative associated with a 0.5 probability of 10 or 0 pellets, as a function of the number of pellets delivered with a probability of 1.0. (Reprinted by permission of the Society for the Experimental Analysis of Behavior.)

with the transition at the point the value of the certain reinforcer exceeded the value of the uncertain reinforcer. Instead, the relation was roughly linear, indicating that preference was probabilistically related to the relative value of the two alternatives. However, matching was not obtained either, as the sensitivity parameter form a fit of Equation (6) to the data was approximately 2.0, substantially exceeding the 1.0 value entailed by perfect matching.

While the results of the experiments just described argue against the notion that matching is the fundamental choice rule, other data are consistent with that possibility. In the study of Williams and Royalty (1989) described above, probe trials between separately trained VI 20-s and VI 80-s schedules produced a mean preference of 0.82, near the 0.80 level predicted by perfect matching. In the discrete-trial version of VI schedules used by Williams (1993), probe tests between scheduled probabilities of 0.20 versus 0.10 yielded a preference of 0.62, only slightly below the predicted value of 0.67 given perfect matching. One possible reason for the disparity between these recent data and the earlier results of Herrnstein and Loveland (1976) and Edmon et al. (1980) was the degree of probe testing. The earlier two studies used a large number of probe trials; the latter studies, in contrast, used minimal probe testing with substantial retraining on the original contingencies between sessions that included probe trials. However, Herrnstein and Loveland (1976) did note that the tendency toward exclusive preference was present even on the first 30 s of probe testing. Thus, the nature of the discrepancy between the different outcomes remains unclear. Given the conflicting nature of the various results that have been described, additional research is needed to determine the outcome of choice procedure when different responses are combined together for the first time.

E. Matching to Interreinforcement Intervals

As noted in the discussion of the melioration theory of matching, the research reported by Williams (1993) shows that obtained probabilities of reinforcement do not control choice directly, because the presence of an alternative choice opportunity somehow devalues a choice alternative from what it would be if the same history of obtained reinforcement probability occurred in the presence of the stimulus in isolation. One possible explanation of this finding is that choice is controlled by the distribution of interreinforcement intervals associated with each stimulus. In other words, the value of an alternative is not calculated with respect to the probability of reinforcement received for actual responding to that alternative; instead, subjects register in some fashion the entire set of interreinforcement times for each stimulus. Rather than the individual responses gaining/losing response strength by individual episodes of reinforcement/nonreinforcement, the subject acquires veridical knowledge of the frequency of reinforcement associated with the different response alternatives, and then on the basis of some choice rule allocates its behavior accordingly.

A choice theory explicitly based on this assumption has been proposed by Gibbon, Church, Fairhurst, and Kacelnik (1988). Much of its explanatory power depends on the scalar theory of timing behavior (see Gibbon, 1991), which is too complex to discuss in detail here (see also Gallistel, Chapter 8, this volume). However, the critical feature of the account can be illustrated by considering its explanation of the results of Williams (1993). The crucial assumption is that the interreinforcement intervals associated with each alternative are calculated only with respect to those periods in which a given stimulus is actually available. Thus, for the subjects with the opportunity to choose between the 0.20 and 0.05 choice probability, the interreinforcement intervals for the 0.05 alternative included any time allocated to the 0.20 alternative because the 0.05 stimulus was physically present at that time, regardless of whether a response to the 0.05 stimulus actually occurred. In contrast, the "0.05" stimulus in the procedure without choice opportunities was present only when a response occurred to it, which meant that the recorded interreinforcement intervals would be substantially less. Such differences in the recorded interreinforcement intervals occurred despite the similar distributions of reinforced and nonreinforced responses to each alternative for the different procedures.

The theory proposed by Gibbon et al. (1988) constitutes a "representational theory" of matching in which the traditional concept of "response strength" is irrelevant. For our present purposes, its validity can be evaluated by considering if choice actually is controlled by the distribution of interreinforcement intervals associated with a response alternative, independent of the obtained distribution of reinforced and nonreinforced responses. The results of Williams (1993) give support for this possibility. However, a related study by Belke (1992) reveals a major difficulty with this approach. In his study pigeons chose between a VI 20-s and a VI 40-s schedule during one component of a multiple schedule, and between a VI 80-s and a second VI 40-s schedule in a second component. The two stimuli correlated with the different VI 40-s schedules were then presented together during probe tests, with the result that the VI 40-s schedule paired with the VI 80-s schedule was strongly preferred over that paired with the VI 20-s schedule. According to the account of Gibbon et al. (1988), no differential preference should have occurred, because the two schedules had identical distributions of interreinforcement intervals. At a more general level, Belke's results cause difficulty for the general position that matching occurs because of the automatic allocation of responding in proportion to the reinforcement rates associated with the different alternatives. It may be that matching is the choice rule given a determination of response strength by something other than reinforcement rate, but such an alternative remains unspecified (but see Gallistel, Chapter 8, this volume).

V. CONCLUSIONS

Despite 30 years of intensive research on choice behavior in free-operant procedures, fundamental theoretical issues remain unresolved. Research in the tradition of the matching law has assumed that all of the major parameters of reinforcement map in similar ways onto the dimension of reinforcer value, and thus are interchangeable in their effects. Accordingly, delay of reinforcement is assumed not to be qualitatively different in its effects than rate, amount, or probability of reinforcement. This assumption faces several empirical challenges and remains open to serious question. Equally at issue is the combination rule by which the different reinforcement parameters combine. The assumptions in previous research that the relationship is multiplicative, and that the ratio of the parameter values, not their absolute values, is the critical variable, continue to be confronted by challenging and often inconsistent empirical findings, which will require further research for clarification.

At a theoretical level, the matching law has inspired a rich assortment of different hypotheses about the underlying processes, but none of the major theories can presently explain all of the available data. Still unresolved is the most fundamental issue of all—whether matching should be regarded as the basic choice rule by which competing response strengths are allocated among the various behavioral alternatives, or whether the basic choice rule is that the response with the greatest strength always occurs ("maximizing"), with matching then derived from how the changing contingencies interact over the course of an experimental session. Assuming that matching is the basic choice rule, the rules determining the individual response strengths are themselves uncertain. It is perhaps disheartening that the intensive amount of research on choice has failed to resolve such fundamental issues. Despite these shortcomings, the arena of choice has served as a fertile ground for inspiring many different levels of theoretical development, of both the molar and molecular varieties, and has inspired several major efforts to apply the principles of animal behavior to other domains of psychology (e.g., Herrnstein, 1990; Rachlin et al., 1986). A deeper understanding of the processes underlying choice can only further that important enterprise.

References

Baum, W. M. (1974). On two types of deviation from the matching law: Bias and undermatching. *Journal of the Experimental Analysis of Behavior, 22,* 231–242.

Baum, W. M. (1979). Matching, undermatching, and overmatching in studies of choice. *Journal of the Experimental Analysis of Behavior, 32,* 269–281.

Baum, W. M. (1981). Optimization and the matching law as accounts of instrumental behavior. *Journal of the Experimental Analysis of Behavior, 36,* 387–403.

Belke, T. W. (1992). Stimulus preference and the transitivity of preference. *Animal Learning & Behavior, 20,* 401–406.

Catania, A. C. (1973). Self-inhibiting effects of reinforcement. *Journal of the Experimental Analysis of Behavior, 19,* 517–526.

Chung, S. H., & Herrnstein, R. J. (1967). Choice and delay of reinforcement. *Journal of the Experimental Analysis of Behavior, 10,* 67–74.

Collier, G. H., Johnson, D. F., Hill, W. L., & Kaufman, L. W. (1986). The economics of the law of effect. *Journal of the Experimental Analysis of Behavior, 46,* 113–136.

Collier, G. H., Johnson, D. F., & Morgan, C. (1992). The magnitude-of-reinforcement function in closed and open economies. *Journal of the Experimental Analysis of Behavior, 57,* 81–89.

Couvillon, P. A., & Bitterman, M. E. (1985). Analysis of choice in honeybees. *Animal Learning & Behavior, 13,* 246–252.

Daly, H. B., & Daly, J. T. (1982). A mathematical model of reward and aversive nonreward: Its application in over 30 appetitive learning situations. *Journal of Experimental Psychology: General, 111,* 441–480.

Davison, M., & McCarthy, D. (1988). *The matching law.* Hillsdale, NJ: Erlbaum.

DeCarlo, L. T. (1985). Matching and maximizing with variable-time schedules. *Journal of the Experimental Analysis of Behavior, 43,* 75–81.

de Villiers, P. A. (1977). Choice in concurrent schedules and a quantitative formulation of the law of effect. In W. K. Honig & J. E. R. Staddon (Eds.), *Handbook of operant behavior* (pp. 233–287). Englewood Cliffs, NJ: Prentice-Hall.

Edmon, E. L., Lucki, I., & Gresham, M. (1980). Choice responding following multiple schedule training. *Animal Learning and Behavior, 8,* 287–292.

Gibbon, J. (1991). Origins of scalar timing. *Learning and Motivation, 22,* 3–38.

Gibbon, J., Church, R. M., Fairhurst, S., & Kacelnik, A. (1988). Scalar expectancy theory and choice between delayed rewards. *Psychological Review, 95,* 102–114.

Graf, V., Bullock, D. H., & Bitterman, M. E. (1964). Further experiments on probability matching in the pigeon. *Journal of the Experimental Analysis of Behavior, 7,* 151–157.

Green, L., Rachlin, H., & Hanson, J. (1983). Matching and maximizing with concurrent ratio-interval schedules. *Journal of the Experimental Analysis of Behavior, 40,* 217–224.

Hall, G. A., & Lattal, K. A. (1990). Variable-interval schedule performance in open and closed economies. *Journal of the Experimental Analysis of Behavior, 54,* 13–22.

Herrnstein, R. J. (1961). Relative and absolute strength of responses as a function of frequency of reinforcement. *Journal of the Experimental Analysis of Behavior, 4,* 267–272.

Herrnstein, R. J. (1970). On the law of effect. *Journal of the Experimental Analysis of Behavior, 13,* 243–266.

Herrnstein, R. J. (1982). Melioration as behavioral dynamism. In M. L. Commons, R. J. Herrnstein, & H. Rachlin (Eds.), *Quantitative analyses of behavior* (Vol. 2, pp. 433–458). Cambridge, MA: Ballinger.

Herrnstein, R. J. (1990). Rational choice theory: Necessary but not sufficient. *American Psychologist, 45,* 356–367.

Herrnstein, R. J., & Heyman, G. M. (1979). Is matching compatible with reinforcement maximization on concurrent variable interval variable ratio? *Journal of the Experimental Analysis of Behavior, 31,* 209–233.

Herrnstein, R. J., & Loveland, D. H. (1976). Matching in a network. *Journal of the Experimental Analysis of Behavior, 26,* 143–153.

Herrnstein, R. J., & Vaughan, W. (1980). Melioration and behavioral allocation. In J. E. R. Staddon (Ed.), *Limits to action: The allocation of individual behavior* (pp. 143–176). San Diego: Academic Press.

Heyman, G. M., & Herrnstein, R. J. (1986). More on concurrent interval-ratio schedules. A replication and review. *Journal of the Experimental Analysis of Behavior, 46,* 331–351.

Hineline, P. N., & Sodetz, F. J. (1987). Appetitive and aversive schedule preferences: Schedule transitions and intervening events. In M. L. Commons, J. E. Mazur, J. A. Nevin, & H. Rachlin (Eds.), *Quantitative analysis of behavior* (Vol 5, pp. 141–157). Hillsdale, NJ: Erlbaum.

Hinson, J. M., & Staddon, J. E. R. (1983). Hill-climbing by pigeons. *Journal of the Experimental Analysis of Behavior, 39,* 25–47.

Hodos, W., & Trumbule, G. H. (1967). Strategies of schedule preference in chimpanzees. *Journal of the Experimental Analysis of Behavior, 10,* 503–514.

Ito, M. (1985). Choice and amount of reinforcement in rats. *Learning and Motivation, 16,* 95–108.

Keller, J. V., & Gollub, L. R. (1977). Duration and rate of reinforcement as determinants of concurrent responding. *Journal of the Experimental Analysis of Behavior, 22,* 179–196.

Killeen, P. R. (1982). Incentive theory: II. Models for choice. *Journal of the Experimental Analysis of Behavior, 38,* 217–232.

Lea, S. E. G. (1981). Correlation and contiguity in foraging behaviour. In P. Harzem & M. D. Zeiler (Eds.), *Advances in analysis of behavior: Vol. 2. Predictability, correlation, and contiguity* (pp. 344–406). New York: Wiley.

Mazur, J. E. (1984). Tests of an equivalence rule for fixed and variable reinforcer delays. *Journal of Experimental Psychology: Animal Behavior Processes, 10,* 426–436.

Mazur, J. E. (1987). An adjusting procedure for studying delayed reinforcement. In M. L. Commons, J. E. Mazur, J. A. Nevin, & H. Rachlin (Eds.), *Quantitative analyses of behavior* (Vol. 5, pp. 55–73). Hillsdale, NJ: Erlbaum.

Mazur, J. E. (1992). Choice behavior in transition: Development of preference with ratio and interval schedules. *Journal of Experimental Psychology: Animal Behavior Processes, 18,* 364–378.

Mazur, J. E., & Vaughan, W., Jr. (1987). Molar optimization versus delayed reinforcement as explanations of choice between fixed-ratio and progressive-ratio schedules. *Journal of the Experimental Analysis of Behavior, 48,* 251–261.

Myerson, J., & Miezin, F. M. (1980). The kinetics of choice: An operant systems analysis. *Psychological Review, 87,* 160–174.

Nevin, J. A. (1979). Reinforcement and response strength. In M. D. Zeiler & P. Harzem (Eds.), *Advances in the analysis of behaviour: Vol. 1. Reinforcement and the organization of behaviour* (pp. 117–158). New York: Wiley.

Nevin, J. A. (1988). Behavioral momentum and the partial reinforcement effect. *Psychological Bulletin, 103,* 44–56.

Nevin, J. A. (1992). An integrative model for the study of behavioral momentum. *Journal of the Experimental Analysis of Behavior, 57,* 301–316.

Nevin, J. A., Mandell, C., & Atak, J. R. (1983). The analysis of behavioral momentum. *Journal of the Experimental Analysis of Behavior, 39,* 49–59.

Rachlin, H. (1971). On the tautology of the matching law. *Journal of the Experimental Analysis of Behavior, 15,* 249–251.

Rachlin, H. (1973). Contrast and matching. *Psychological Review, 80,* 217–234.

Rachlin, H., Battalio, R., Kagel, J., & Green, L. (1981). Maximization theory in behavioral psychology. *Behavioral and Brain Sciences, 4,* 371–388.

Rachlin, H., & Green, L. (1972). Commitment, choice, and self-control. *Journal of the Experimental Analysis of Behavior, 17,* 15–22.

Rachlin, H., Green, L., & Tormey, B. (1988). Is there a decisive test between matching and maximizing? *Journal of the Experimental Analysis of Behavior, 50*, 113–123.

Rachlin, H., Logue, A. W., Gibbon, J., & Frankel, M. (1986). Cognition and behavior in studies of choice. *Psychological Review, 93*, 33–45.

Reed, P. (1991). Multiple determinants of the effects of reinforcement magnitude on free-operant response rates. *Journal of the Experimental Analysis of Behavior, 55*, 109–123.

Shimp, C. P. (1966). Probabilistically reinforced choice behavior in pigeons. *Journal of the Experimental Analysis of Behavior, 9*, 433–455.

Shimp, C. P. (1976). Short-term memory in the pigeon: The previously reinforced response. *Journal of the Experimental Analysis of Behavior, 26*, 487–493.

Shimp, C. P. (1984). Timing, learning, and forgetting. In J. Gibbon & L. Allan (Eds.), *Timing and time perception* (pp. 346–360). New York: New York Academy of Sciences.

Silberberg, A., Hamilton, B., Ziriax, J. M., & Casey, J. (1978). The structure of choice. *Journal of Experimental Psychology: Animal Behavior Processes, 4*, 368–398.

Skinner, B. F. (1986). Some thoughts about the future. *Journal of the Experimental Analysis of Behavior, 45*, 229–235.

Spence, K. W. (1936). The nature of discrimination learning in animals. *Psychological Review, 43*, 427–449.

Staddon, J. E. R. (1992). Rationality, melioration, and law-of-effect models for choice. *Psychological Science, 3*, 136–141.

Staddon, J. E. R., & Motheral, S. (1978). On matching and maximizing in operant choice experiments. *Psychological Review, 85*, 436–444.

Taylor, R., & Davison, M. (1983). Sensitivity to reinforcement in concurrent arithmetic and exponential schedules. *Journal of the Experimental Analysis of Behavior, 39*, 191–198.

Timberlake, W., & Peden, B. F. (1987). On the distinction between open and closed economies. *Journal of the Experimental Analysis of Behavior, 48*, 35–60.

Vaughan, W., Jr. (1985). Choice: A local analysis. *Journal of the Experimental Analysis of Behavior, 42*, 383–405.

Vom Saal, W. (1972). Choice between stimuli previously presented separately. *Learning and Motivation, 3*, 209–222.

Vyse, S. A., & Belke, T. W. (1992). Maximizing versus matching on concurrent variable-interval schedules. *Journal of the Experimental Analysis of Behavior, 58*, 325–334.

White, K. G., & Pipe, M. E. (1987). Sensitivity to reinforcer duration in a self-control procedure. *Journal of the Experimental Analysis of Behavior, 48*, 235–250.

Williams, B. A. (1985). Choice behavior in discrete-trial concurrent VI VR: A test of maximizing theories of matching. *Learning and Motivation, 16*, 423–443.

Williams, B. A. (1988). Reinforcement, choice, and response strength. In R. C. Atkinson, R. J. Herrnstein, G. Lindzey, & R. D. Luce (Eds.), *Stevens' Handbook of Experimental Psychology* (2nd ed., pp. 167–244). New York: Wiley.

Williams, B. A. (1991). Choice as a function of local versus molar contingencies of reinforcement. *Journal of the Experimental Analysis of Behavior, 56*, 455–473.

Williams, B. A. (1992). Dissociation of theories of choice via temporal spacing of choice opportunities. *Journal of Experimental Psychology: Animal Behavior Processes, 18*, 287–297.

Williams, B. A. (1993). Molar versus local reinforcement probability as determinants of stimulus value. *Journal of the Experimental Analysis of Behavior, 59*, 163–172.

Williams, B. A., & Fantino, E. J. (1978). Effects on choice of reinforcement delay and conditioned reinforcement. *Journal of the Experimental Analysis of Behavior, 29*, 77–86.

Williams, B. A., & Royalty, P. (1989). A test of the melioration theory of matching. *Journal of Experimental Psychology: Animal Behavior Processes, 15*, 99–113.

Young, J. S. (1981). Discrete-trial choice in pigeons: Effects of reinforcer magnitude. *Journal of the Experimental Analysis of Behavior, 35*, 23–29.

CHAPTER 5

Discrimination and Categorization

John M. Pearce

When different stimuli signal the occurrence of different events to an animal, then it will typically behave differently in their presence. Such a capacity for discrimination learning can be found throughout most, if not all, of the animal kingdom. Hennessey, Rucker, and McDiarmid (1979), for example, maintain that even the complex protozoa, *Paramecium,* can be trained to respond differentially to two auditory stimuli when only one of them signals an electric shock. The purpose of the first part of this chapter is to review some of the theories that have been developed to account for the way in which animals solve discriminations.

Studies of discrimination learning normally employ simple stimuli, such as tones or colored lights, but this is not always the case. Herrnstein, Loveland, and Cable (1976), for instance, presented pigeons in each experimental session with a sequence of 80 different slides, half of which contained pictures of trees. Subjects were rewarded for pecking a response key in the presence of slides showing a tree, whereas responses in the presence of the remaining slides were ineffective. The slides were selected from a pool containing more than 600 pictures, and eventually a considerably higher rate of responding was recorded during pictures showing the presence rather than the absence of trees. The second part of the chapter examines if an ability to solve this type of category discrimination is a consequence of the

Animal Learning and Cognition

mechanisms that are believed to govern the acquisition of simpler discriminations.

I. THEORIES OF DISCRIMINATION LEARNING

A. Pavlov

The origins of theorizing concerning discrimination learning rest with the work of Pavlov (1927). In one simple experiment (Pavlov, 1927, p. 121), a hungry dog received a discrimination in which an illuminated circle signaled food and an illuminated square was followed by nothing. Initially, a reasonably strong conditioned response (CR) was observed on both types of trial, but as training progressed this response occurred predominantly during the circle. The decline of responding during the square was attributed to it arousing inhibition, which counteracted any excitatory tendency to perform the response.

Evidence in support of this role of inhibition in discrimination learning can be found in another study by Pavlov (1927, pp. 75–76). The experiment involved a flashing light that signaled food, and a compound of the flashing light with a tactile stimulus, which was followed by nothing. As a result of this training, a CR was eventually observed only when the light was presented by itself. Pavlov (1927) proposed that this discrimination depended on inhibition aroused by the tactile stimulus suppressing any tendency to perform a CR when it was accompanied by the light. In keeping with this analysis, a further stage of the experiment demonstrated that the tactile stimulus also suppressed responding during a third stimulus, which had previously been paired with food but which had not taken part in a discrimination.

B. Spence

The idea that inhibition plays an important role in discrimination learning was adopted by Spence (1936), who was concerned with the way in which animals solve simultaneous discriminations between, say, the black and white doors of a jumping stand. In essence, he proposed that rewarding animals for responding to one stimulus (S+) will increase the excitatory tendency to repeat the response in the presence of S+. But when a reward is not delivered for responses to a different stimulus (S−), there will be an increment in a negative or inhibitory tendency not to perform the response to S−. At first the animal may perform on the basis of the position of the stimuli (e.g., always going left regardless of whether this is toward the black or the white door). Despite being incorrect, this strategy will ensure a gradual acquisition of excitatory tendencies to S+ and inhibitory tendencies to S−. As these tendencies develop, a preference for S+ will emerge and the discrimination will be solved.

Spence (1937) developed these proposals to take account of the fact that when S+ and S− are similar, there will be a measure of generalization between them. Figure 1 shows a stimulus dimension from which two stimuli have been selected for a discrimination. Responding in the presence of S+ and S− is assumed to result in the acquisition, respectively, of excitatory and inhibitory tendencies to these stimuli. The larger curve in Figure 1 depicts the excitatory generalization gradient that is assumed to develop after a number of rewarded trials in the presence of S+, and the smaller curve depicts the generalization gradient of inhibition around S−. A tendency to approach one stimulus rather than the other is determined by the difference between these gradients. Because the magnitude of this difference is greater for S+ than for S−, it follows that animals are more likely to select the former rather than the latter stimulus.

Further inspection of Figure 1 reveals that an interesting prediction can be derived from this attractively simple analysis. For the stimulus S′ on the dimension to the left of S+, the difference between the excitatory and inhibitory gradients is greater than for S+ itself. The theory thus predicts

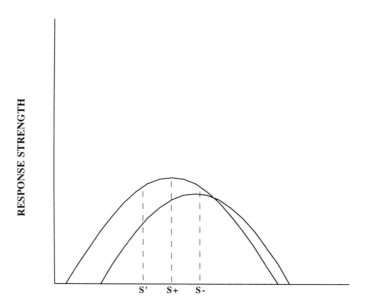

STIMULUS DIMENSION

FIGURE 1   The gradients of excitation and inhibition that are predicted to develop by the theory of Spence during a discrimination in which reward is presented in the presence of one stimulus (S+) but not in the presence of another (S−).

that after discrimination training with S+ and S−, if animals are given a choice between S′ and S+, they should prefer the former even though they have always been rewarded for approaching the latter. Although this prediction may seem surprising, it has received ample support from transposition studies (e.g., Kohler, 1918). A related test of this aspect of Spence's (1937) theory was conducted by Hanson (1959), who rewarded pigeons for pecking a response key when it was illuminated by light of 550 nm (S+) but not when it was illuminated by light of 590 nm (S−). Once the discrimination had been mastered, test trials were administered with light ranging in wavelength from 480 to 620 nm. The success of the original training was revealed by a higher level of responding on trials with 550-nm than with 590-nm light. It is most interesting, however, that the highest level of responding was shifted away from S+ at 540 nm. Such a peak shift has been reported on a number of occasions (see Rilling, 1977, for a review), and provides further forceful support for Spence's gradient interaction theory.

Spence (1936, 1937) assumed that the level of excitation or inhibition acquired by a stimulus was determined principally by the number of trials on which it had been paired with either the presence or the absence of reward. There is now, however, clear evidence that this factor alone does not determine the associative properties of a stimulus. These properties also appear to be governed by the context in which a stimulus is presented for conditioning. Consider a feature-positive (AB+ A0) discrimination in which a compound, AB, is paired with food, and A by itself is paired with nothing. According to the theory of Spence (1936, 1937), the intermittent pairings of A with food will result in this stimulus becoming excitatory and it will elicit a response whenever it is presented. Discriminations of this sort, however, are generally found to leave A with essentially no influence at all on responding (Wagner, 1969). In other words, the presence of B on reinforced trials prevents A from acquiring the excitatory properties that would develop if it were by itself paired with food on the same reinforcement schedule. Perhaps this outcome should not be surprising because B, rather than A, accurately signals when food will be delivered. It is unfortunately that Spence's theory lacks any mechanism that would allow the associative properties of a stimulus to be determined by its relative value as a predictor of reinforcement. Largely because of its inability to explain such stimulus selection effects, the theory is no longer popular. These effects can, however, be readily explained by the theory considered next.

C. Rescorla and Wagner

The Rescorla–Wagner (1972) theory is described in detail elsewhere in this volume (see Hall, Chapter 3, this volume), so the emphasis here will be on its application to discrimination learning. Equation (1) shows, for any conditioning trial, the change in associative strength of a CS, ΔV, that is predicted by this theory.

$$\Delta V = \alpha \beta (\lambda - V_\Sigma) \tag{1}$$

In the equation α and β are learning rate parameters (with values between 0 and 1), λ is the asymptote of conditioning, and V_Σ is the sum of the associative strengths of all the stimuli present on the trial in question. The term V_Σ is also important because its value determines the strength of responding on any trial.

As far as a feature-positive discrimination (AB+ A0) is concerned, Equation (1) predicts that on the initial compound trials both A and B will gain associative strength. On the nonreinforced trials with A the value of λ will be zero and this stimulus will lose some of the associative strength it has just acquired. As training progresses, therefore, the associative strength of B will grow to λ while that of A will decline to zero and the discrimination will be solved. Hence, in contrast to the theory of Spence (1936, 1937), this theory predicts that the nonreinforced component of a feature-positive discrimination will eventually elicit a negligible level of responding.

Consider, now, a feature-negative discrimination, A+ AB0, such as the one conducted by Pavlov (1927), in which an unconditioned stimulus (US) is presented after the element A, but not after the compound AB. On reinforced trials, Equation (1) predicts that A will gain positive associative strength, whereas on nonreinforced trials, B will acquire negative, or inhibitory, associative strength. At asymptote, the associative strength of A will be λ, and that of B $-\lambda$, which will permit a CR of asymptotic strength during A alone, and no response at all during AB. By predicting that B will possess negative associative strength, the Rescorla and Wagner (1972) theory implies that this stimulus will function as a conditioned inhibitor. For example, if B should be presented in compound with another stimulus, C, that has been paired with the US, then responding during BC is predicted to be weaker than when C is presented by itself. The theory can therefore explain all the findings by Pavlov (1927) that were considered earlier.

As it has been presented thus far, the Rescorla–Wagner (1972) model is unable to explain why a discrimination between two stimuli from a single dimension will be harder when the stimuli are close together on that dimension rather than when they are far apart. However, if it is assumed that a stimulus is composed of many different elements, some of which are shared by other stimuli, then the model is readily able to explain this finding (Blough, 1975; Rescorla, 1976). Indeed, Blough (1975) has shown how such an assumption permits the model to explain a variety of phenomena associated with discrimination learning, including peak shift.

The observation has been made on more than one occasion that the Rescorla–Wagner (1972) model is formally equivalent to a number of single-layer network theories of human learning and categorization (Gluck and Bower, 1988; McClelland & Rummelhart, 1985). Thus, changes in associative strength between a CS and a US, in terms of the Rescorla–Wagner (1972) model, correspond to changes in connection weights between the input and output units of network theories. Furthermore, the Rescorla–Wagner equation has been shown to be formally equivalent to the Woodrow–Hoff rule, or delta rule, which determines the course of learning in these connectionist networks (Gluck & Bower, 1988; Maki & Abunawass, 1991; Sutton & Barto, 1981). As a consequence of this equivalence, the Rescorla–Wagner theory can be regarded as possessing all the positive features that are attributed to these network theories. At the same time, the theory also shares their shortcomings.

One shortcoming of many single-layer network theories is that they are incapable of solving what is known as the *exclusive-or* problem in which an outcome is signaled by either of two events when they are presented separately, but not when they are presented together. An example of such an experimental design is *negative patterning*, in which reward is presented on trials with A or B alone, but not when they are presented in compound, A+ AB0 B+. Although animals can readily solve this type of discrimination (e.g., Woodbury, 1943), as it has been presented the Rescorla–Wagner (1972) model predicts that responding will always be greater to the compound than to its components. To overcome this problem, Rescorla and Wagner (1972; see also Gluck, 1991) proposed that combinations of stimuli create unique, or configural, cues, which can take part in associative learning as if they were normal stimuli. In the case of the negative patterning problem, a configural cue would be generated only on AB trials and thus, according to Equation (1), will acquire negative associative strength. As this cue gains in negative strength, responding on the trials with the compound will diminish, and the discrimination will eventually be solved.

Despite the foregoing modification, there remain certain discriminations for which the Rescorla–Wagner (1972) model predicts the incorrect outcome. Consider an A+ B+ C+ AB+ AC+ BC+ ABC0 discrimination in which reward is signaled by three stimuli when they are presented independently, or in pairs, but not when they are all presented together. This discrimination can be regarded as a complex version of negative patterning, and the Rescorla–Wagner model thus requires that configural cues gain associative strength if the discrimination is to be solved. However, because responding during the pairs of stimuli is predicted to be determined by the sum of the associative strengths of their components, it follows that during the acquisition of the discrimination the compounds AB, AC, and BC will elicit a higher rate of responding than will A, B, and C. The effect of this pattern of responding will be to ensure that the discrimination between the single elements and ABC will be acquired with greater difficulty than that encountered between the pairs of elements and ABC.

The foregoing prediction might seem surprising. It implies that a discrimination between two relatively different types of trial (ABC vs. A, B, or C) will be acquired more slowly than between two relatively similar types of trial (ABC vs. AB, AC, or BC). In order to demonstrate that this unlikely prediction has been correctly derived from the Rescorla–Wagner (1972) model, a series of computer simulations for the A+ B+ C+ AB+ AC+ BC+ ABC0 discrimination was conducted.[1] In all of the simulations, the predicted outcome of the discrimination was the same as that derived above, and was similar to the results of one simulation, which are shown in Figure 2A. For this simulation, four configural stimuli were employed, α for all the stimuli was .4, and β was .4 on reinforced, or .2 on nonreinforced, trials.

Figure 2B portrays the results of an experiment by Redhead and Pearce (1994), which employed the A+ B+ C+ AB+ AC+ BC+ ABC0 discrimination. For this experiment a single group of pigeons received autoshaping with stimuli that consisted of numerous randomly located dots of different colors. The figure makes it quite clear that in contrast to the pattern of results predicted by the Rescorla–Wagner (1972) theory, the discrimination between ABC and A, B, or C alone was, on a number of sessions, more pronounced than between ABC and AB, AC, or BC.

There are a number of other experimental designs for which the Rescorla–Wagner (1972) model makes unlikely and, as it turns out, incorrect predictions concerning the ease with which a discrimination will be solved. For example, Pearce and Redhead (1993) have shown the theory predicts that adding a common cue to an A+ AB0 discrimination, to create an AC+ ABC0 discrimination, will facilitate the acquisition of the discrimination. This prediction is surprising because it implies that a discrimination between two relatively similar signals for reward and nonreward (AC and ABC) will progress more readily than when the signals are relatively different (A and AB). In fact, the results from the experiment indicated that the A+ AB0 discrimination was acquired more rapidly than the AC+ ABC0 discrimination.

One shortcoming of many network theories of learning is that they overpredict the effects of retroactive interference (McCloskey & Cohen, 1989). This criticism can also be leveled at the Rescorla–Wagner (1972) model. Pearce and Wilson (1991) presented rats with a feature-negative discrimination, A+ AB0, which was followed by excitatory conditioning

---

[1] For these simulations a wide range of values for α and β were employed. In addition, the values of α were occasionally different for the experimental and configural stimuli, and the values of β were occasionally different for the reinforced and nonreinforced trials. In some simulations only three configural cues were employed, each corresponding to a different pair of elements; and in other simulations a fourth configural cue was used, which was effective only on trials with ABC.

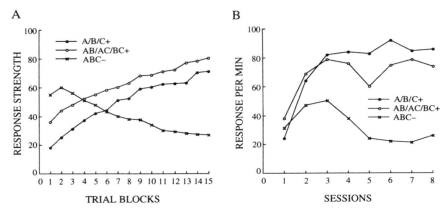

FIGURE 2   The pattern of responding predicted by the Rescorla–Wagner (1972) model for an A+ B+ C+ AB+ AC+ BC+ ABC0 discrimination (A) and the results from an experiment by Redhead and Pearce (1993) based on this design (B). The results for trials of the same type have been combined.

with B before the A+ AB0 discrimination was again presented for testing. According to the Rescorla–Wagner (1972) model, the excitatory conditioning with B will first erase the negative associative strength initially acquired by this stimulus, and then replace it with positive associative strength. As a consequence, when AB is presented for testing not only should the original suppression of responding during this compound be eliminated, it should be replaced by an unusually high level of responding because of the summation of the associative strengths of A and B. That is, the training in the second stage of the experiment is predicted to produce the opposite pattern of responding of that observed in the original stage. In fact, subjects revealed substantial savings from the original discrimination training, and in the test phase reacquired the A+ AB0 discrimination significantly more rapidly than did a control group. Pearce and Wilson (1991) have shown in detail that it is difficult for the Rescorla–Wagner model to account for these results, even if appeal is made to the influence of configural cues.

The Rescorla–Wagner (1972) model has provided an extremely influential account of the way in which discriminations are solved. By assuming the existence of configural cues, and by assuming that stimuli are composed of elements, the theory has been able to account for a far wider range of phenomena than have any of its predecessors. In keeping with other network theories to which it is closely related, however, the theory may be criticized for making unrealistic predictions concerning the effects of retroactive interference. In addition, there are occasions when the theory makes incorrect, and perhaps counterintuitive, predictions concerning the outcome of feature-negative and negative patterning discriminations.

D. Configural Theory

The Rescorla–Wagner (1972) model can be regarded as an *elemental* theory of conditioning because it assumes that compound conditioning provides the opportunity for each element of the compound to become connected with the US. On the other hand, certain authors (e.g., Gulliksen & Wolfle, 1938; Pearce, 1987, 1994) have argued that a more satisfactory analysis of discrimination learning might stem from a *configural* theory. Indeed, such a theory is able to overcome the obstacles that have just been shown to confront the Rescorla–Wagner theory. According to configural theory (e.g., Pearce, 1987), any compound conditioning trial will result in the formation of a configural representation of the pattern of stimulation on that trial, and this representation in its entirety will then enter into an association with the outcome of the trial. In a manner similar to that proposed by Rescorla and Wagner (1972), this association is assumed to develop over trials and its strength will determine the vigor of the CR that occurs to that particular pattern of stimulation. If the pattern of stimulation should change in any way, a weaker CR will be performed with a vigor that is related to the similarity between the training and test stimuli.

As far as an A+ AB0 discrimination is concerned, configural theory predicts that an excitatory association will develop between A and the US. Whenever AB is presented, there will then be a measure of generalization from A to AB and the compound will elicit a CR. The absence of the US after the compound will then result in AB entering into an inhibitory association, which will counteract the excitation generalizing from A to AB. There will also be a measure of generalization of inhibition from AB to A, which will necessitate further excitatory conditioning with A to ensure that a CR of asymptotic strength occurs during A.

The above account may seem complex, but it is really no more than a modification of the analysis of discrimination learning developed by Spence (1937). Figure 3 shows the way in which the ideas developed in the previous paragraph can be accommodated within Spence's (1937) framework. In this figure, distance along the abscissa does not reflect the difference between two stimuli from the same dimension. This distance instead reflects the difference between configurations of stimuli measured in some more complex manner.[2] The figure demonstrates that an A+ AB0 discrimination will result in an inhibitory generalization gradient around AB (small curve) and an excitatory gradient around A (large curve). If the difference between

---

[2] There are a variety of ways in which similarity may be computed in these circumstances. According to Pearce (1987), the similarity between two configurations is related to the number of common elements they share. Alternatively, Shepard (1987) has suggested that similarity is determined by distance measured in a multidimensional space.

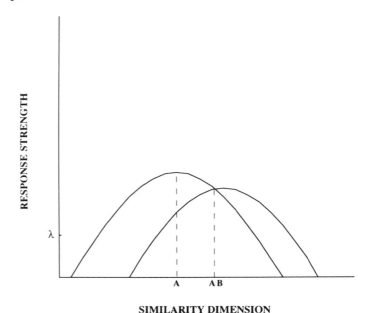

SIMILARITY DIMENSION

FIGURE 3   One way in which the proposals of Spence (1937) can be developed to account for the solution of an A+ AB0 discrimination.

these gradients determines the strength of responding, then it follows that the element A, but not the compound AB, will elicit a CR.

One advantage of the foregoing analysis is that it correctly predicts that once the A+ AB0 discrimination has been mastered, excitatory conditioning with B will only moderately disrupt the original discrimination (Pearce & Wilson, 1991; Wilson & Pearce, 1992). Recall that in these circumstances the Rescorla–Wagner (1972) model predicts that excitatory conditioning with B will reverse the effects of the original training. However, in terms of configural theory, if the A+ AB0 discrimination is reintroduced once conditioning with B is complete, then excitation associated with B will generalize to AB and some responding will now occur during AB. But this generalization will be incomplete and result in AB eliciting weaker responding than either A or B alone. In other words, for this particular example, configural theory predicts that excitatory conditioning with B will interfere retroactively with the effects of the original discrimination, but to a considerably lesser extent than that predicted by the Rescorla–Wagner theory.

Figure 3 can also be used to explain the finding by Redhead and Pearce (1993) that an AC+ ABC0 discrimination was acquired more slowly than an A+ AB0 discrimination. According to Pearce (1987), the close similarity of AC and ABC will ensure that these compounds are nearer each other on the abscissa than are A and AB. As a consequence, there will be substantial generalization of excitation from AC to ABC, and many training trials will be needed before the inhibition associated with ABC has developed to such an extent that no response occurs on the nonreinforced trials. This theoretical account thus readily predicts that any manipulation that enhances the similarity of the signals for reward and nonreward will make the discrimination between them more difficult.

The complex version of negative patterning, A+ B+ C+ AB+ AC+ BC+ ABC0, is difficult to interpret using Figure 3. In general terms, however, configural theory predicts that the individual stimuli and the compounds composed of pairs of stimuli will gain excitatory strength and ABC will gain inhibitory strength. This inhibitory strength will generalize to a greater extent to the two-element compounds than to the single elements and disrupt responding during the pairs of stimuli more than during the individual stimuli. A more formal configural explanation for these results can be found in Redhead and Pearce (1994).

This discussion of configural theory would be incomplete if it did not acknowledge that certain findings are difficult to explain with this type of theory. For example, if two stimuli have separately been paired with a US, it is often found that responding during a compound composed of these stimuli is stronger than during either stimulus alone (e.g., Kehoe, 1986; Wagner, 1971). According to a formal version of configural theory developed by Pearce (1987), this summation effect should not occur because the generalization decrement that results from presenting each element in a compound will ensure that responding during the compound is approximately equal to that during either element. It is possible for configural theory to explain summation (see Pearce, 1994), but whether or not this explanation is justified remains to be determined.

In summary, configural theory differs from the other theories that have been considered because of the assumption that configurations of stimuli can enter into individual associations. By making the additional assumption that generalization will occur between similar configurations, configural theory can explain all the findings from the discrimination studies that have been described above.

II. RELATIONAL DISCRIMINATIONS

A common assumption of all the theories considered thus far is that subjects choose between discriminative stimuli on the basis of their absolute physical properties. For example, if reward is signaled by a bright but not a dim visual stimulus, then according to all of the above theories, this discrimination will be solved when the absolute brightness of each stimulus is associated with its respective outcome. Ever since the work of Kinnaman (1902), however, it has been acknowledged that there is an alternative strategy open to animals when confronted with this type of problem. They may rely more on the relative than the absolute properties of the stimuli and, in the case of the present example, select the brighter of the two. The purpose of this section is to evaluate the findings from experiments that have been conducted to test this relational analysis of discrimination learning.

One method that has been employed to show that animals use relational information when solving certain discriminations is the transposition design discussed earlier. The successful transposition of a discrimination to new values on a stimulus continuum is consistent with the idea that the original discrimination was based on the relationship between S+ and S−. However, it will be recalled that the theory of Spence (1937) is also able to explain this effect without making reference to relational learning. Despite this success of Spence's theory, it is by no means certain that his theory can explain all transposition effects. For example, Gonzalez, Gentry, and Bitterman (1954) trained chimpanzees on a discrimination involving three stimuli that varied in size, with food being signaled by only the intermediate stimulus. In these circumstances, the theory of Spence (1937) predicts that if testing is conducted with three new stimuli that differ in size to those used originally, then transposition will not occur. In contrast to this prediction, Gonzalez et al. (1954) found clear evidence of transposition on the test trials. Further evidence that animals are able to detect relations between stimuli can be found in experiments by Lawrence and DeRivera (1954) and Saldanha and Bitterman (1951).

A rather different methodology to transposition for studying whether or not animals can detect the relationship between two stimuli is provided by experiments on matching and oddity. In these experiments a subject is typically required to respond to a single, *sample* stimulus whereupon two *comparison* stimuli are presented, one of which will be the same as the sample. To gain reward the subject must choose the comparison stimulus that is either the same as the sample (matching), or different to the sample (oddity). At first sight, successful performance on this task might be taken to indicate that the selection of the correct comparison stimulus is determined by its relationship (of sameness or difference) with the sample. However, convincing evidence to support this conclusion has proved elusive. If the experiment involves only a limited number of sample stimuli, then successful performance could be based on the absolute properties of the stimuli concerned. That is, the different samples could elicit responses directed toward specific comparison stimuli, and the fact that there was a consistent relationship between the two would be of no consequence. A direct way of refuting this possibility is to train subjects with a restricted set of stimuli, and then to employ novel stimuli for testing. If subjects show appropriate matching, or oddity, performance on the test trials, then this could not be explained by the account that has just been developed. Experiments based on this design have revealed successful transfer with some animals: corvids (Wilson, Mackintosh, & Boakes, 1985), chimpanzees (Oden, Thompson, & Premack, 1988), and dolphins (Herman & Gordon, 1974), but not pigeons (Wilson et al., 1985).[3]

However, even successful transfer of matching or oddity provides questionable proof of the ability of animals to solve relational discriminations. Premack (1983) has suggested that it is possible to solve either type of problem by relying on information about novelty and familiarity, rather than about relationships. In the case of a matching experiment, once a response has been directed to the sample, the subject will experience a stronger reaction of familiarity toward one comparison stimulus than toward the other. By selecting the stimulus that elicits the stronger reaction of familiarity, the subject will then be able to solve the discrimination, even on trials when the sample is novel. A reaction of familiarity might be difficult to specify, but quite clearly it does not depend on the subject perceiving the relationship of sameness between the sample and comparison stimulus in order for it to occur. In view of Premack's suggestion, therefore, the results from all matching and oddity experiments must be viewed as providing equivocal evidence that animals can solve discriminations on the basis of relational information.

Evidence that at least one species, the African Grey parrot, can use information about sameness and difference to solve discriminations comes from a study by Pepperberg (1987). The single subject in the experiment, Alex, had been trained over many years to use vocal English labels to refer to a variety of objects in response to questions posed in spoken English by his trainers. He could also use such categorical labels as "color" and "shape." For the experiment under consideration, Alex was presented with two objects, such as a red triangle and a red square, and was asked, "What's same?" or "What's different?" He was then expected to respond by choosing the appropriate category label: "color" for the first question and "shape" for the second. Performance on these tests, even when the objects had not previously been presented for testing, was considerably better than that expected on the basis of chance.

In this experiment it is extremely difficult to understand how a reaction

---

[3] Further evidence that pigeons experience considerable difficulty with relational discriminations is provided by Pearce (1988, 1991) and Herrnstein, Vaughan, Mumford, and Kosslyn (1989). Many sessions of training were required before any indication was given that subjects could judge whether or not two adjacent vertical bars were of the same or different heights (Pearce, 1988, 1991); or whether they could discriminate between loops with a dot located on either the inside or the outside (Herrnstein et al., 1989). Moreover, because of the nature of the training, it is possible that this success was due entirely to an ability to remember the individual patterns.

of familiarity would help Alex solve the problem with which he was confronted. Instead, the results suggest that he was able to perceive the relationship between two test objects for a given attribute (e.g., shape) and use this information to respond correctly. Such a finding clearly encourages the conclusion that one species, at least, is capable of appreciating relational information.

Evidence that another species, the chimpanzee, can also utilize relational information to solve discriminations can be found in a study by Gillan, Premack, and Woodruff (1981). A single chimpanzee, Sarah, was given a variety of analogical reasoning problems, of which one example is presented in Figure 4. An array of shapes was placed in front of her (see Figure 4A). To earn reward she had to place one of the shapes from below the line in the space in the array above the line (see Figure 4B). This task thus required her to perceive the relationship between the shapes in the left-hand column above the line in 4A, and re-create it in the right-hand column, 4B, using different shapes. Not only was she able to perform this task accurately with a variety of geometric shapes, she was also successful when the items presented to her were everyday objects, for example, keys, locks, paint pots, and brushes. It would seem, therefore, that Sarah was able to perceive a wide range of relationships between the many different types of stimuli with which she was presented.

The findings with Alex and Sarah provide strong evidence for believing that at least some animals can use relational information to solve discriminations. Both animals, however, had received many years of specialized training before they demonstrated this impressive ability, and it is not clear

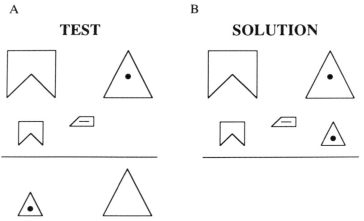

FIGURE 4   An example of an analogical reasoning problem presented to a chimpanzee.

which aspects of their training were responsible for the success of the experiments.

III. CATEGORIZATION

The remainder of this chapter is concerned with understanding the way in which animals solve categorization problems. We shall see that the theories developed to account for the solution of simple discriminations can be used with considerable success to account for the results from experiments with more complex stimuli.

Typically, a categorization experiment involves a discrimination in which reward is not signalled by a single stimulus, but, instead, is signaled by a variety of stimuli that share some common characteristic. In the case of the experiment by Herrnstein et al. (1976), mentioned earlier, the common characteristic was trees. Whenever the subject was shown a photograph of a tree, responding resulted in the occasional delivery of food. It is important to note that not only were pigeons able to solve this problem with familiar stimuli, but, once they had been successfully trained, they were able to categorize correctly novel stimuli. Thus, categorization does not depend merely on remembering each exemplar and its significance (Bhatt, Wasserman, Reynolds, & Knauss, 1988; Schrier & Brady, 1987).

There is now ample evidence showing that a wide range of species are capable of solving categorization problems. Indeed, after reviewing some of this evidence, Herrnstein (1990) was led to conclude that categorization has "turned up at every level of the animal kingdom where it has been competently sought" (p. 138). Studies of categorization have involved monkeys (e.g., Schrier, Angarella, & Povar, 1984), chinchillas (Burdick & Miller, 1975), pigeons (e.g, Herrnstein et al., 1976), chickens (Ryan, 1982), quail (Kluender, Diehl, & Killeen, 1987), blue jays (Pietrewicz & Kamil, 1977), and an African Grey parrot (Pepperberg, 1983).

A wide variety of categories have been studied in these experiments. Some categories have been based on natural objects: trees, water, people, cats, and flowers (e.g., Bhatt et al., 1988; Herrnstein et al., 1976). Other categories have involved human-made objects: cars, chairs, the letter A in various typescripts, and the cartoon character Charlie Brown (e.g., Bhatt et al., 1988; Cerella, 1980; Morgan, Fitch, Holman, & Lea, 1976). Yet other categories have involved auditory information such as phonemes and music (Burdick & Miller, 1975; Kluender et al., 1987; Porter & Neuringer, 1984). Finally, some studies have involved more abstract categories such as color, shape, sameness, and tools (Oden et al., 1988; Pepperberg, 1983; Savage-Rumbaugh, Rumbaugh, Smith, & Lawson; 1980).

A. Feature Theory

Humans have been said to assign individual stimuli to categories on the basis of the features of which they are composed. According to certain theorists, membership of a category is determined by whether or not an individual instance possesses some necessary set of defining features (see Smith & Medin, 1981). More recently, categorization has been said to depend on the strength of connections, between individual features and the appropriate label, in a connectionist network (Gluck & Bower, 1988; McClelland & Rummelhart, 1985). A feature analysis has also been developed to account for categorization by animals (e.g., D'Amato & Van Sant, 1988; Lea, 1984).

The stimuli employed for animal studies of categorization are often pictures of natural scenes, which means that they are complex and involve a great deal of redundant information. Any satisfactory feature theory must, therefore, specify how the relevant rather than the irrelevant features in a succession of exemplars acquire their influence. One solution to this problem is to look to a theory such as the Rescorla–Wagner (1972) model. Suppose that animals receive trials in which photographs of trees signal food, whereas nontree slides are followed by nothing. Whenever a photograph is presented, its features can be expected to undergo a change in associative strength in the manner specified by Equation (1). The features that belong to trees will, in general, be present only on reinforced trials and they will steadily acquire associative strength. In contrast, the features that are irrelevant to the solution of the discrimination, for example, patches of sky, will occur on both reinforced and nonreinforced trials and the manner in which they acquire associative strength will be erratic. Eventually,, it follows from Equation (1) that the features that are reliably present on reinforced trials, specifically those belonging to trees, will gain considerably more associative strength than will the irrelevant features. Once this has occurred, then the discrimination will be solved and even novel photographs will be classified correctly.

Evidence consistent with this analysis can be found in a study by D'Amato and Van Sant (1988), who trained monkeys with a categorization problem using photographs showing the presence or absence of humans. Although the monkeys were successful on this problem, test trials revealed a number of interesting errors. For example, any nonperson slide that contained a red patch, such as half a watermelon, or a dead flamingo being carried by a jackal, was likely to be classified as belonging to the person category. One straightforward interpretation of this finding is that the feature of red patch acquired considerable associative strength, presumably because it is common to many faces, and resulted in a high level of responding whenever it was presented.

An obvious problem with experiments involving photographs of natural scenes is that their complexity makes it difficult to derive precise theoretical conclusions. There have, therefore, been several attempts to understand how animals solve categorization problems by using artificial stimuli. Figure 5 shows three examples of such stimuli from an experiment by Huber and Lenz (1993). Inspection of these figures will reveal that the faces differ on four dimensions: the area above the eyes, the distance between the eyes, the length of the nose, and the area below the mouth. The figures also show, progressively from left to right, the three values that were used to represent each dimension. The features in the left-hand figure were assigned a value of -1, in the center figure, 0, and in the right-hand figure, $+1$.

In the experiment by Huber and Lenz (1993), pigeons were shown 62 different faces, and were rewarded for pecking a response key in the presence of any one of them for which the sum of the feature values was greater than zero. A substantially greater level of responding was eventually recorded in the presence of those faces that signaled the availability rather than the absence of food. The advantage of using artificially created stimuli in this study is that the experimenter knows which features the subjects must use in order to solve the discrimination. It is also possible to study performance in the presence of different faces, to gain an understanding of the control exerted by these features. In fact, Huber and Lenz (1993) found that there was an extremely orderly relationship between the sum of the feature values of a particular face and the rate of responding that it elicited. In other words, the more features a face had in common with the right-hand face of Figure 5, the faster subjects would respond in its presence.

This experiment demonstrates that not all members of a category are treated equally, with some being classified more readily than others. Furthermore, this conclusion is entirely consistent with a feature analysis of

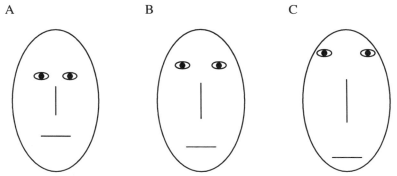

FIGURE 5   Examples of three of the faces employed in a categorization study by Huber and Lenz (1993).

categorization (Rescorla & Wagner, 1972). According to this theory the discrimination employed by Huber and Lenz (1993) will result in each of the positive features gaining positive associative strength and each of the negative features gaining negative associative strength. Since the theory assumes that the level of responding during a compound is determined by the sum of the associative strengths of its components, there will be a direct relationship between the rate of responding during a particular face and the sum of the values of its component features.

B. Exemplar Theory

A number of authors have argued that the ability of both humans (Hintzman, 1986; Kruschke, 1992; Medin & Schaffer, 1978) and animals (Astley & Wasserman, 1992; Pearce, 1988, 1989, 1991) to categorize objects depends on them remembering each instance or exemplar and the category to which it belongs. Such an account can clearly explain the ability to categorize familiar exemplars, but some additional mechanism is required if this account is to explain the successful categorization of novel stimuli. Both Pearce (1988, 1991) and Astley and Wasserman (1992) have proposed that this additional mechanism is based on the principles of stimulus generalization as envisaged by Spence (1937). That is, when a stimulus is presented for the first time in a categorization problem, there will be generalization of both excitation and inhibition to it as a result of the training with the previous exemplars. If the new stimulus belongs to the category that signals food, then it is likely that it will bear a strong similarity to many other stimuli that have already been shown and which belong to the same category. There will, as a consequence, be a considerable generalization of excitation to this new stimulus. In contrast, it is unlikely that the new pattern will bear a close resemblance to the members of the nonreinforced category and the generalization of inhibition that will result from the nonreinforced trials will be slight. The interaction of generalized excitation and inhibition will thus leave the new stimulus with a net level of excitation, and it will elicit a response appropriate to the category to which it belongs.

Two points merit emphasis concerning this account of categorization. The first concerns an argument that may occur to many readers: the account is implausible because it requires that animals remember an unreasonably large number of individual stimuli. In the case of the study by Herrnstein et al. (1976), this number would have to be a sizable proportion of 600. There is, however, good evidence that at least some animals are capable of retaining information about a large number of photographs. In an experiment by Vaughan and Greene (1984), pigeons were exposed to 320 photographs that were allocated at random to two groups. They were trained in a manner similar to that of many categorization studies, except that the availability of food was signaled by 160 photographs that were unrelated rather than from the same category. Despite the large number of stimuli employed in this experiment, the subjects mastered the discrimination. This finding provides at least some justification for believing that animals remember the various exemplars of a category that they encounter.

The second point concerns the finding that when the same pool of photographs is used for training with a category discrimination, then performance in the presence of these photographs is often better than with novel photographs belonging to the same category (Bhatt et al., 1988; Schrier et al., 1984). This result has also been found in experiments with humans where it is referred to as the exemplar effect (Homa, Dunbar, & Nohre, 1991). The implication of this result is that animals can learn about individual training stimuli in a categorization task, and that this information is at least partially responsible for their successful performance.

Despite the different assumptions on which they are based, it is often extremely difficult to differentiate between a feature and an exemplar theory in terms of the predictions they make concerning experimental findings. Consider, for example, the finding by Huber and Lenz (1993), that faces with a net feature sum of, say, 3 elicited faster responding than those with a net feature sum of 1, even though they belonged to the same category. At first sight this finding may appear to be more compatible with a feature analysis of categorization, but it can be readily explained by an exemplar analysis. Any face with a net feature sum of 1 will be more similar to the members of the negative category than an instance with a net feature sum of 3. The generalization of inhibition from members of the nonreinforced category will thus be greater to the former instance than to the latter, and result in faces with a feature sum of 3 eliciting a higher rate of responding than those with a feature sum of 1. In other words, the reason for the high rate of responding to a +3 face is essentially the same as that offered by Spence's (1937) theory of discrimination learning for the peak shift effect.

The results from certain studies of discrimination learning reviewed in the first part of this chapter were interpreted as being more consistent with a configural than with an elemental theory of conditioning. The experiments employed relatively simple designs and their findings may be of little relevance where the evaluation of theories of categorization is concerned. When complex stimuli are employed, animals may adopt different strategies to solve discriminations than those adopted when simple stimuli are employed. In addition, certain categorization tasks may encourage animals to focus on the components of the stimuli, as feature theory predicts; whereas in other tasks they may focus on more global aspects of the stimuli, as exemplar theory predicts (see Huber & Lenz, 1993). Because of these possibilities, it is difficult to derive any firm conclusions concerning the relative merits of feature and exemplar accounts of categorization by animals. Instead, we shall conclude for the present that the mechanisms that are believed to be responsible for the way animals solve relatively simple discriminations are also likely to be responsible for the way they solve categorization problems. We turn now to examine whether any additional processes are involved in categorization by animals.

C. Prototype Theory

The results from experimental studies of categorization by humans have led a number of investigators to suggest that exposure to the members of a category results in the formation of a prototype (e.g., Posner & Keele, 1968; see Shanks, Chapter 12, this volume). The prototype of a category is supposed to be a summary representation that corresponds to the average, or central tendency, of all the exemplars that have been experienced. Once a prototype has been formed, it is assumed to be activated whenever an exemplar is presented, and once activated it will elicit the response that is appropriate for the category. The likelihood of the response is assumed to be determined by the extent to which the prototype is activated, which, in turn, is assumed to be related to the degree of similarity between the exemplar and the prototype.

Evidence in support of these proposals comes from the finding that exemplars that bear a close resemblance to the prototype are classified more easily than exemplars that are rather different than the prototype (Posner & Keele, 1968). Although initial attempts to find a similar prototype effect with animals were unsuccessful (see, e.g., Lea & Harrison, 1978; Pearce, 1987; Watanabe, 1988), more recent experiments have been successful (von Fersen & Lea, 1990). But even these successes do not necessarily mean that a prototypical representation is responsible for successful categorization by animals. It now seems that demonstrations of such an effect with humans can be explained by either a feature (McClelland & Rummelhart, 1985) or an exemplar (Hintzman, 1986; Shin & Nosofsky, 1992) theory of categorization (see Shanks, Chapter 12, this volume); and the same may well be true for experiments with animals.

In terms of feature theory, for instance, a stimulus that corresponds to the average of all the members of a category is likely to be composed of elements that occur frequently in that category. If the category in question has been used to signal food, then each of these frequently occurring features will have considerable associative strength, and their combined influence will result in a high level of responding during the prototypical stimulus. Furthermore, this level of responding is likely to be higher than that during stimuli that are only distantly related to the prototype, and which can be expected to possess only a few features with high associative strength.

D. Categorization as Concept Formation

It is tempting to say that animals are able to categorize because they possess a concept. Schrier and Brady (1987), for instance, have proposed that monkeys can categorize photographs of people because they possess a "concept *humans* as we commonly understand it" (p. 142). Although it is easy to talk about concepts when they are used by humans, it is in fact rather more difficult to specify what this term means when it applies to animals. Lea (1984) has considered this matter in some detail. He concludes that if animals acquire concepts, then successful categorization need not depend on the physical similarity of the members of the category. Thus far, successful categorization of a novel exemplar has been attributed to it bearing some similarity, or sharing common elements, with at least one training stimulus. But if categorization is based on learning about the significance of concepts, then it should be successful even when the exemplars bear no physical similarity to each other. To help clarify this point, consider the following experiment by Savage-Rumbaugh et al. (1980). Chimpanzees were required to sort a mixed pile of tools and food into two separate piles, based on these categories, in order to obtain reward. Initially they were trained with the same objects, but when test trials were given with new objects they were able to categorize these successfully. It is not easy to argue that this problem was solved on the basis of the physical similarity of the test items to the training items, since it is difficult to identify a set of physical features that an object must possess in order to be classified as a tool, or as food. Instead, these objects may have been categorized successfully because the subjects possessed the concepts "food" and "tool." In fact, the chimpanzees had received considerable language training (see Rumbaugh and Savage-Rumbaugh, Chapter 11, this volume) prior to this experiment and it is conceivable that this was responsible for their ability to acquire these concepts. There is, however, an alternative, theoretically less exciting explanation for the results of this experiment.

When a chimpanzee picks up an item of food it may react in some consistent way, such as by salivating. In order to solve the discrimination, therefore, all the animal has to learn is that any object that elicits this reaction must be treated in one way, and objects that do not elicit this reaction must be treated in another way. Such a "mediated generalization" account is, admittedly, cumbersome, but it has a long history (e.g., Osgood, 1953, p. 353) and there are good reasons for believing it to be true for both simple (Honey & Hall, 1989) and complex (Wasserman, DeVolder, & Coppage, 1992) discriminations. In view of this possible role of mediated generalization, it currently remains an open question as to whether or not success by animals on any categorization problems involves the possession of a concept. However, the success of certain species in solving relational discriminations suggests that some animals, at least, should be able to categorize according to more abstract criteria than those based on physical similarity.

IV. CONCLUDING COMMENTS

In closing this discussion of discrimination learning and categorization by animals, the direction in which future research on these topics may progress will be briefly considered.

The debate is likely to continue as to whether configural or elemental associations form the foundations of what is learned during simple discriminations. Configural and elemental theories are likely to be refined and developed to take account of evidence that at first sight might appear to lend more support to one than the other. Eventually we can hope that the differences between the theories are reconciled and that the mechanisms that are fundamental to discrimination learning and categorization will be fully understood.

A comparison of the theories considered in this chapter with many of those discussed by Shanks in Chapter 12 on human associative learning reveals a considerable overlap between these different areas of research. The formal similarity between the Rescorla–Wagner (1972) model and network theories of human categorization (Gluck & Bower, 1988; McClelland & Rummelhart, 1985) has already been mentioned. There are also a number of important commonalities between the configural theory of Pearce (1987) and exemplar theories of human categorization (e.g., Kruschke, 1992). In the future, perhaps, these links may develop to such an extent that the same theory is able to encompass findings with both animals and humans.

Finally, the slender evidence available suggests that some animals are capable of using relational information more readily than others. If future research should confirm this suggestion, this will provide a rare indication of a difference between the intelligence of different nonhuman vertebrates (Mackintosh, Wilson, & Boakes, 1985; Macphail, 1982).

References

Astley, S. L., & Wasserman, E. A. (1992). Categorical discrimination and generalization in pigeons: All negative stimuli are not created equal. *Journal of Experimental Psychology: Animal Behavior Processes, 18,* 193–207.

Bhatt, R. S., Wasserman, E. A., Reynolds, W. F., & Knauss, K. S. (1988). Conceptual behavior in pigeons: Categorization of both familiar and novel examples from four classes of natural and artificial stimuli. *Journal of Experimental Psychology: Animal Behavior Processes, 14,* 219–234.

Blough, D. S. (1975). Steady state data and a quantitative model of operant generalization and discrimination. *Journal of Experimental Psychology: Animal Behavior Processes, 1*, 3–21.

Burdick, C. K., & Miller, J. D. (1975). Speech perception by the chinchilla: Discrimination of sustained /a/ and /i/. *Journal of the Acoustical Society of America, 58*, 415–427.

Cerella, J. (1980). The pigeon's analysis of pictures. *Pattern Recognition, 12*, 1–6.

D'Amato, M. R., & Van Sant, P. (1988). The person concept in monkeys (*Cebus apella*). *Journal of Experimental Psychology: Animal Behavior Processes, 14*, 43–55.

Gillan, D. J., Premack, D., & Woodruff, G. (1981). Reasoning in the chimpanzee: I. Analogical reasoning. *Journal of Experimental Psychology: Animal Behavior Processes, 7*, 1–17.

Gluck, M. A. (1991). Stimulus generalization and representation in adaptive network models of category learning. *Psychological Science, 2*, 50–55.

Gluck, M. A., & Bower, G. H. (1988). Evaluating an adaptive network model of human learning. *Journal of Memory and Language, 27*, 166–195.

Gonzalez, R. C., Gentry, G. V., & Bitterman, M. E. (1954). Relational discrimination of intermediate size in the chimpanzee. *Journal of Comparative and Physiological Psychology, 47*, 385–388.

Gulliksen, H., & Wolfle, D. L. (1938). A theory of learning and transfer: I. *Psychometrika, 3*, 127–149.

Hanson, H. M. (1959). Effect of discrimination training on stimulus generalization. *Journal of Experimental Psychology, 58*, 321–334.

Hennesey, T. M., Rucker, W. B., & McDiarmid, C. G. (1979). Classical conditioning in paramecia. *Animal Learning and Behavior, 7*, 417–423.

Herman, L. M., & Gordon, J. A. (1974). Auditory delayed matching in the bottlenose dolphin. *Journal of Experimental Analysis of Behavior, 21*, 19–29.

Herrnstein, R. J. (1990). Levels of stimulus control. *Cognition, 37*, 133–166.

Herrnstein, R. J., Loveland, D. H., & Cable, C. (1976). Natural concepts in pigeons. *Journal of Experimental Psychology: Animal Behavior Processes, 2*, 285–301.

Herrnstein, R. J., Vaughan, W., Jr., Mumford, D. B., & Kosslyn, S. M. (1989). Teaching pigeons an abstract relational rule: Insideness. *Perception and Psychophysics, 46*, 56–64.

Hintzman, D. L. (1986). 'Schema abstraction' in a multiple trace memory. *Psychological Review, 93*, 411–428.

Homa, D., Dunbar, S., & Nohre, L. (1991). Instance frequency, categorization, and the modulating effect of experience. *Journal of Experimental Psychology: Learning, Memory, and Cognition, 17*, 444–458.

Honey, R. C., & Hall, G. (1989). Acquired equivalence and distinctiveness of cues. *Journal of Experimental Psychology: Animal Behavior Processes, 15*, 338–446.

Huber, L., & Lenz, R. (1993). A test of the linear feature model of polymorphous concept discrimination with pigeons. *Quarterly Journal of Experimental Psychology, 46B*, 1–18.

Kehoe, E. J. (1986). Summation and configuration in conditioning of the rabbit's nictitating membrane response to compound stimuli. *Journal of Experimental Psychology: Animal Behavior Processes, 12*, 186–195.

Kinnaman, A. J. (1902). Mental life of two macacus rhesus monkeys in captivity. *American Journal of Psychology, 13*, 98–148.

Kluender, K. R., Diehl, R. L., & Killeen, P. R. (1987). Japanese quail can learn phonetic categories. *Science, 237*, 1195–1197.

Kohler, W. (1918). Nachweis einfacher structurfunktionen beim schimpansen und beim haus-huhn. *Abhandlungen der Preussischen Akademie der Wissenschaften, Physikalisch-Mathematische Klasse, 2*, 1–101. (Translated and condensed as Simple structural functions in chimpanzee and chicken. In W. D. Ellis, *A source book of gestalt psychology*. London: Routledge & Kegan Paul, 1969).

Kruschke, J. K. (1992). ALCOVE: An exemplar-based connectionist model of category learning. *Psychological Review, 99,* 22–44.

Lawrence, D. H., & DeRivera, J. (1954). Evidence for relational transposition. *Journal of Comparative and Physiological Psychology, 47,* 465–471.

Lea, S. E. G. (1984). In what sense do pigeons learn concepts? In H. L. Roitblat, T. G. Bever, & H. S. Terrace (Eds.), *Animal cognition* (pp. 263–276). Hillsdale, NJ: Erlbaum.

Lea, S. E. G., & Harrison, S. N. (1978). Discrimination of polymorphous stimulus sets by pigeons. *Quarterly Journal of Experimental Psychology, 30,* 521–537.

Mackintosh, N. J., Wilson, B., & Boakes, R. A. (1985). Differences in mechanisms of intelligence among vertebrates. *Philosophical Transactions of the Royal Society of London Series, B 308,* 53–65.

Macphail, E. M. (1982). *Brain and intelligence in vertebrates.* Oxford: Clarendon.

Maki, W. S., & Abunawass, A. M. (1991). A connectionist approach to conditional discriminations: Learning, short-term memory, and attention. In M. L. Commons, S. Grossberg,, & J. E. R. Staddon (Eds.), *Neural network models of conditioning and action* (pp. 241–278). Hillsdale, NJ: Erlbaum.

McClelland, J. L., & Rummelhart, D. E. (1985). Distributed memory and the representation of general and specific information. *Journal of Experimental Psychology: General, 114,* 159–188.

McClosky, M., & Cohen, N. J. (1989). Catastrophic interference in connectionist networks: The sequential learning problem. In G. Bower (Ed.), *The psychology of learning and motivation* (Vol. 24, pp. 109–165). San Diego: Academic Press.

Medin, D. L., & Schaffer, M. M. (1978). Context theory of classification learning. *Psychological Review, 85,* 207–238.

Morgan, M. J., Fitch, M. D., Holman, J. G., & Lea, S. E. G. (1976). Pigeons learn the concept of an "A". *Perception, 5,* 57–66.

Oden, D. L., Thompson, R. K. R., & Premack, D. (1988). Spontaneous transfer of matching by infant chimpanzees (*Pan troglodytes*). *Journal of Experimental Psychology: Animal Behavior Processes, 14,* 140–145.

Osgood, C. E. (1953). *Method and theory in experimental psychology.* New York: Oxford University Press.

Pavlov, I. P. (1927). *Conditioned reflexes.* New York: Oxford University Press.

Pearce, J. M. (1987). A model of stimulus generalization for Pavlovian conditioning. *Psychological Review, 94,* 61–73.

Pearce, J. M. (1988). Stimulus generalization and the acquisition of categories by pigeons. In L. Weiskrantz (Ed.), *Thought without language* (pp. 132–152). Oxford: Oxford University Press.

Pearce, J. M. (1989). The acquisition of an artificial category by pigeons. *Quarterly Journal of Experimental Psychology, 41B,* 381–406.

Pearce, J. M. (1991). The acquisition of abstract and concrete categories by pigeons. In L. Dachowski & C. Flaherty (Eds.), *Current topics in animal learning: Brain, emotion, and cognition* (pp. 141–164). Hillsdale, Erlbaum.

Pearce, J. M. (1994). Similarity and discrimination: A selective review and a connectionist model. *Psychological Review,* in press.

Pearce, J. M., & Redhead, E. S. (1993). The influence of an irrelevant stimulus on two discriminations. *Journal of Experimental Psychology: Animal Behavior Processes, 19,* 180–190.

Pearce, J. M., & Wilson, P. N. (1991). Failure of excitatory conditioning to extinguish the influence of a conditioned inhibitor. *Journal of Experimental Psychology: Animal Behavior Processes. 17,* 519–529.

Pepperberg, I. M. (1983). Cognition in the African grey parrot: Preliminary evidence for auditory/vocal comprehension of the class concept. *Animal Learning and Behavior, 11,* 179–185.

Pepperberg, I. M. (1987). Acquisition of the same/different concept by an African Grey parrot (*Psittacus erithacus*): Learning with respect to color, shape and material. *Animal Learning and Behavior, 15,* 423–432.

Pietrewicz, A. T., & Kamil, A. C. (1977). Visual detection of cryptic prey by blue jays. *Science, 195,* 580–582.

Porter, D., & Neuringer, A. (1984). Music discrimination by pigeons. *Journal of Experimental Psychology: Animal Behavior Processes, 10,* 138–148.

Posner, M. I., & Keele, S. W. (1968). On the genesis of abstract ideas. *Journal of Experimental Psychology, 77,* 353–363.

Premack, D. (1983). The codes of man and beasts. *Behavioral Brain Sciences, 6,* 125–167.

Redhead, E. S., & Pearce, J. M. (1994). Similarity and discrimination learning. *Quarterly Journal of Experimental Psychology,* in press.

Rescorla, R. A. (1976). Stimulus generalization: Some predictions from a model of Pavlovian conditioning. *Journal of Experimental Psychology: Animal Behavior Processes, 2,* 88–96.

Rescorla, R. A., & Wagner, A. R. (1972). A theory of Pavlovian conditioning: Variations in the effectiveness of reinforcement and nonreinforcement. In A. H. Black & W. F. Prokasy (Eds.), *Classical conditioning II: Current research and theory* (pp. 64–99). New York: Appleton-Century-Crofts.

Rilling, M. (1977). Stimulus control and inhibitory processes. In W. K. Honig & J. E. R. Staddon (Eds.), *Handbook of operant behavior* (pp. 432–480). Englewood Cliffs, NJ: Prentice-Hall.

Ryan, C. W. E. (1982). Concept formation and individual recognition in the domestic chicken (*Gallus gallus*). *Behavior Analysis Letters, 2,* 213–220.

Saldanha, E. L., & Bitterman, M. E. (1951). Rleational learning in the rat. *American Journal of Psychology, 64,* 37–53.

Savage-Rumbaugh, E. S., Rumbaugh, D. M., Smith, S. T., & Lawson, J. (1980). Reference— the linguistic essential. *Science, 210,* 922–925.

Schrier, A. M., Angarella, R., & Povar, M. L. (1984). Studies of concept formation by stumptailed monkeys: Concepts humans, monkeys, and letter *A. Journal of Experimental Psychology: Animal Behavior Processes, 10,* 564–584.

Schrier, A. M., & Brady, P. M. (1987). Categorization of natural stimuli by monkeys (*Macaca mulatta*): Effects of stimulus set size and modification of exemplars. *Journal of Experimental Psychology: Animal Behavior Processes, 13,* 136–143.

Shepard, R. N. (1987). Toward a universal law of generalization for psychological science. *Science, 237,* 1317–1323.

Shin, H. J., & Nosofsky, R. M. (1992). Similarity-scaling of dot-pattern classification and recognition. *Journal of Experimental Psychology: General, 121,* 278–304.

Smith, E. E., & Medin, D. L. (1981). *Categories and concepts.* Cambridge, MA: Harvard University Press.

Spence, K. W. (1936). The nature of discrimination learning in animals. *Psychological Review, 43,* 427–449.

Spence, K. W. (1937). The differential response in animals to stimuli varying within a single dimension. *Psychological Review, 44,* 430–444.

Sutton,, R. S., & Barto, A. G. (1981). Toward a modern theory of adaptive networks: Expectation and prediction. *Psychological Review, 88,* 135–170.

Vaughan, W., Jr., & Greene, S. L. (1984). Pigeon visual memory capacity. *Journal of Experimental Psychology: Animal Behavior Processes, 10,* 256–271.

von Fersen, L., & Lea, S. E. G. (1990). Category discriminations by pigeons using five polymorphous features. *Journal of the Experimental Analysis of Behavior, 54,* 69–84.

Wagner, A. R. (1969). Stimulus validity and stimulus selection in associative learning. In N. J. Mackintosh & W. K. Honig (Eds.), *Fundamental issues in associative learning* (pp. 90–122). Halifax: Dalhousie University Press.

Wagner, A. R. (1971). Elementary associations. In H. H. Kendler & J. T. Spence (Eds.), *Essays in neobehaviorism: A memorial volume to Kenneth W. Spence* (pp. 187–213). New York: Appleton-Century-Crofts.

Wasserman, E. A., DeVolder, C. L., & Coppage, D. J. (1992). Non similarity-based conceptualization in pigeons via secondary or mediated generalization. *Psychological Science, 3,* 374–379.

Watanabe, S. (1988). Failure of visual prototype learning in the pigeon. *Animal Learning and Behavior, 16,* 147–152.

Wilson, B., Mackintosh, N. J., & Boakes, R. A. (1985). Transfer of relational rules in matching and oddity learning by pigeons and corvids. *Quarterly Journal of Experimental Psychology, 37B,* 313–332.

Wilson, P. N., & Pearce, J. M. (1992). A configural analysis for feature-negative discrimination learning. *Journal of Experimental Psychology: Animal Behavior Processes, 18,* 265–272.

Woodbury, C. B. (1943). The learning of stimulus patterns in dogs. *Journal of Comparative Psychology, 35,* 29–40.

CHAPTER 6

The Neural Basis of Learning with Particular Reference to the Role of Synaptic Plasticity

Where Are We a Century after Cajal's Speculations?*

Richard G. M. Morris

I. INTRODUCTION

The study of animal learning and cognition is concerned with the psychological processes that subserve various types of learning and memory in animals. There has been, both historically and in contemporary work dating from roughly the mid-1960s, particular emphasis on the supposition that associative processes underlie many types of learning including several that appear, at least to initial inspection (e.g., taste aversion, spatial learning), to lie outside its explanatory domain. The reason for this wide scope is that associative learning is now recognized—and it is a recurring theme of several of the chapters in this handbook—as consisting of a much richer array of associative processes than the textbook description usually offered of classical and instrumental conditioning. If this generalized or "enriched" view of associative learning is correct, then one might reasonably expect the brain to possess one or more neural mechanisms through which stimuli, actions, and/or ideas can be "associated" in various ways. As such information is presumably represented in the brain as spatiotemporal patterns of neuronal firing, it is likely that associative processes will depend, at least in part, on neural mechanisms activated at the point of contact between neu-

* This chapter is based on a Plenary Lecture given by the author at the 25th anniversary meeting of the *European Brain and Behaviour Society* held in Madrid in September, 1993.

Animal Learning and Cognition

rones, that is, at synapses. Further, it also seems reasonable to suppose that neural mechanisms exist that enable the long-term storage of associations in such a manner that they may later be retrieved for the purposes of controlling behavior (including behavior that may not have occurred at the time of learning), or retrieved for combining flexibly with other learned associations in inferential ways.

Relatively few physiological psychologists interested in the neural basis of behavior have made a direct attack on such putative neural mechanisms. Instead, they have preoccupied themselves with the interface between psychology and neuroanatomy, as in work on the functions of the hippocampal formation, amygdala, and cerebellum; or with the interface between psychology and neuropharmacology, as in study of the mesolimbic dopamine system of the ascending cholinergic pathways of the brain. However, recent developments in neuroscience have led to significant conceptual discoveries about neuronal plasticity and a range of useful techniques with which the task or identifying the underlying neural mechanisms of associative learning might be tackled head on. In fact, research on the neural basis of plasticity has advanced to the point where it is realistic to consider whether there exist mechanisms, specifically mechanisms at the synaptic level, that might enable psychological theories of associative learning (such as those described by Hall in Chapter 2, this volume) to be implemented by neuronal machinery.

Persistent changes in synaptic efficacy have long been thought to be involved in the neural mechanisms of learning and memory. The idea has its roots in the writings of such figures as Ramón y Cajál, Konorski, and Hebb, and, today, finds expression in work ranging from experimental studies of the role of activity–dependent synaptic changes in certain kinds of learning through to neural network modeling in which such changes are held to occur according to specific learning rules. A synaptic locus has the merit of affording, at least in principle, greater storage capacity than a mechanism based on, for example, somatic changes in excitability, because the many thousands of synapses on an individual neuron (>10,000 on a hippocampal pyramidal cell) can participate on many different memories. Indeed, if information is represented in the vertebrate brain as distributed patterns of neural activity, which seems most likely, individual synapses can participate in many memories.

The purpose of this chapter is to consider experimental evidence bearing on whether synaptic plasticity is *causally* involved in some or, indeed, any form of learning. Given both the richer theoretical view we now have of associative learning processes and the existence of neural network models of learning and memory that incorporate similar learning rules, it would seem valuable to establish whether their fundamental assumptions—such as the idea that associative mechanisms are engaged in information processing—

have any neurobiological validity. Irrespective of the finer details and differential predictions of particular models, a common *motif* is that learning takes place through modifications of connections according to activity-dependent rules. It follows, first, that we should be able to observe such synaptic changes in vivo during learning and, second, that were we able to interfere with or enhance their expression, learning should proceed more slowly or more quickly, respectively. These are key predictions amenable to experimental test. The empirical issue at stake is therefore: *Does activity-dependent synaptic plasticity occur during learning and is its occurrence a necessary condition for learning?* I shall concentrate in the later sections of the chapter on one well-characterized form of synaptic plasticity—hippocampal long-term potentiation—but I shall first set out the broader intellectual context and some general issues.

II. HISTORICAL OVERVIEW

It is now 100 years since Ramón y Cajál gave his Croonian Lecture to the Royal Society in 1894 in which he first proposed that the growth of contacts between neurones was the likely basis through which the nervous system would alter in response to experience. This idea was discussed more fully in a well-known passage in the Spanish (1899, p. 1150) and French (1911, p. 887) editions of his treatise on the vertebrate nervous system:

> To become a pianist, an orator, a mathematician or a thinker one needs, does one not, not only a suitable physical constitution but also many years of exercise of mind and muscles? How might one understand this transformation and its slowness? One may do so by proposing, first, that pre-existing connections are reinforced by the exercise and then by adding the supposition that new pathways become established by ramification and progressive increase in the size of arborisations, both dendritic and axonal. (translation, I. M. L. Donaldson)

While, in this passage, Cajál distinguishes between the reinforcement of existing contacts and the growth of new ones, he sets his authority firmly behind the notion that experience, rather than the mere passage of time, causes localized anatomical changes in the nervous system and that one locus at which these would occur is at the contacts between neurones. These contacts are now called *synapses,* a term based on the Greek "to clasp," and coined by Sherrington around the same time as Ramón y Cajál's lecture (1897). Synapses are the points at which the axons of one neurone terminate in a specialized or *presynaptic* ending from which transmitter is released across a small gap or *synaptic cleft* to act on receptors in the cell membrane of specialized *postsynaptic* terminals of the target neurons (Figure 1). These pre- and postsynaptic elements of the synapse were first discovered by Cajál.

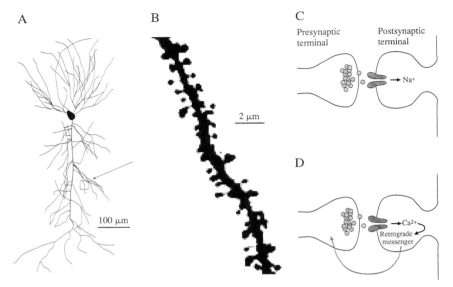

FIGURE 1   The synapse, conventional fast neurotransmission, and the possibility of bidirectional communication. (A) Camera lucida drawing of lucifer yellow–filled hippocampal CA1 pyramidal cell from a confocal microscope image (40× objective) showing ramifications of dendritic arborization. (B) Section of apical dendrite from same cell as shown in (A) (area of arrowed box) from confocal microscope image (100× objective). Drawings courtesy of Per Andersen (University of Oslo). (C) Conventional fast excitatory neurotransmission. L-glutamate is released from presynaptic vesicles, crosses the synaptic cleft, and causes sodium (Na^+) influx in the postsynaptic dendritic spine. (D) Possibility of bidirectional communication via intercellular or retrograde messenger from postsynaptic to presynaptic terminal. Drawings based on Figure 2 of Jessell and Kandel (1993).

Anatomical and physiological research over the past century has revealed the brain to possess a myriad of different types of neurons, various types of synaptic connections between them, and several patterns of synaptic transmission. The basic principles of fast and slow synaptic transmission have long been established, along with the cardinal principle that cell–cell communication is mediated by the release of neurotransmitters from the presynaptic terminal that then cross the synaptic cleft and bind onto receptors of the postsynaptic neuron. This unidirectional principle of communication is, however, being challenged by recent cell-biological research pointing to the existence of *intercellular* or *retrograde messengers* capable of passing from the postsynaptic to presynaptic terminal and there influencing transmitter release (Figure 1D; Jessell & Kandel, 1993).

A key set of issues in contemporary work on the neural mechanisms of learning is to understand the circumstances in which synapses alter (or, in modern parlance, show *plasticity*), how they change, and whether there is any causal connection between synaptic plasticity and learning. These issues are of importance to the readers of this volume because the concepts of synaptic plasticity, although cast in a language different from that of animal learning, echo ideas that have long been discussed by experimental psychologists. An early illustration of their relevance are the theoretical propositions made independently by Konorksi (1948) and by Hebb (1949) about the circumstances in which associative learning and synaptic growth should occur:

> . . . the optimal precondition for the formation of the conditioned reflex is the coincidence of excitation in the emitting center with a sufficiently abrupt rise of activity in the receiving center. In other words . . . the impulses set up by the center to be conditioned must reach the unconditioned centre in the stage when the latter's activation is growing. (Konorski, 1948, p. 106)
>
> . . . when an axon of cell A is near enough to excite cell B and repeatedly or persistently takes part in firing it, some growth process or metabolic change takes place in one or both cells such that A's efficiency, as one of the cells firing B, is increased. (Hebb, 1949)

Konorski's and Hebb's proposals differed in detail, and have been shown to give rise to quite different formal learning rules with only those of the former anticipating rules such as the Rescorla–Wagner equation (McClaren & Dickinson, 1990), but they share the honor of being the first statements of a circumstance in which synaptic change might take place. Konorski was also unique in anticipating the possible need for some statement of the circumstances in which *decreases* in neuronal connectivity might occur, a point later developed in the context of neural development by Stent (1973). But Konorski's book was not widely distributed in America and only Hebb acquired the dubious privilege of becoming an adjective. Neither Hebb nor Konorski were able to follow up their proposals with relevant experiments, but it was not long before their ideas were pursued—initially by neuroanatomists and later by neurophysiologists. Only now has the opportunity arisen for linking their findings to behavioral studies.

III. EXPERIMENTAL STRATEGIES FOR EXPLORING THE RELATION BETWEEN SYNAPTIC PLASTICITY AND LEARNING

There have been, since the early 1950s, several attempts to explore the possible relation between learning and neuronal or synaptic plasticity. An early and well-known example is the work of Rosenzweig, Bennett, and Diamond (1972), which was the first to establish experimentally that differential rearing of rats (isolation, social housing, and enriched environments) was associated with striking changes in cortical thickness and dendritic branching. This work is much in what may be described as the "Cajál tradition," in as much as its emphasis was on identifying the alterations in neuronal architecture that occurred as a function of differential experience. However, the environmental manipulations used were so global that it is impossible to know what psychological process or correlate was most important in relation to the anatomical changes observed (differential sensory input? stress-induced effects? motor activity? learning?). More recent studies summarized by Greenough and Bailey (1988) have refined this approach considerably. Thus, rats given extensive training to use only their left paw to retrieve food show anatomical changes specific to the topographically appropriate part of the right motor cortex; and *Aplysia* subject to long-term habituation or sensitization of specific sensorimotor reflex pathways show several long-lasting changes in synaptic morphology (e.g., up- and down-regulation, respectively, of the numbers of nerve terminals and size of the synaptic apposition zone). These neuroanatomical changes can sometimes be quite dramatic. For example, even experiences as short as a single trial, such as 1-trial avoidance learning in the chick (avoidance of a bead covered with methyl anthranylate), appear to give rise to striking changes in synaptic morphology in areas of the brain implicated in the learning process on other grounds (Doubell & Stewart, 1993).

Interesting as these findings may be to the neuroscientist, they have had little impact on the field of animal learning. The reason for this is clear enough. The character of the work is strictly *top-down* and the behavioral paradigms used and the neural observations made are not of a kind that would be likely to inform or constrain psychological theories in a *bottom-up* fashion. However, not all work on the neural basis of learning is of this character. Work on the neural correlates of filial imprinting (see Shettleworth, Chapter 7, this volume), for example, has been analyzed from the perspective of contemporary animal learning theory (e.g., Bolhuis, 1991) and revealed to show a number of changes that might influence psychological accounts of the phenomenon. To take one example, the efficacy of excitatory synaptic receptors thought to participate in information storage shows slow time-dependent changes over a period of several hours that might reflect the process of memory consolidation (McCabe & Horn, 1991). If this interpretation is correct, one may predict that the performance in a two-alternative, forced-choice preference test following imprinting would improve with the passage of time after the end of training (and in the absence of further training). There is, of course, no reason in principle why such a finding could not emerge purely on the basis of behavioral studies alone. However, while logic dictates that we can understand what neuronal mechanisms do only once we have a satisfactory account of the overlying psychological processes they implement, the *path of discovery* in the mapping of process to mechanism is more of a two-way street. In this instance, thinking about how a change in receptor efficacy comes about physically brings to mind biochemical processes that might require the passage of time in a way that thinking about changes in ΔV in the Rescorla–Wagner equation does not.

This line of thought is the appropriate point to introduce a quite different research strategy that has been pursued in many contemporary studies investigating the role of synaptic plasticity in learning. The work described so far has the description of a type of learning as its starting point and only then attempts to identify the underlying neural mechanisms—the work on filial imprinting being a classic example in that years of careful behavioral studies by Bateson and his colleagues (Bateson, 1976) preceded the subsequent research by Horn, Bateson, and Rose (see Horn, 1985) to identify the neural substrate. The alternative strategy begins with a specific form of synaptic plasticity and the attempts to utilize discoveries about its neural mechanisms for the purposes of functional studies. Clearly, there is a gamble here in as much as it may turn out that the particular type of synaptic plasticity chosen is, by analogy with the spandrels of San Marco (Gould & Lewontin, 1979), of no particular functional significance whatsoever. However, the advantages of this research strategy are, in my view, more than enough to offset the risk. For neuroscientists have identified a small number of distinct types of synaptic plasticity occurring at the major excitatory synapses of the brain, the physiological circumstances in which such changes in connectivity come about, and are well on the way to revealing the underlying neuronal mechanisms at the cell-biological and molecular-biological level. This research effort, by literally hundreds of scientists, has thrown up an arsenal of electrophysiological, pharmacological, biochemical, and molecular-biological tools that can be exploited in behavioral studies. Thus, by taking the neuron as one's conceptual starting point (an article of faith among neuroscientists!) and then considering the plasticity of neuronal connectivity, one can move *up* to behavioral studies, *down* to more exacting neuroscientific studies, or *jump* between levels as appropriate (Figure 2). If this chapter carries any take-home messages to students of animal learning, I hope these will be: (1) the sense of excitement prevalent within contemporary neuroscience about our emerging understanding of the neural mechanisms of synaptic plasticity; and (2) the conceptual contribution that students of animal learning could make to unraveling its functional significance. It will become apparent, as the chapter unfolds, that we have only scratched the surface of the problems inherent in the mapping of psychological process to neural mechanism.

For this alternative strategy to be viable, we need a well-understood form

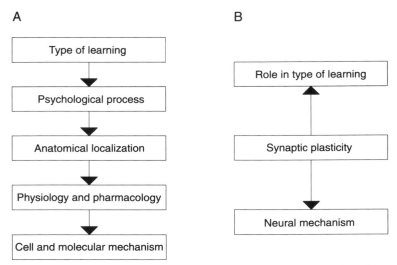

FIGURE 2   Two different research strategies for investigating the neural mechanisms of learning. (A) Traditional reductionist framework beginning with phenomena of learning. (B) Interactive strategy involving a targeted analysis of a specific form of synaptic plasticity.

of synaptic plasticity as a starting point. As we shall see below, several different forms of long-lasting, activity-dependent synaptic plasticity have been identified, the best characterized of these being hippocampal long-term potentiation (Bliss & Lømo, 1973). That the synaptic mechanisms on which it depends converge with those identified from the research on filial imprinting (pursuing the first strategy) is some grounds for thinking that long-term potentiation may be a phenomenon whose underlying mechanisms are of exquisite relevance to various types of learning.

IV. HIPPOCAMPAL LONG-TERM POTENTIATION

A. The Phenomenon of Long-Term Potentiation and Its Principal Characteristics

Hippocampal long-term potentiation was discovered in 1966 and first reported in detail by Bliss and Lømo (1973). They implanted electrodes into the brains of anesthetized rabbits such that the stimulating electrode could activate a pathway entering the hippocampal formation (called the "perforant path") while the recording electrode monitored the consequent electrical changes in a region called the hilus or polymorphic zone of the dentate

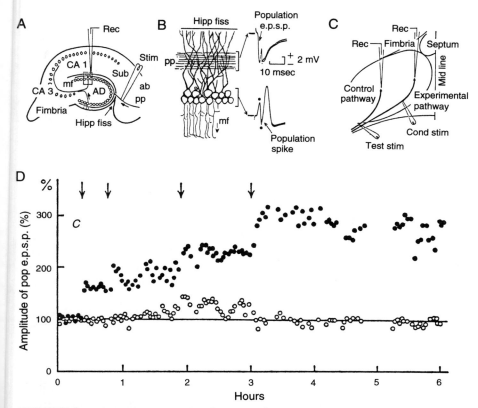

FIGURE 3   Bliss and Lømo's (1973) discovery of long-term potentiation. (A) Cross-sectional drawing of hippocampus showing sites of stimulating (stim) and recording (rec) electrodes. (B) Drawing showing negative-going field EPSPs recorded from the molecular layer of the fascia dentata above the granule cells, and positive-going fEPSPs, including population spike, from the cell-body layer and hilus. (C) Drawing showing ipsilateral two-pathway pathway experiment, with two stimulating and recording electrodes. (D) Growth in amplitude of the fEPSP of the experimental pathway (solid circles) following successive bursts of high-frequency stimulation. Adapted from Bliss and Lømo (1973). Courtesy of the *Journal of Physiology*.

gyrus (Figure 3). They found that if periods of low-frequency stimulation were interspersed with brief trains of high-frequency stimulation, the size of the evoked potentials became larger. The increase was observed in both the synaptic component of the field potential (usually referred to as the field excitatory postsynaptic potential, or fEPSP) and a second component of the wave form reflecting the firing of dentate granule cells (referred to as the population spike). This potentiation of the fEPSP and population spike occurred shortly after the high-frequency stimulation and lasted a long time. Bliss and Lømo observed that the magnitude of the potentiation of the fEPSP could account for only a component of the potentiation of the population spike, indicating that this long-lasting change, generally referred to as long-term potentiation (LTP), involved both synaptic and somatic components.

Several characteristics of LTP were apparent at the outset, while others have emerged in the course of research on the phenomenon over the past twenty years (see Bliss & Collingridge, 1993, for a recent detailed review). Those that have attracted most interest as properties desirable of a candidate learning mechanism are persistence, synapse specificity, and associativity (Figure 4). *Persistence* refers to the fact that, while LTP sometimes lasts only a few hours, certain methods of inducing it result in an increase of synaptic efficacy lasting days or weeks (Barnes, 1979; Bliss & Gardner-Medwin, 1973; Racine, Milgram, & Hafner, 1983). *Synapse specificity* refers to the fact that not all the synapses on a neuron, or set of neurons, will demonstrate the increase—only those activated by high-frequency presynaptic activity (Andersen, Sundberg, Sveen, & Wigström, 1977; Lynch, Dunwiddie, & Gribkoff, 1977). *Associativity* refers to the fact that synapses that are stimulated too weakly to cause an increase in their own right can, if stimulated at the same time as other stronger afferents to the same neuron, nevertheless show potentiation (Barrionuevo & Brown, 1983; Levy & Steward, 1979; McNaughton, Douglas, & Goddard, 1978). Taken together, these properties of LTP are suggestive of a possible underlying neural mechanism that could be useful for associative learning. Whatever the mechanism is in detail, it enables the temporal conjunction of a weak stimulus (w) with a strong one (S) to be detected by a neuron (N) and then expressed as an alteration in connectivity in a stimulus-specific way (i.e., w can more easily activate N). Such a cellular mechanism of detection and expression cannot, on its own, enable a neural representation of S to be evoked by that of w, but, embedded into an appropriate neural architecture, this simple property of associative conditioning might yet be realized (see McNaughton & Morris, 1987). The cell-biological machinery can be thought of as the bit that does the job of detecting and signaling stimulus contingencies, while the neural circuit and neural coding of information determine the nature of the representation that is realized.

B. The Neural Mechanisms of Long-Term Potentiation

While some students of animal learning may prefer to be spared the details of how LTP is thought to work physiologically (they should skip to Section V), others may be interested in a beautiful and still-unfolding story. The aim

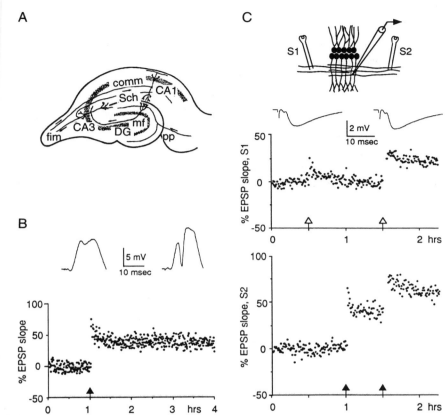

FIGURE 4   Characteristics of hippocampal long-term potentiation. (A) Cross-sectional view of hippocampus orthogonal to longitudinal axis showing both perforant path, dentate gyrus synapse (commonly studied in vivo) and Schaffer collateral, CA1 synapse (commonly studied in vitro). (B) *Persistence.* Following the high-frequency train, field EPSPs remain elevated at a new stable level for several hours. (C) *Synapse specificity.* Strong stimulation (solid arrow) given after 1 hour to pathway S2 (see drawing) gives LTP specific to pathway S2. Pathway S1 does not change; *Associativity.* Weak stimulation (open arrows) to pathway S1 does not give LTP unless paired with simultaneous strong stimulation (solid arrows) to an independent pathway (S2). From Bliss and Collingridge (1993). Courtesy of Macmillan Publishers.

has been to find a satisfactory reductionist account of persistence, synapse specificity, and associativity. In the process of unraveling such an account, several useful tools for investigating the functional significance (if any) of LTP have become apparent. One could not have a clearer illustration of the fact that the path of discovery in science, if not the logic of explanation, sometimes works *bottom-up* as well as *top-down*.

1. Induction of LTP

At an excitatory central synapse, transmission is mediated by the release from the presynaptic terminal of packets of a neurotransmitter, L-glutamate, which cross the synaptic cleft and have the potential to act on an array of postsynaptic glutamate receptors embedded in the membrane of the target neuron. Fast synaptic transmission, at least in the hippocampus, is mediated by two molecules of L-glutamate binding to the ionotropic AMPA receptor, which then opens an ion channel through which sodium ions (Na$^+$) move to enter the target neuron (Figure 5A). This influx of positively charged ions increases the voltage of the dendritic spine of the target neuron, on which the postsynaptic AMPA receptors are located, giving rise to

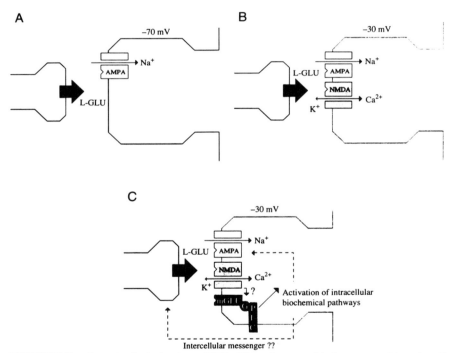

FIGURE 5   Steps in the induction of LTP. (A) L-glutamate binds to recognition site of AMPA receptor giving rise to Na$^+$ influx via its associated ion channel and, thus, fast synaptic transmission. (B) L-glutamate also binds to the NMDA receptor. The ion channel associated with the NMDA receptor is, ordinarily, blocked by the divalent cation Mg^{2+}. However, under conditions of tetanic stimulation or where the postsynaptic cell has otherwise become depolarized for a sustained period (note depolarization to -30 mV), the Mg^{2+} block is released and Ca^{2+} can enter the spine. K$^+$ may leave by the same channel. (C) Activation of both NMDA and mGLU receptors results in activation of intracellular biochemical pathways resulting in possible synthesis of intercellular messenger (see text for discussion).

what is called a "miniature" excitatory postsynaptic potential (mEPSP). Their amplitude is typically in the region of about 250 μV (depending on the method of measurement and the state of the synapse). The flow of current across several synapses of a single neuron summate and, if the summated potential is sufficient to raise the resting potential from about -65 mV to about -50 mV, will collectively fire the cell. The action potential so generated can then pass down the target neuron's axon, cause release of transmitter at its terminals, and so on.

One way of changing the probability with which a cell will fire is, therefore, to change the efficacy of synapses. This is where the other ionotropic glutamate receptor comes into the story (Figure 5B); for L-glutamate will bind not only to the ligand-gated AMPA receptor (a ligand is a neurotransmitter), but also to the ligand- and voltage-gated NMDA receptor. Ordinarily (i.e., when the resting potential is at -65 mV), the binding of L-glutamate to NMDA receptors is without further consequence because the ion channel associated with these receptors is blocked by the divalent cation magnesium (Mg^{2+}). However, if the voltage of the postsynaptic cell is raised to about -30 mV for a sustained period, Mg^{2+} is released from the ion channel ("voltage gating") and the L-glutamate bound to the receptor's recognition site ("ligand gating") results in opening of the ion channel. NMDA receptor ion channels are more permeable to Ca^{2+} than to Na^+. Thus, successful activation of NMDA receptors selectively results in Ca^{2+} entering the dendritic spine. This Ca^{2+} signal will occur only if there has been presynaptic activity (causing the release of L-glutamate) and sustained postsynaptic depolarization (causing the release of the Mg^{2+} block). Thus, NMDA receptors serve as conjunction devices—recognizing the very conditions anticipated so long ago by Konorski and Hebb as likely conditions for associative learning. We are, however, still some way from "explaining" LTP in reductionist terms.

It used to be thought that the Ca^{2+} signal entering via the NMDA receptor might be sufficient to trigger various Ca^{2+}-dependent enzymes whose activity would, somehow, alter the efficiency of AMPA receptors or otherwise bring about changes that would be manifest as a change in synaptic weight. The key discovery behind this supposition was Collingridge, Kehl, and McLennan's (1983) finding that a selective antagonist of NMDA receptors, 2-amino-5-phosphonopentanoic acid (AP5), blocks LTP. However, while this observation has been amply confirmed on numerous occasions, the role of NMDA receptors in providing the necessary Ca^{2+} transient is now thought to be an oversimplification; for there is a third type of glutamate receptor to consider as well (Figure 5C). This is the recently described metabotropic glutamate receptor (mGLUR; Schoepp, Bockaert, & Sladaczek, 1990), which, unlike ionotropic AMPA or NMDA receptors, does not have an associated ion channel. It is a member of a large family of so-called metabotropic receptors that are coupled to membrane-bound "G-proteins" whose activation can trigger a cascade of intracellular biochemical pathways. Their relevance to the LTP story is that it has recently been discovered that sustained activation of mGLURs by the selective agonist 1S,3R-ACPD can induce LTP and that blockade of mGLURs by the selective antagonist MCPG inhibits the induction of LTP, respectively (Bashir, et al., 1993; Bortolotto & Collingridge, 1993). These findings are still controversial, but it now seems possible that Ca^{2+} entry through the NMDA receptor interacts in some as yet ill-defined way with the mGLUR and that this interaction gives rise to a massive release of Ca^{2+} from intracellular stores. Thus, speaking metaphorically, the Ca^{2+} entry through the NMDA receptor is like the push of a boulder over a cliff, the boulder gathering momentum is the release of Ca^{2+} from intracellular stores, and its eventual impact into the sea below is the triggering of enzymes that give rise to the expression of LTP.

2. Expression of LTP

So, how is LTP expressed? There are several possibilities. The simplest is that the ion channel associated with the AMPA receptor is altered, perhaps by phosphorylation, such that a greater Na^+ current flows through it upon subsequent activation. It might even be the case that individual synapses could have an array of functional and nonfunctional AMPA receptors and that LTP is a Markov-like conversion of "silent" AMPA receptors from a nonfunctional to a functional state. In either case, a change in the efficacy of AMPA receptors would constitute a *postsynaptic* mode of expression. Lynch and Baudry (1984) were the first to articulate such a hypothesis about the locus of expression of LTP, and others since have taken the same stance.

An alternative (or additional) possibility is that a greater amount of transmitter is released upon subsequent activation of the presynaptic terminal—a *presynaptic* mode of expression. This requires an interestingly more complex scheme, for it is clear that the initial steps of LTP induction are definitively postsynaptic and thus some message must be passed back from the postsynaptic terminal to the presynaptic terminal until the conditions for the induction of LTP have been met. There are several candidate messengers, which, like so many messengers of news, do not have an easy life. They include arachidonic acid (Bliss, Douglas, Errington, & Lynch, 1986), the small, membrane-permeable molecule nitric oxide (Böhme, Bon, Stutzmann, Doble, & Blanchard, 1991; Garthwaite, Charles, & Chess-Williams, 1988; O'Dell, Hawkins, Kandel, & Arancio, 1991; Schumann & Madison, 1991), and K^+ permeating out of the NMDA receptor at the time that Ca^{2+} itself enters (Collingridge, 1992). There is evidence supporting and against each of these proposals. Perhaps more important than the physical identity of the messenger is the principle it opens up, namely that the synapse that detects associative conjunctions need not be the only one to be influenced by the conjunctive event. Strictly speaking, the synapse specificity of LTP means that the only synapses to be potentiated are those at which presynaptic activity is coupled to postsynaptic depolarization. But the existence of an intercellular or retrograde messenger opens up the possibility that nearby synapses could (so to speak) "listen in" on an exciting conjunction of events nearby and make their own decisions about what to do. Whether the nervous system is quite so liberal minded is not yet clear, but evidence that something like this may occur has been reported by Bonhoeffer, Staiger, and Aertsen (1989) and has formed the basis of an intriguing proposal by Edelman and his colleagues about the developmental formation of cortical columns with common functional properties (e.g., directional selectivity in area V1) in the absence of interneuronal circuitry (Gally, Montague, Reeke, & Edelman, 1990).

Finally, a third possibility is that LTP involves *both* enhanced presynaptic release of transmitter *and* enhanced postsynaptic receptor sensitivity, with the time course of the transition between these components still a matter of debate. This is a mixed *pre- and postsynaptic* hypothesis. One version of this idea, defended by Collingridge (see Bliss & Collingridge, 1993), holds that tetanic stimulation gives rise to a short-term form of potentiation (STP), which decays within 60 minutes and is largely dependent on presynaptic changes. STP then gives rise to a longer lasting LTP, which is characterized by postsynaptic changes in receptor sensitivity. An alternative view (Larkman, Stratford, & Jack, 1991) holds that whether LTP is expressed pre- or postsynaptically depends on the initial probability (p) of transmitter release. If p is small, LTP tends to be presynaptic and is characterized by an increase in p; whereas, if p is large, LTP tends to be postsynaptic and is characterized by a change in quantal size (generally taken to imply an increase in postsynaptic receptor sensitivity).

This is not the place to review critically all the physiological evidence relevant to the concepts discussed above—the interested reader should consult Bliss and Collingridge's (1993) review. What is important is to see that the physiological properties of persistence, synapse specificity, and associativity can be understood (if not yet fully explained) within a framework based around the biophysical properties of excitatory glutaminergic neurotransmission. *Associativity* falls out of the dual ligand- and voltage-gated properties of the NMDA receptor, amplified by the actions of mGLUR activation; *synapse specificity* arises because the intradendritic Ca^{2+} transient triggers enzymes that have local rather than neuronewide action (but the locality of their action may not be exclusive to the particular synapse at which these enzymes are triggered); and *persistence* occurs because the changes brought about, be they alterations of transmitter release or postsynaptic receptor efficacy, are believed to be long lasting. In short, LTP has several desirable properties of a candidate learning mechanism and, although our understanding of its mechanisms is incomplete, enough is already known for its induction and expression to be experimentally controlled for the purpose of functional studies.

V. OTHER TYPES OF ACTIVITY-DEPENDENT SYNAPTIC PLASTICITY AND THE NATURE OF SYNAPTIC LEARNING RULES

While associative LTP is the most prominent and best studied type of long-lasting synaptic plasticity, it is important to recognize that it is not the only type of activity-dependent change seen in the central nervous system. The existence of other forms of synaptic plasticity has considerable implications for thinking about synaptic learning rules and, with it, the extent to which different forms of plasticity encode the variance or invariance of events. First, there are a family of short-lasting changes such as facilitation, augmentation, and post-tetanic potentiation, which last for periods ranging from seconds through, at most, a few minutes. I shall not consider these further (although they may be relevant to short-term memory). Second, there are several forms of non-NMDA-dependent LTP, one of which is expressed at the terminals of mossy fibers in the hippocampus. Third, there are various forms of activity-dependent, long-term depression (LTD), which enable synapses to decrease in weight.

A. Mossy Fiber Potentiation

A qualitatively different form of long-term potentiation is found on one of the intrinsic pathways of the hippocampus—the mossy fibers that innervate CA3 from the dentate gyrus. Its determinants remain a matter of controversy. It may (Zalutsky & Nicoll, 1990) or may not (Johnston, Williams, Jaffe, & Gray, 1992) be associative, and may or may not be Ca^{2+} dependent. What is intriguing is that it occurs at the largest synapses in the brain and on cells, the CA3 pyramidal neurones of the hippocampus, whose neuronal architecture makes them particularly well suited to serve as elements of an episodic or event-based memory system (Rolls, 1990). As Treves and Rolls (1992) have emphasized, CA3 pyramidal neurones are unusual in having two sets of inputs that terminate at synapses *with* NMDA receptors (the perforant path and their own recurrent collaterals) and one set of inputs *without* NMDA receptors (the mossy fibers). This circuit arrangement may be to facilitate different forms of neuronal processing at the storage and retrieval phases of learning. Specifically, by arranging for one set of inputs to terminate near the cell body at strong synapses without NMDA receptors, the pattern of presynaptic activity on these afferents will dominate whether or not that neuron is depolarized to $-30\,mV$ for a sustained period irrespective of the pattern of co-occurrent input from terminals at the weaker synapses with NMDA receptors. However, the pattern of input at the terminals ending in NMDA receptors will determine which of these synapses get potentiated. In this way, it seems that activity at mossy fibers can control whether or not an event is "stored" in memory but may be unnecessary for its subsequent retrieval. This intriguing prediction remains to be tested. Mossy fiber potentiation could then come into the story as a mechanism for boosting the efficacy of specific mossy fiber terminals during the storage phase.

B. Long-Term Depression

Not all synaptic change is "growth" as Ramón y Cajál or Hebb envisaged it. Synaptic *depression* may also occur. The simplest form of synaptic depression would be what amounts to "forgetting," whereby terminals decay back to their baseline state over time. When LTP is induced in vivo in animals prepared with chronic indwelling electrodes, decay of fEPSPs back to baseline is always observed with a time course that can vary from hours to weeks (Barnes, 1979; Bliss & Gardner-Medwin, 1973; Racine et al., 1983). It is possible, indeed likely, that a major component of this long-term decay is strictly time dependent. However, neural activity continues throughout and, thus, it is possible that this decay is partly activity dependent. The question then arises: What are the circumstances under which activity-dependent LTD occurs? Are these the mirror image of those giving rise to LTP?

   Neurophysiologists distinguish conceptually between two types of LTD—heterosynaptic LTD and homosynaptic LTD. Heterosynaptic LTD means depression occurring at *other* synapses than those that trigger it, while homosynaptic LTD means depression occurring at the *same* synapses as those that trigger it. Recasting this into the language of the Rescorla–Wagner equation, heterosynaptic LTD would mean that the presentation of a stimulus A can, under certain conditions (specifically when V_A increases), give rise to negative values of ΔV_B and ΔV_C on the *same* learning trial and in the *absence* of presentations of stimuli B and C (a change which would not normally occur according to the Rescorla–Wagner rule); while homosynaptic LTD would mean, more conventionally, that certain conditions of stimulus presentation can give rise to negative values for ΔV for any given stimulus. The latter is unproblematic, but the former may need a word of explanation because it is clearly unconventional. Heterosynaptic changes on a neurone mean that some triggering event at a synapse, be it for LTP or LTD, gives rise to diffuse changes that affect neighboring synapses. One way of rationalizing this might be to suppose that, to protect itself from injury through overexcitation, a neurone has machinery to adjust synaptic weights (ω) such that the sum of its synaptic weights ($\Sigma\omega$) always adds roughly to a constant. If this were the case, the conditions for inducing LTP on one subset of a neurone's afferents would, simultaneously, be the very conditions for reducing the efficacy of neighboring synapses that did not participate in the LTP so induced. To describe this, as above, as analogous to decreasing V_B or V_C upon increasing V_A is an oversimplification because it assumes that the neural code in which stimuli are represented involves a one-to-one mapping between stimulus X and synapse X—and this is at best unlikely. A better analogy would, therefore, be in terms of interference. However, the spirit of the earlier point is correct in as much as thinking about decreases in synaptic weights in neuronal terms and, by inference, decreases in associative strength in neuronal terms, encourage one to consider possibilities that may not emerge from a purely psychological analysis.

What is the empirical evidence for LTD? Heterosynaptic LTD, it turns out, can indeed occur when LTP is induced. For example, Levy and Steward (1979) showed that potentiation of the ipsilateral perforant path in vivo resulted in depression of the weaker contralateral perforant path, provided the contralateral pathway was inactive at the time of LTP induction. Abraham and Goddard (1983) showed a similar but symmetrically reciprocal relationship between the medial and lateral components of the perforant path. However, the effect is weak and there is some doubt about whether it is long lasting. There is also general agreement that, at least in vitro, where stimulus patterns can be more precisely controlled, LTP can be induced without concomitant LTD. That is, LTD on other synapses is not an inevitable correlate of LTP. Homosynaptic LTP is equally controversial. Stanton and Sejnowski (1989) claimed that particular patterns of stimulation gave rise to an associative homosynaptic LTD. Using in vitro hippocampal slices, they potentiated one synaptic pathway with repeated bursts of 5 high-frequency pulses at 200-ms intervals. They then stimulated a second and weaker pathway to the same set of CA1 neurons with single pulses timed to arrive at the midpoint of the 200-ms interval. This would be the peak of neuronal after-hyperpolarization when the target neurones' intracellular potential would be most negative. They claimed that this stimulation protocol induced LTD on the second pathway. If true, this is an important result (see Morris & Willshaw, 1989) because it implies that associative activity-dependent LTD really occurs and that the precise timing of stimulus presentation to the hippocampus is critical. It would also imply that LTD is triggered when there is presynaptic activity but no postsynaptic activity (the opposite of heterosynaptic LTD). However, several laboratories have attempted and failed to document the phenomenon (e.g., Paulsen, Li, Hvalby, Andersen, & Bliss, 1993). Artola and Singer (1990) have described a form of LTD in neocortex in vitro, and Dudek and Bear (1992) have introduced a protocol, also in vitro, involving the delivery of 900 weak stimuli at 1 Hz interspersed between two phases of conventional low-frequency test pulses at 1 per 30 s. This produces a reliable LTD that can be seen upon recommencing the second phase of test pulses. Dudek and Bear's LTD is AP5 sensitive. The phenomenon has been reproduced independently (Mulkey & Malenka, 1992) and is widely agreed by other laboratories to be robust. As weak stimuli would not be expected to have much effect postsynaptically, the conditions for inducing this form of LTD can also be said to involve presynaptic activity but no postsynaptic activity. An outstanding difficulty is that the Dudek and Bear form of LTD is most prominent in brain tissue taken from young animals (and may therefore represent a form of synaptic pruning) and has not yet been thoroughly documented in adult animals in vivo.

C. Synaptic Learning Rules

The empirical study of LTP and LTD is accompanied by a range of theoretical proposals about so-called "synaptic learning rules." The simplest of these is the Hebbian scheme, which supposes that the increase in synaptic weight (Δw) is a function of the degree of correlation between pre- and postsynaptic activity (Figure 6). Sejnowski's (1977) modification of this scheme—the covariance rule—allows Δw to become negative when presynaptic activity is high but postsynaptic activity is low. Biennenstock, Cooper, and Munro (1982) have proposed a rule (the BCM rule) comprising both potentiation and depression and a sliding threshold (θ). This also requires Δw to be negative when postsynaptic activity (c) is low, but because Δw is proportional to the product of ϕc and the amount of presynaptic activity (d), Δw goes to zero when there is neither presynaptic nor postsynaptic activity (i.e., when both d and $c = 0$). Artola, Brocher, and Singer (1990) have also proposed a variant of this latter rule.

Study of the circumstances in which LTP and LTD occur is currently of great interest among neuroscientists (Malenka, 1993). Theoretical work by Willshaw and Dayan (1990) has established that having synaptic weights that both increase and decrease according to an activity-dependent rule is not only necessary to prevent catastrophic interference through saturation, but also improves the "signal-to-noise" ratio of a distributed associative matrix and, in turn, its memory storage capacity.

Students of animal learning theory will recognize many points of similarity between these rules and those of formal conditioning theories. It would, then, seem natural if research on the functional significance of LTP and LTD made fairly direct contact with modern research on classical conditioning. Curiously, this has not happened—much of the work to date

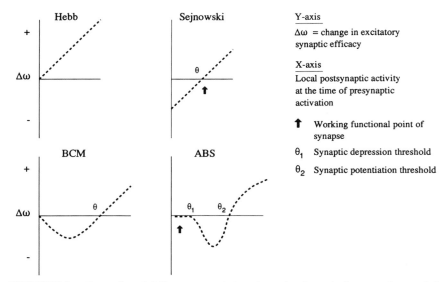

FIGURE 6   Examples of different synaptic learning rules. In each diagram, the y-axis is the change ($\Delta\omega$) in synaptic efficacy and the x-axis is the level of postsynaptic activity at the time of presynaptic activation. The rules shown include the traditional "Hebbian" scheme in which only increases in ω can occur; Sejnowski's (1977) covariance rule in which both increases and decreases occur, linearly related to the level of postsynaptic activation; the Biennenstock, Cooper, and Munro (BCM) (1982) rule in which a nonlinear function (ϕc) determines the change in ω such that $\omega = d \times \phi c$, where d is the level of presynaptic activity—the point where $\phi c = 0$ is determined by an adjustable threshold θ; and Artola, Brocher, and Singer's (1990) scheme (the ABS rule), which also has a nonlinear function, but a region of stability when the levels of pre- and postsynaptic activity are both low.

(including my own) having taken ideas about the functions of the hippocampal formation as its starting point and the primary determinant of the choice of behavioral training paradigm. This mismatch between experimental work on the functions of synaptic plasticity and theoretical work in animal learning is unfortunate. The point will be developed further in the conclusion to this chapter. For the present, it is necessary to add only one final point.

D. Encoding Variance and Invariance

Current theories of classical conditioning have been described, in a nice metaphor, as being concerned with representing "the causal structure of the world" (Dickinson & Mackintosh, 1980). To do this, they attempt to capture the regularities or conditional probabilities of one event given another.

They are, therefore, theories in which increases and decreases of associative strength come to represent the *invariance* of the world. To be implemented in neural hardware one would (probably) need both activity-dependent synaptic potentiation and activity-dependent synaptic depression. Suppose, however, one wished to have, *in addition,* a system at the brain's disposal for encoding recent events. Such a system might be a time–limited episodic memory upon which the brain could draw in order to remember events that occurred up to a few days before and which might, in their conditional relations with other events, temporarily contradict the invariant associations stored in long-term memory. Such a system would be immensely useful for encoding the unusual and often inconsequential events of everyday life but would also need to have the potential to alter long-term memory. Holding onto information for longer than the usual span of short-term or working memory (Baddeley, 1992), it might be adequately implemented with rapid ("1-trial") synaptic potentiation together with strictly time-dependent decay (Figure 7). It would not need activity-dependent LTD because saturation of synaptic weights would never occur—the storage limit of the system being never seriously approached given the usual rates at which events occur and, in general, are conveniently forgotten. Such a system could include the hippocampal formation. Indeed, the evidence to date indicates that the hippocampus shows robust LTP, which decays over time but, as yet, there is no strong indication of activity-dependent homosynaptic LTD in adult animals in vivo.

Encoding Invariance	Encoding Variance
Long-term associative memory system for encoding reliable relationships between stimuli and events.	Intermediate memory system for encoding recent events and relationships.
LTP and LTD ?	*LTP plus time-dependent decay ?*

FIGURE 7   Encoding invariance and variance. (A) Modern theories of animal learning have learning rules incorporating an independence of path assumption whereby the associative strength (V) is a single one–dimensional parameter that can be increased or decreased slowly as a function of experience. Such systems are suitable for encoding associations in long-term memory and are relatively immune from disproportionate influence by recent events, that is, useful for encoding invariance. (B) An intermediate memory system useful for keeping track of recent events might not need activity-dependent depression. Synaptic enhancement and time-dependent decay may suffice.

VI. EXPLORING THE EMPIRICAL RELATION BETWEEN HIPPOCAMPAL LONG-TERM POTENTIATION AND LEARNING

Having described the most prominent form of long-term, activity-dependent synaptic potentiation (LTP), and recent experimental and theoretical research on synaptic depression (LTD), the question arises: Are the underlying mechanisms of these physiological phenomena of any functional significance? As no systematic work to date has been done on LTD, I shall focus on research examining whether associative hippocampal LTP occurs during, and is necessary for, certain kinds of learning. Several predictions follow from the analysis presented so far:

1. There should be a statistical *correlation* between physiological indices of LTP and behavioral measures of learning.
2. *Saturation* of LTP should interfere with existing memories and occlude subsequent learning.
3. *Erasure* of LTP should cause forgetting of certain types of memory.
4. *Blockade* of LTP should impair certain types of learning and the formation of long-term memories.

A. Do Long-Term Potentiation-like Changes Occur in Association with Learning and Does Their Magnitude Correlate with the Extent of Learning?

The correlational strategy is, in principle, the most direct attack on the question and the experimental design is straightforward. Recordings of hippocampal fEPSPs (or transmitter release, postsynaptic receptor efficacy, etc.) are taken before and after a variety of different learning experiences. The simple prediction is that a lasting increase in fEPSP (or other measures) should be observed following learning if LTP has been involved. Types of learning that depend on an LTP-like mechanism should show such changes, while others would not.

However, the problem with this "simple" prediction is that changes in synaptic efficacy after an individual learning experience might be difficult to detect because, if a brain structure has significant storage capacity, an individual training experience would not be expected to cause a change in fEPSPs across a large array of synapses. Although changes in neuronal excitability, transmitter release, and second-messenger activation have each been reported (Laroche, Doyère, Rédini-Del Negro, & Burette, in press), no long-lasting alteration in fEPSPs has yet been found. Laroche et al. (in press) describe the time course of changes in fEPSPs that occur differentially during the course of classical conditioning (relative to a truly random control condition), but these changes are short lived.

Considerable excitement surrounded the discovery by Sharp,

McNaughton and Barnes (1989) of what appeared to be a striking short-term modulation of hippocampal fEPSPs (dentate gyrus) occurring during spatial exploration. Rats were implanted with stimulating and recording electrodes and kept undisturbed in environmentally well-controlled housing conditions. They were then placed in a laboratory on a triangular table top containing a number of unusual objects. The animals soon began exploring the objects and, as they did so, a highly consistent growth of the fEPSP was observed. Upon return to their home cages, the fEPSP gradually returned to baseline over a period of 20 to 30 minutes. This phenomenon clearly resembles an STP-like induced change, which, as we have seen, may be a step on the way toward LTP. However, in addition to the *increase* in fEPSP, Sharp et al. (1989) found that the population spike *decreased* in amplitude and then returned to baseline according to a roughly comparable time course. This is clearly a different pattern to that of STP/LTP in as much as both measures normally increase together. It might have been explicable in terms of a gradual hyperpolarization so hippocampal neurons (which would tend to make fEPSPs larger but the cells harder to fire—the very pattern observed), were it not for the further observation that the latency of the population spike also declined (i.e., the spike becomes smaller but, paradoxically, *easier* to fire). Green, McNaughton, and Barnes (1990) analyzed this short-term modulation phenomenon in greater detail and offered the tentative hypothesis that the changes might reflect a potentiation of both direct excitatory perforant path connections and feed-forward inhibitory connections as well. It is now known, however, that the major cause of the short-lasting change lies elsewhere.

Moser, Mathieson, and Andersen (1993) have recently suggested that this unusual pattern of electrophysiological changes may be explained in terms of a muscular activity–induced alteration in brain temperature. They prepared animals of chronic recording in a similar way to Sharp et al. (1989), but also implanted miniature thermistors in the brain at approximately the same dorsoventral level to that of the hippocampal recording electrode. While repeating all of Sharp et al.'s observations (fEPSP *increase*, population spike *decrease*, spike latency *decrease*), they found that these changes occurred in association with an increase in brain temperature as the animals explored (Figure 8). If the rats were instead placed into a pool of cold water, brain temperature declined and each of the above measures reversed sign (i.e., fEPSP *decreased*, etc.). Further evidence that this bidirectional pattern is really associated with changes in brain temperature came from experiments showing that similar changes could be induced by the application of radiant heat (from a heating lamp) and could be blocked by "temperature clamping" the animal's brain as close as possible to a constant temperature (by alternately applying and withdrawing the heating lamp as required). Moser et al.'s (1993) experiment raises the fascinating issue of why the brain is so

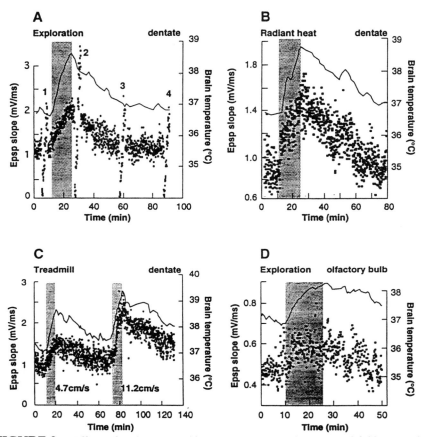

FIGURE 8   Effects of exploration and brain temperature on hippocampal field potentials. (A) Exploration (15-min period shown with gray hatching) was associated with an increase in fEPSP slope (left-hand y-axis) and brain temperature (right-hand y-axis). (B) Radiant heat is sufficient to induce the changes in fEPSPs. (C) Running on a treadmill, rather than exploration and new learning, is also sufficient to induce changes in fEPSPs and brain temperature. (D) The electrophysiological changes are not confined to the hippocampus. They can, for example, also be recorded in the olfactory bulb. From Moser, Mathiesen, and Andersen (1993). Courtesy of American Association for the Advancement of Science.

much less homeothermic than had been thought previously, but it also presents the procedural complication that, in any subsequent attempt to pursue the issue of learning-associated changes in synaptic efficacy, it will be necessary to establish that any changes in fEPSPs are independent of a temperature-induced change. As one pair of commentators aptly put it, now is the time to buy stock (shares) in companies that make miniature thermistors (Eichenbaum & Otto, 1993).

Moser, Moser, and Andersen (1993) have themselves taken an important next step in the direction of unconfounding temperature-associated changes from other changes in fEPSPs. They first worked out the calibration functions for individual animals between brain temperature and fEPSP under conditions in which the rats sat quietly in a corner. They then placed them into an environment containing four landmarks that the animals were allowed to explore. After complete cessation of exploration (about 30 min), one of the objects was replaced by a new one. This triggered renewed exploration, largely directed at the novel object. Recordings of fEPSPs were taken prior to, during, and after this exploration period. The second period of exploration did, of course, trigger a muscular activity–associated increase in brain temperature, but Moser and his colleagues were able to use the calibration functions to identify a temperature-independent component of exploration. This component increased rapidly at the start of exploration and declined gradually to baseline over approximately 15 min. This change *may* be a true learning-related alteration in synaptic efficacy. However, it could also be a movement-associated change in fEPSP (Winson & Abzug, 1977) that is associated with a release of neuromodulators in the hippocampus, such as noradrenalin. The key next step is, therefore, to dissociate these two alternative explanations by examining whether repeated exploration episodes can bring about a cumulative and sustained potentiation of fEPSPs that is independent of concurrent motor activity.

As a final *coda* on the issue of detecting changes in synaptic efficacy in association with learning, it should be noted that changes in fEPSPs are not necessarily the most suitable measure. Given the argument presented at the start of this section to the effect that synaptic changes associated with an individual training episode are not likely to be a large fraction of the available synaptic resource in a structure with significant storage capacity (see also Morris, 1990), it might be better to try to record from identified pairs of neurons known to be activated during a learning task. Hippocampal "place cells" (O'Keefe, 1976) are such a possibility. A *gedanken* experiment might, then, be to record simultaneously from a group of neighboring hippocampal CA3 cells and a group of neighboring CA1 cells (focusing the electrode array at the CA3/CA1 border where the probability of finding connected pairs of cells is highest). In the event of finding pairs of cells where the cross-correlation of firing patterns is nonrandom (i.e., there is some evidence of synaptic connectivity), one would then examine whether the learning of a simple spatial task is associated with a change in the cross-correlation function. It would be a mistake to underestimate the difficulty of such an experiment. The chances of finding connected pairs of CA3 and CA1 cells is remote, given the cell numbers involved (roughly 500,000 of each), although it may be easier at the border between CA3 and CA1. In addition, there is the possibility that a large component of LTP is the conversion of near-silent synapses (in which the probability of transmitter release is near zero or there is a high preponderance of nonfunctional AMPA receptors) into active ones (in which case the experiment is well-nigh impossible).

B. Does Saturation of Long-Term Potentiation Interfere with Stored Information or Occlude Subsequent Learning?

If information were temporarily stored in the hippocampus as distributed patterns of synaptic weights, direct physiological activation of as many synapses as possible with LTP inducing patterns of stimulation should have two effects. First, it should disrupt information already stored and so cause retrograde amnesia. Second, if the LTP induced, approached, or reached its physiological asymptote (i.e., was saturated), it should occlude subsequent learning. The logic underlying the first prediction relies on rather specific assumptions about how information is represented and on the unlikely supposition that the hippocampus is, even temporarily, the only site in the brain at which information is stored. The logic implicit in the second prediction seems less ambiguous—namely, that when an animal's brain attempts to alter synaptic weights in the hippocampus during learning, it would be unable to do so because the prior experimentally induced LTP had either "used up" all the synaptic resource available or had saturated the biochemical machinery that mediates such changes. I shall consider only the second of these predictions.

Research on whether saturation of LTP causes a learning impairment has recently been fraught with a number of failures to replicate earlier findings documenting the phenomenon (Cain, Hargreaves, Boon, & Dennison, 1993; Jeffery & Morris, 1993; Korol, Abel, Church, Barnes, & McNaughton, 1993; Sutherland, Dringenberg, & Hoesing, 1993). These studies were each an attempt to replicate the observations of Castro, Silbert, McNaughton, and Barnes (1989), namely, that rats are unable to learn a spatial water maze task following 14 days of high-frequency stimulation of the perforant path, which was demonstrated to have induced cumulative LTP to asymptote; on the other hand, animals that had received either identical high-frequency stimulation and then had been left for 3 weeks until the fEPSPs declined to baseline, or had been given low frequency stimulation, which does not induce LTP, were each able to learn satisfactorily. Our own study (Jeffery & Morris, 1993) included both as near an exact replication as we could achieve of part of Castro et al.'s (1989) experiment and a second experiment using somewhat different experimental conditions. In neither experiment was a learning deficit observed (Figure 9). Several other laboratories have also failed to find a positive effect of saturation, and it is noteworthy that one of these recent papers documenting a failure to replicate Castro et al.'s results

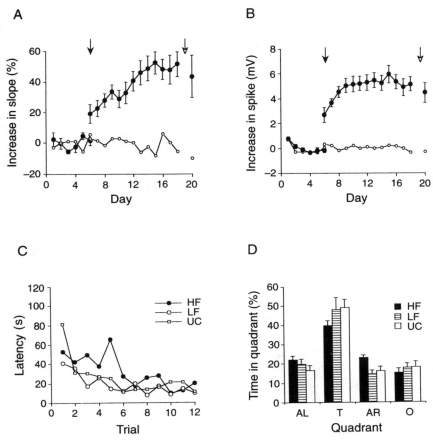

FIGURE 9   Failure of cumulative LTP to influence subsequent learning in the water maze. (A) Gradual increase in fEPSP in animals given high-frequency (HF) stimulation on each of 14 days (solid symbols). Control animals received low-frequency (LF) stimulation only (open symbols). (B) Gradual increase and possible saturation of population spike over the same period. (C) Subsequent performance of rats in a standard reference memory spatial learning task in the water maze. There is a slight but non significant trend toward poorer performance by the HF rats than by the two control groups (LF and unoperated control). (D) Equivalent transfer test performance by all groups. Redrawn from Jeffery and Morris (1993).

(Korol et al., 1993) comes from the same laboratory as that of the original findings.

The reasons for the irreproducability of results is unclear (see Bliss & Richter-Levin, 1993, for a commentary). It may be that LTP is not involved in learning and that its saturation is therefore without effect. Alternatively, it could be (1) because only perforant path terminals were sufficiently saturated but not those of other extrinsic or intrinsic hippocampal pathways; or

(2) due to a failure to saturate the full septotemporal axis of the hippocampus. With respect to the second of these possibilities, if saturation of LTP were to be restricted to a small part of the dentate gyrus, a substantial "resource" of modifiable synapses would still be available. It would not then be surprising that learning proceeds normally. The extent of potentiation measured could, rather than impair learning, serve instead as an "index" of synaptic modifiability in individual animals. In this vein, while Jeffery and Morris (1993) found no difference in rate of learning in either of their two experiments between animals given LTP-inducing or non-LTP-inducing stimulation, they did find, again in both experiments, a significant correlation *within the high-frequency group* between the magnitude of LTP and performance in the water maze (Figure 10). Animals showing good LTP tended to persist in searching in the correct quadrant of the pool during a post-training retention test, whereas animals showing weak LTP performed more poorly. This result extends earlier ex vivo findings by Deupree, Turner, and Watters (1991) and suggests that it may be necessary to have two separate indices of the LTP displayed by individual animals. These are: (1) the absolute amount of LTP displayed by an individual animal; and (2) the proportion of that animal's maximum this level of LTP represents (Figure 11). If an animal displays good LTP (i.e., has readily modifiable synapses), but LTP that does not reach asymptote, that animal should subsequently learn well. Conversely, if either little LTP occurs, or the LTP observed (irrespective of its magnitude) reaches an asymptote throughout the hippocampus, that animal should subsequently learn poorly.

This analysis of the correlation between LTP and learning makes two predictions. First, the correlations should still be present if the order of

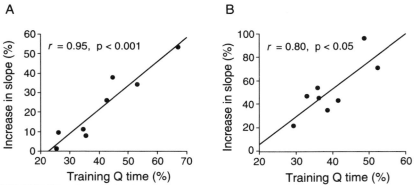

FIGURE 10   Correlation between extent of LTP and spatial performance. In each of two experiments (A and B), there was a positive correlation *within* the high-frequency group between the extent of the increase in fEPSP slope and time spent searching in the correct quadrant of the water maze. Redrawn from Jeffery and Morris (1993).

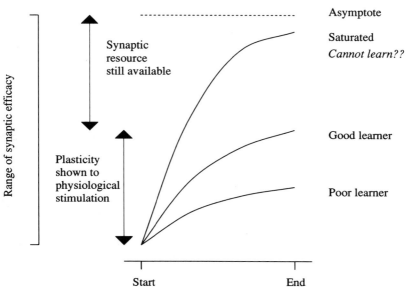

FIGURE 11   Diagram indicating that two parameters concerning the LTP shown by individual animals may need to be known to predict performance in a "saturation" experiment: (1) the plasticity shown to physiological stimulation; and (2) the proximity of this change in synaptic efficacy to the true asymptote. "Good learners" may be those that show relatively greater LTP until, and if, they reach the asymptote.

behavioral training and LTP induction were to be reversed such that the former preceded the latter rather than vice versa. Jeffery (1993) has completed such an experiment and finds the prediction upheld. However, she also ran into the complication that the correlation between magnitude of LTP and extent of spatial learning is strongly *positive* at high intensities of test stimuli to the perforant path, but strongly *negative* at weak intensities. The reason for this paradoxical reversal of the correlation with variation in the strength of perforant path stimulation is unclear. A second prediction is that the correlation should break down and the saturation–induced learning impairment be observed in a "reduced preparation" in which the synaptic resource is independently made smaller. Such a preparation might be an animal in which lesions have been made to many parts of the hippocampus leaving only an island or "lamella" (Andersen, 1975) of intact tissue. Preliminary evidence from a study by Mumby, Weisand, Barela, and Sutherland (1993) indicates that a saturation–induced learning impairment in the water maze may occur in animals that have been given unilateral lesions of the hippocampus prior to implantation of recording electrodes in the contralateral hippocampus.

As a *coda* to this section, it is worth noting that the fundamental assumption on which the saturation approach is based might be criticized on a priori grounds. The problem is whether it is reasonable to suppose that saturation of LTP could *ever* occur without the animal becoming epileptic. It is thought that activation of about 50 of the 10,000+ synapses of a pyramidal cell is necessary to fire that cell. It follows, as has been noted by Storm (1993), that most of the synapses must be either nonfunctional or temporarily silent. If LTP were to cause the up-regulation of a significant proportion of these synapses, the cell would clearly become at risk of firing too frequently. One way in which it might protect itself, as we have seen (Section V.B), would be to combine up-regulation of some synapses (LTP) with down-regulation of others (heterosynaptic LTD). If the former synapses were those preferentially activated by the stimulating electrode, and the latter those of fibers distal from the stimulating electrode, *apparent* saturation of LTP would be observed that, in practice, reflected nothing more than a redistribution of synaptic weights on an individual cell's dendrites such that $\Sigma\omega$ was roughly constant. Saturation of LTP would then remain, like the horizon at sea, always just beyond our grasp.

C. Does Erasure of Long-Term Potentiation Cause Forgetting?

If information about a recent learning experience were temporarily stored within the hippocampus, erasing LTP should cause forgetting. To my knowledge, this intriguing prediction has not yet been tested. Two ways in which LTP might be erased are: (1) the application of long trains of 1-Hz stimulation to bring about depotentiation following the procedure deployed by Dudek and Bear (1992) in their discovery of homosynaptic LTD; and (2) the application of drugs or enzyme inhibitors that interrupt the expression of LTP when given shortly after its induction.

The problem with the first proposal is, as we have seen, that no reliable method of inducing homosynaptic LTD in adult animals in vivo has yet been reported (outside of the cerebellum) and it would be premature to suppose that a procedure that works in vitro would necessarily work in vivo. A related difficulty is that LTD is more prominent in hippocampal slices prepared from the brains of young animals than those from adults (Malenka, 1993), indicating that its normal function may be in the developmental process of synapse elimination rather than in the down-regulation of synapses whose neural activity is without consequence. A third problem is that it may be difficult to induce LTD on all the relevant pathways of the hippocampal formation, that is, on all those actually used to store the events and associations formed during a recent learning experience. Nonetheless, the experiment is surely worth trying.

The alternative pharmacological approach has the advantage that it is easier to target all the relevant synapses with a drug, but at the cost of unknown side effects. Stevens and Wang (1993) have recently reported that application of Zinc Protoporphyrin IX (ZnPP), an inhibitor of heme-oxygenase (the enzyme that makes carbon monoxide in the brain), could bring a potentiated pathway back to its pre-LTP baseline without effect on a second unpotentiated pathway. This pathway specificity of LTP erasure is important as it indicates that, however ZnPP acts (about which there is some doubt; see Morris & Collingridge, 1993), it would appear to be doing something other than merely causing a nonspecific down-regulation of synaptic function. A *gedanken* experiment would, therefore, be to investigate whether ZnPP (or other related compound) could erase LTP in vivo within a specific time period after its induction and, in parallel, see if it could also erase recently formed memories for tasks dependent on hippocampal function within the same time window. Conducting both parts of the experiment successfully in the same set of animals would add force to the findings.

D. Does Blockade of Long-Term Potentiation Impair Certain Types of Learning?

To date, the most persuasive evidence that the neural mechanisms of LTP might participate in associative learning has come from experiments in which LTP is blocked and learning has been shown to be compromised. There are three ways in which associative LTP has been or might be blocked in vivo: (1) application of NMDA receptor antagonists (such as AP5); (2) inhibition of enzymes putatively involved in the expression of LTP, such as the synthesis of candidate intercellular messengers (e.g., nitric oxide synthase); (3) targeted deletion of genes encoding receptors or enzymes involved in the expression of LTP (gene "knock-outs").

1. Application of NMDA Receptor Antagonists

The first indication that blocking NMDA receptors might affect learning was Morris, Anderson, Lynch, and Baudry's (1986) finding that chronic infusion of the selective NMDA receptor antagonist, AP5, impaired the acquisition of spatial learning in the water maze.

In this and some later experiments, AP5 was infused continuously over a two-week period into the lateral ventricle of the brain via subcutaneously implanted minipumps (this drug does not easily cross the blood–brain barrier). These pumps, which work by osmosis, cause the slow, steady infusion (0.5 μl/hour) of a small quantity of the drug (or artificial cerebrospinal fluid vehicle solution) into the ventricular space, from which it can readily spread throughout the forebrain. The advantage of this method of applying the drug is that its concentration is relatively well restricted to the forebrain (i.e., it does not infiltrate the spinal cord to an appreciable degree) and relatively uniform both spatially and temporally. Using this technique, and at a drug concentration found in separate animals to be sufficient to block hippocampal LTP without affecting baseline fEPSPs mediated by AMPA receptors, learning the location of a single, hidden platform in a water maze was impaired, while acquisition of a visual discrimination task was un-affected (Figure 12). The spatial task involved searching for an escape plat-form hidden at one place in the pool, while the visual discrimination task involved distinguishing between a visible black–and–white platform that provided escape from the water and a visible gray platform that did not (or vice versa). The apparent selectivity of the learning impairment induced by AP5 is important for at least two reasons: (1) it suggests that the drug is not impeding spatial performance by virtue of some gross sensorimotor distur-bance (e.g., impaired vision); and (2) it points to a functional parallel be-tween the effects if blocking hippocampal synaptic plasticity and those of lesions (Morris, Garrud, Rawlins, & O'Keefe, 1982; Morris, Hagan, & Rawlins, 1986; Morris, Schenk, Tweedie, & Jarrard, 1990; Sutherland, Whishaw, & Kolb, 1983).

These findings have since been followed up in a number of ways. The possibility that the AP5-induced impairment of spatial learning is an artifact of disturbed sensorimotor function was further diminished by the discovery that the drug does not affect performance in the water maze if learning precedes application of the drug (Morris, 1989). Thus, under the influence of AP5, rats can see distal cues and can search at the site of the normally hidden platform as accurately as vehicle-treated controls—provided the drug was not present during initial learning. The dissociation between spa-tial learning and visual discrimination learning has also been substantiated by the findings that (1) intracortical infusion of AP5 into a site just anterior to the visual cortex does not impair visual discrimination learning despite giving rise to a cortical concentration of the drug exactly comparable to that which in hippocampus is associated with a spatial learning impairment (Butcher, Hamberger, & Morris, 1991); and (2) acute intrahippocampal infusion of AP5 in a manner shown to restrict the drug to the hippocampus is sufficient to cause the impairment of spatial learning (Morris, Halliwell, & Bowery, 1989). Finally, the link to hippocampal LTP has been refined by the demonstration that the dose-response profile of the learning impairment exactly parallels the dose-response profile of the blockade of LTP in vivo and, strikingly, that the measured intrahippocampal concentrations of AP5 that cause these effects are comparable to those concentrations that block LTP in the in vitro hippocampal slice (Davis, Butcher, & Morris, 1992). In addition, work by others has established that the similarities between the deleterious effects of AP5 and hippocampal lesions extend to various non-spatial tasks such as DRL 18-s schedules (Tonkiss, Morris, & Rawlins, 1989) and delayed matching to sample with a restricted set of stimuli (Lyford, Gutnikov, Clark, & Rawlins, 1993).

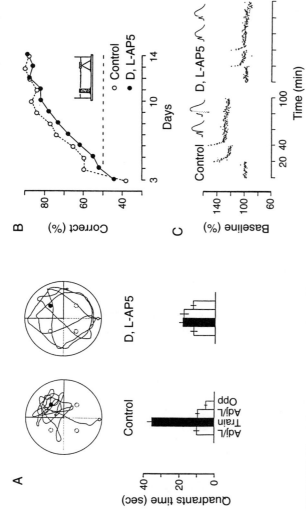

FIGURE 12 Selective impairment of spatial learning by i.c.v. infusion of AP5. (A) Spatial learning is impaired by AP5. The paths taken by the median animal during a final transfer test are shown for the control (vehicle infusion) and AP5-treated groups. (B) AP5 does not impair visual discrimination learning plotted as choice performance over days. (C) This behavioral dissociation occurs at a rate of drug infusion that is sufficient to block LTP in vivo. Redrawn from Morris, Anderson, Lynch, and Baudry (1986). Courtesy of Macmillan Publishers.

Taken together, these findings suggest that blocking hippocampal synaptic plasticity impairs types of learning, both spatial and nonspatial, that depend on the integrity of hippocampal function. A more general conclusion about the role of LTP might be possible of blocking NMDA receptors in other brain areas also impairs types of learning dependent on the integrity of these brain areas. One such example is the elegant work by Michael Davis and his colleagues showing that intra-amygdala infusions of nanomolar quantities of AP5 block the acquisition and extinction, but not the expression, of fear-potentiated startle (Falls, Miserendino, & Davis, 1992; Miserendino, Sananes, Melia, & Davis, 1990). These latter studies deployed a wide range of controls to demonstrate that, at the drug concentrations used, AP5 did not affect processing of either the CS or the US, while nonetheless impairing learning. For example, Falls et al. (1992) implanted rats with bilateral cannulas aimed at the basolateral nucleus of the amygdala and then trained them in a standard fear-potentiated startle paradigm. One week later, they were given a short test session to enable them to be matched into four groups displaying equivalent levels of fear-potentiated startle. The next day, half the animals were infused with AP5 and the remainder with vehicle (aCSF). A few minutes later, half of each of these groups were presented with CS-alone extinction trials during which shock presentations were omitted. The remaining animals were placed in a test chamber but no CS presentations were scheduled. The following day, all animals were tested for fear-potentiated startle. As shown in Figure 13, animals that were not given CS-alone extinction trials had levels of potentiated startle equivalent to that observed in their initial test, that is, no measurable extinction had occurred. Animals infused with aCSF immediately before the CS-alone trials exhibited little potentiated startle—indicating that extinction had occurred. In contrast, animals infused with AP5 just prior to their CS-alone trials displayed levels of potentiated startle significantly greater than those of the aCSF-treated group and not significantly different from their initial test levels. These data indicate that infusions of AP5 into the amygdala block the extinction of fear-potentiated startle. The more general implication of my own and M. Davis's work is that associative NMDA receptor–dependent plasticity might be involved in a number of different brain-region-specific types of associative learning.

There remain, however, a number of problems with the strategy of blocking NMDA receptors. One is that the intraventricular route of administration of AP5 in my earlier experiments does not restrict the drug to the hippocampus and thus NMDA receptor function is likely to be compromised in a number of brain areas. These would include such areas as the striatum, thalamus, and neocortex where NMDA receptors participate in diverse aspects of brain function including, but not restricted to, synaptic plasticity (Daw, Stein, & Fox, 1992). It is, therefore, an oversimplification

A

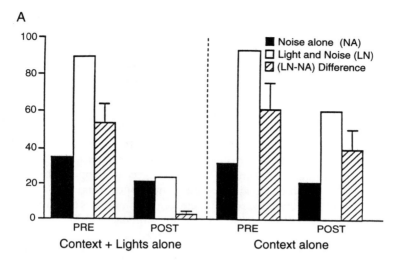

Context + Lights alone   Context alone

B

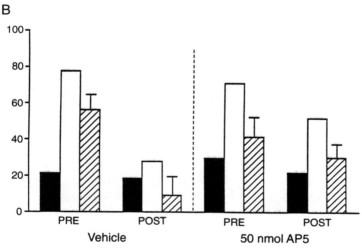

Vehicle   50 nmol AP5

FIGURE 13   Impairment of extinction of fear-potentiated startle by AP5. (A) Lights presented in the absence of shock are required to produce extinction. Left-hand panel shows that the difference (hatched bar) in mean amplitude of startle in the Noise-Alone (solid bar) and Light & Noise (open bar) conditions was significantly reduced following presentations of the light in the absence of shock. No change was observed (right-hand panel) following a Context-Alone condition in the absence of light presentations. (B) Comparison of the panels shows that extinction of the light was reduced by AP5. From Falls, Miserendino, and Davis (1992). Courtesy of the Society for Neuroscience.

to suppose that the major, still less the only, effect of AP5 administration is to block synaptic plasticity. This problem can, in part, be obviated by using acute intrastructure drug administration, which, as we have seen, is apparently sufficient to cause a spatial learning impairment in the water maze following intrahippocampal infusion, and an impairment in the learning and extinction of fear-potentiated startle after intra-amygdala infusions.

The second problem is that even if the only (or major) effect of acute AP5 administration in a given experiment is to block NMDA receptors in a regionally specific way, there can be no guarantee that doing this would not disrupt aspects of the dynamic system properties of the hippocampus or amygdala in addition to blocking LTP. If this were to happen, the fact that AP5 blocks LTP could be beside the point; the hippocampus (or amygdala) may simply not be working properly and region-dependent learning could be impaired for that reason alone. This is an immensely difficult ambiguity to break apart experimentally. One route to tackling it is to attempt to block LTP at a point biochemically downstream of the NMDA receptor such that those system properties of the hippocampus that depend on NMDA receptor function, excluding synaptic plasticity, are ostensibly unaffected.

2. Inhibition of Enzymes Putatively Involved in the Synthesis of Candidate Intercellular Messengers Mediating the Expression of LTP

Given the strictures raised in the immediately preceding discussion, an ideal experiment might be one exploring the behavioral effects of a regionally specific blockade of LTP downstream of the NMDA receptor and in the absence of other neuropharmacological "side effects." Achieving a regionally specific blockade of the synthesis of a key intercellular messenger that communicates from the postsynaptic side of the synapse to the presynaptic side would be one such case. However, the pharmacokinetics of doing this is fairly exotic and not yet within our grasp. As ever, we must work up to the ideal experiment in stages.

To that end, Bannerman, Chapman, Kelly, and Morris (1993a,b) have recently completed a series of experiments exploring the effects of inhibiting nitric oxide synthase (NOS) in vivo using intraperitoneal administration of the selective NOS inhibitor L-nitro-arginine methyl ester (L-NAME). Nitric oxide is a small membrane molecule with a half-life of about 4 s, which, on the basis of studies in vitro, was shown to be released in response to glutamate receptor activation (Garthwaite et al., 1988) and has been implicated in a number of neuronal functions including LTP. Thus, inhibition of NOS in hippocampal slices blocks LTP (Böhme et al., 1991; Haley, Wilcox, & Chapman, 1992; O'Dell, Hawkins, Kandel, & Arancio, 1991; Schumann and Madison, 1991), while pairing NO (or NO donors) with subthreshold presynaptic stimulation can induce LTP (Zhuo, Small, Kandel, & Hawkins, 1993). However, as Bannerman et al., were to discover, the fact that these findings can be obtained in vitro is no guarantee that they will work in vivo. The long and short of it is that, in our hands, acute administration of L-NAME causes a time-dependent disturbance of fast synaptic transmission but does not block LTP in vivo (Figure 14A,B). This lack of an effect on LTP occurred despite use of a drug dose that brought about a demonstrated 93% inhibition of NOS in the brain. Bannerman et al. (1993b) found that L-NAME can cause an impairment in the water maze, but variation of the training protocol across a series of experiments revealed this to be a disturbance of performance rather than a selective learning effect (L-NAME also impaired the visual discrimination task, and its deleterious effects on spatial learning in the water maze could be obviated by extending training over many days by running only one trial per day; Figure 14C,D). This performance effect, and the acute effects of L-NAME on baseline fEPSPs, may be secondary to a range of cardiovascular effects that inhibiting NOS is also known to cause (NO has also been definitively implicated in the relaxation of blood vessels in response to cardiovascular demand; Moncada, 1992).

The failure to block LTP in vivo with inhibition of an enzyme involved in the synthesis of a candidate intercellular messenger was a disappointment, but it is neither reason to abandon the general strategy nor grounds for casting doubt on the hypothesis that LTP might play a role in learning. As Bannerman et al. (1993b) put it, their studies indicate that "a compound which does *not* block LTP does *not* impair spatial learning." It does, however, illustrate an important point about trying to tie together neuropharmacological and behavioral results: *Never assume the brain is just a test tube.* The judgment required in pursuing this strategy is to guess which of the things that neuropharmacologists get to work in vitro will also turn out to work in vivo.

3. Targeted Deletion of Genes Encoding Receptors or Enzymes Involved in the Induction or Expression of LTP

A radically new approach is provided by the use of homologous recombination to make mutant mice deficient in the alpha subunit of $Ca^{2+}/$ calmodulin-dependent protein kinase type II (αCaMKII; Silva, Paylor, Wehner, & Tonegawa, 1992; Silva, Stevens, Tonegawa, & Wang, 1992) or deficient in *fyn* tyrosine kinase (Grant et al., 1992). Publicized in *The New York Times* as soon as they were in *Science* magazine, these remarkable animals jumped onto the LTP scene with a splash. They don't show normal LTP and (apparently) cannot learn the water maze.

The making of such mice is a difficult matter. Briefly, it involves a series

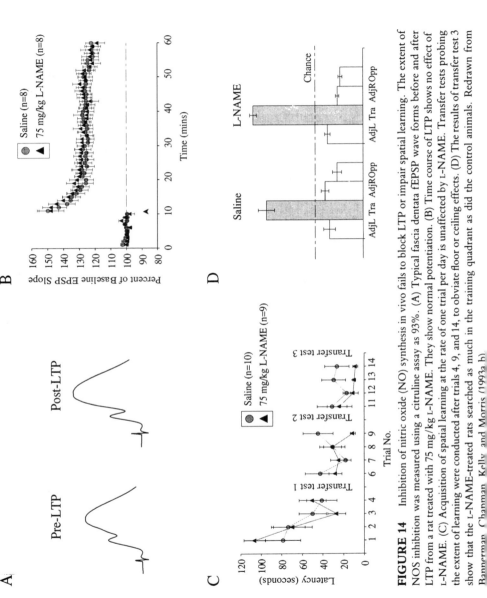

FIGURE 14 Inhibition of nitric oxide (NO) synthesis in vivo fails to block LTP or impair spatial learning. The extent of NOS inhibition was measured using a citruline assay as 93%. (A) Typical fascia dentata fEPSP wave forms before and after LTP from a rat treated with 75 mg/kg L-NAME. They show normal potentiation. (B) Time course of LTP shows no effect of L-NAME. (C) Acquisition of spatial learning at the rate of one trial per day is unaffected by L-NAME. Transfer tests probing the extent of learning were conducted after trials 4, 9, and 14, to obviate floor or ceiling effects. (D) The results of transfer test 3 show that the L-NAME-treated rats searched as much in the training quadrant as did the control animals. Redrawn from Bannerman, Chapman, Kelly, and Morris (1993a,b).

of four stages shown in Figure 15. The first stage is the culturing of embryonic stem (ES) cells derived from blastocysts. The second stage involves engineering the deletion of the gene of interest, called the gene "knockout." The third stage involves the injection of the mutant cells into host blastocysts. The progeny of these cells are chimeric because they are derived from both mutant and normal cells. The fourth and final stage involves selective breeding and the eventual derivation of homozygous mice bearing two copies of the gene of interest $(-/-)$.

Animals made in this way grow up, from birth, missing an enzyme (e.g., αCaMKII) thought to be crucial for the expression of LTP. Behavioral tests were made using adult animals exploring their capacity for learning and, using slices of hippocampal tissue, their capacity to display LTP or other forms of synaptic plasticity. Electrophysiological observations on slices made from the αCaMKII mutant mice indicated that both AMPA and NMDA currents were normal. Thus, unlike direct application of AP5 to the hippocampus, there is no reason to suppose that the dynamic system properties of the hippocampus will be in any way abnormal. Nonetheless, slices from these animals displayed partially compromised post-tetanic potentiation and a near-complete blockade of LTP in almost all slices that were tested. While different animals had to be used for the behavioral experiments (preventing the possibility of looking for interexperiment correlations), these showed that the mutant mice could learn to navigate to a visible platform in the water maze without difficulty but had an enduring deficit on the spatial or hidden-platform version of the task. They were, nonetheless, able to learn a simple spatial cross-maze from any starting point. What are we to make of these findings?

Clearly, the technology is little short of remarkable and it would be short sighted to belittle the potential impact of this approach. But the interpretation offered by the results to date can and has been questioned (Deutsch, 1992; Morris & Kennedy, 1992). There are several problems: first, the alpha-rich CaMKII holoenzyme is known to be quite abundant (Kennedy, et al., 1990), comprising approximately 2% of total protein in hippocampus, 1% in cortex and a far greater proportion in so-called postsynaptic densities (the electron-dense region of synapses where postsynaptic receptors are thought to be located). Given this abundance, it seems ironic that its deletion from birth has such apparently subtle effects. There is an element of suspension of disbelief about the claims to date, which make one wonder whether such a gene deletion could have so exclusive an effect as blocking long-lasting synaptic plasticity without influence on fast synaptic transmission and other neuronal mechanisms. Second, it is not yet possible to delete αCaMKII in a specific region of the brain. The enzyme is particularly high in the forebrain, but is in no way restricted to the hippocampus or, indeed, to any other brain structure. Third, there are grounds for suspecting that the learn-

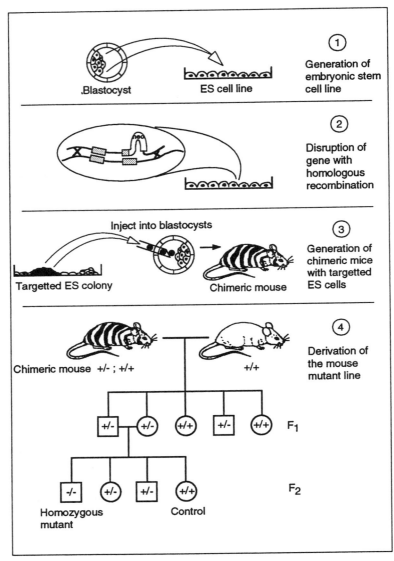

FIGURE 15   Stages in the construction of mutant mice in which a targeted gene, such as the gene putatively involved in LTP, has been deleted by homologous recombination. See text for explanation. Diagram courtesy of Seth Grant.

ing capacity of the mutant mice has been underestimated. It is true that the mice showed an enduring latency impairment in the spatial version of the task, but this phase of training was relatively short. Silva, Paylor et al. (1992) emphasize, however, that it is their chance levels of performance in the subsequent absent-platform retention test that persuades them that their mutant mice cannot learn. That is, not only were they significantly worse than the wild-type controls, they were also *at chance,* displaying no evidence of having learned anything about where the platform had been located during training. This is a good point, but the force of it is compromised by the fact that the spatial learning part of the experiment was conducted using the same animals as those that had participated in the nonspatial visual task immediately beforehand. The reason for caution is that the visual task entailed moving the visible platform around between several places in the pool, training that would have taught the animals that the platform could be anywhere. The levels of performance of *all* the mice might, therefore, be expected to be quite poor during the subsequent spatial training. This was indeed the case (the controls were only 10% above chance in the retention test). Thus, the chance performance of the mutants is against what can fairly be described as a weak baseline. This is not to deny or belittle the fact that a significant impairment between mutants and wild-type nice was observed.

Transgenesis and homologous recombination are each proving valuable in tackling a wide range of problems in biology—there is no reason why they should not also help unravel the neural mechanisms of learning. Once techniques for achieving *reversible, local,* and *conditional* enzyme ablation are realized (i.e., animals that grow up normally but have neural enzymes reversibly inhibited in a specific brain area at a time point under the control of the experimenter), the approach will have enormous power—far exceeding that ever possible with pharmacological techniques alone. For example, the functional significance of LTP would be greatly illuminated by an experiment in which the capacity to display LTP was temporarily blocked and then later restored. Would such animals become amnesic, but then later get up from their beds and remember? It would be a miracle indeed but, in science, miracles *do* happen! In the meantime, we must be cautious. The "small industrial revolution in the construction of mice with mutated neural genes" that is now upon us (Morris & Kennedy, 1992, p. 514) carries with it the potential to mislead as well as to illuminate.

VII. CONCLUSIONS

One purpose of this chapter was to assess the current status of the hypothesis—which has its origins in Ramón y Cajál's speculations a century ago—that activity-dependent synaptic plasticity occurs during, and is a necessary condition for, certain kinds of learning. This hypothesis was discussed particularly in relation to a long-lasting form of plasticity known as hippocampal long-term potentiation. What conclusions can be drawn?

First, there are yet no convincing indications that LTP or LTP-like changes in synaptic potential occur during any of the forms of learning with which it has been studied. Anatomical changes in synaptic number or structure have certainly been observed, particularly following learning in young animals, but it is unclear whether these changes imply any change in functional connectivity. Short-lasting electrophysiological changes occur during spatial exploration and classical conditioning, but there are reasons to suspect these are primarily temperature-dependent alterations in fEPSPs rather than true synaptic changes. Second, while the prediction that saturation of LTP should impair learning has proved hard to uphold, there are grounds for thinking the experiment is simply difficult to do properly. An unexpected consequence of one study investigating the effects of saturation of LTP was a highly significant correlation between the cumulative extent of LTP in individual animals and their behavioral performance in a spatial learning task. Understanding this correlation will, however, require a great deal of further work. Third, the effects of erasing LTP are yet to be investigated. Fourth, blocking LTP by either of two methods (NMDA receptor blockade, gene knock-outs) impairs spatial learning as predicted, but there are several grounds to be guarded about the interpretation of both classes of experiment. In summary, the evidence collected to date neither warrants rejection of the "LTP and learning" hypothesis, nor offers grounds for its unequivocal acceptance.

Numerous outstanding problems remain in both the neurobiological and psychological domains. One is that no amount of work on whether synaptic plasticity is necessary for learning—be it obtained using physiological (e.g., saturation), pharmacological (e.g., receptor blockade), or molecular-biological (e.g., gene knock-out) methods—will ever be persuasive in the absence of studies definitively demonstrating that such plasticity occurs naturally during learning. Uncomfortable as it is, there is no alternative but to face up to the complications of chronic recording experiments with all the necessary in vivo temperature and other controls. As noted in Section VI.A, looking for changes in the cross-correlation functions of simultaneously recorded hippocampal neurons might be one heroic strategy. But progress will be slow because such experiments would be technically exceedingly difficult and few laboratories have, to my knowledge, the necessary expertise for isolated single-cell recording. In addition, interpreting single-unit recording data requires a better understanding of the functional significance of the network architecture of the hippocampus than we possess at present.

A second weakness of the approach offered here is the exclusive concentration on hippocampal LTP as the exemplar of long-lasting synaptic plasticity. LTP occurs in other brain structures, and other plastic changes, such as LTD, occur in hippocampus and elsewhere (at least in vitro). The emphasis on LTP can be defended on grounds that it constrains the boundaries of a huge issue to tractable limits, but efforts to link even hippocampal-dependent learning exclusively to LTP can justly be criticized as an oversimplification. This criticism is all the more forceful if there is any merit in the analogy between, on one hand, potentiation and depression of synaptic strength and, on the other, up- and down-regulation of associative strengths, as in the Rescorla and Wagner (1972) equation.

A third and quite different lacuna relates to the second purpose of this chapter—namely to look at synaptic plasticity in the context of contemporary animal learning theory. One problem is the paucity of learning paradigms that have yet been studied in relation to LTP, particularly paradigms that have been studied by students of animal learning. Because of the additional complications of doing electrophysiology, often in the same animals, experimenters are naturally keen to use simple and rapid learning paradigms—such as spatial learning in the water maze. This task has the additional advantage that it is reliably impaired by hippocampal lesions and there is, therefore, some basis for thinking that neural activity within the hippocampal formation, including the capacity to display LTP, may be involved in spatial learning. But students of animal learning will be well aware that the stimulus control of the water maze is ill defined, and this makes it difficult, at least as it is usually run, to establish precisely what animals are learning as they perform the task. Operationally, it displays features of both classical and instrumental conditioning. Indeed, both may be involved—learning about the associative relations of the extramaze stimuli, and learning to navigate an appropriate path to the hidden platform, respectively. Because of this ambiguity, it is difficult to know exactly what psychological process has been disrupted when changes in rate of learning are observed as a result of some treatment (e.g., blocking LTP with an NMDA antagonist or a gene knock-out). This state of affairs means that the water maze is a useful but, frankly, rather "gross" assay of hippocampal dysfunction, and it bodes ill for its use in analytic studies of what psychological process LTP may be implementing. Either new spatial tasks must be developed with better cue control (e.g., the Manhattan Maze; Biegler & Morris, 1993), or investigators must turn to other paradigms. In this context, M. Davis's work on fear-potentiated startle takes on particular significance, as the cue control realizable in this paradigm is sufficient for it to have already been demonstrated that local infusion of AP5 has no effect on either CS processing or US processing. Thus, the NMDA antagonist–induced impairment of both learning and extinction of fear-potentiated startle looks, at present, to be the best documented instance of a definitive learning impairment that might be linked to NMDA-dependent synaptic plasticity.

However, it is unfortunate that electrophysiological work on the amygdala has progressed relatively slowly and has not been conducted in the same animals as those participating in behavioral experiments. The general point, however, is clear enough: *new learning paradigms for investigating the function of synaptic plasticity are required.* These should exploit the enormous developments in animal learning theory that have taken place over the past few years as documented in other chapters of this book. Only then can the ambition of realizing a mapping of psychological process onto neurobiological mechanism—a true "neuroscience of learning"—be eventually achieved.

Acknowledgments

Preparation of this chapter was made possible by Programme and Project Grants from the Medical Research Council and by a grant from the Human Frontiers Science Panel. I am grateful to Per Andersen, David Bannerman, Tim Bliss, Graham Collingridge, Mark Good, Seth Grant, Kate Jeffery, Rick Lathe, and John O'Keefe for discussion and criticism of the arguments presented in this chapter. Iain Donaldson kindly provided the translation of Ramón y Cajál, Per Andersen the confocal microscope pictures of a hippocampal cell and its dendritic spines (Figure 1), and Seth Grant the diagram comprising Figure 15.

References

Abraham, W. C., & Goddard, G. V. (1983). Asymmetric relationships between homosynaptic long-term potentiation and heterosynaptic long-term depression. *Nature (London), 305,* 717–719.

Andersen, P. (1975). Organization of hippocampal interneurons and their interconnections. In R. L. Isaacson & K. H. Pribram (eds.), *The hippocampus* (Vol. 1). New York: Plenum.

Andersen, P., Sundberg, S. H., Sveen, O., & Wigström, H. (1977). Specific long-lasting potentiation of synaptic transmission in hippocampal slices. *Nature (London), 266,* 736–737.

Artola, A., Brocher, S., & Singer, W. (1990). Different voltage-dependent thresholds for inducing long-term depression and long-term potentiation in slices of rat visual cortex. *Nature (London), 347,* 69–72.

Artola, A., & Singer, W. (1990). The involvement of N-methyl-D-aspartate receptors in induction and maintenance of long-term potentiation in rat visual cortex. *European Journal of Neuroscience, 2,* 254–269.

Baddeley, A. D. (1992). Working memory. *Science, 255,* 556–559.

Bannerman, D. M., Chapman, P. F., Kelly, P. A. T., & Morris, R. G. M. (1993a). Does inhibition of nitric oxide synthase affect the induction of long-term potentiation in vivo? *Journal of Neuroscience,* submitted for publication.

Bannerman, D. M., Chapman, P. F., Kelly, P. A. T., & Morris, R. G. M. (1993b). Does inhibition of nitric oxide synthase impair spatial learning? *Journal of Neuroscience,* submitted for publication.

Barnes, C. A. (1979). Memory deficits associated with senescence: A neurophysiological and behavioural study in the rat. *Journal of Comparative and Physiological Psychology, 93,* 74–104.

Barrionuevo, G., & Brown, T. H. (1983). Associative long-term potentiation in hippocampal slices. *Proceedings of the National Academy of Sciences of the U.S.A., 80,* 7347–7351.

Bashir, Z. I., Bortolotto, Z. A., Davies, C. H., Berretta, N., Irving, A. J., Seal, A. J., Henley, J. M., Jane, D. E., Watkins, J. C., & Collingridge, G. L. (1993). Induction of LTP in the hippocampus needs synaptic activation of glutamate metabotropic receptors. *Nature (London)*, *363*, 347–350.

Bateson, P. P. G. (1976). The specificity and the origins of behaviour. *Advances in the Study of Behavior*, *6*, 1–20.

Biegler, R., & Morris, R. S. M. (1993). Landmark stability is a prerequisite for spatial but not discrimination learning. *Nature*, *361*, 631–633.

Biennenstock, E. L., Cooper, L. N., & Munro, P. W. (1982). Theory for the development of neuron selectivity: Orientation specificity and binocular interaction in visual cortex. *Journal of Neuroscience*, *2*, 32–48.

Bliss, T. V. P., & Collingridge, G. L. (1993). A synaptic model of memory: Long term potentiation in the hippocampus. *Nature (London)*, *361*, 31–39.

Bliss, T. V. P., Douglas, R. M., Errington, M. L., & Lynch, M. A. (1986). Correlation between long-term potentiation and release of endogenous amino acids from dentate gyrus of anaesthetized rats. *Journal of Physiology (London)*, *377*, 391–408.

Bliss, T. V. P., & Gardner-Medwin, A. R. (1973). Long-lasting potentiation of synaptic transmission in the dentate area of the unanaesthetized rabbit following stimulation of the perforant path. *Journal of Physiology (London)*, *232*, 357–374.

Bliss, T. V. P., & Lømo, T. (1973). Long-lasting potentiation of synaptic transmission in the dentate area of the anaesthetized rabbit following stimulation of the perforant path. *Journal of Physiology (London)*, *232*, 331–356.

Bliss, T. V. P., & Richter-Levin, G. (1993). Spatial learning and the saturation of long-term potentiation. *Hippocampus*, *3*, 123–126.

Böhme, G. A., Bon, C., Stutzmann, J.-M., Doble, A., & Blanchard, J. C. (1991). Possible involvement of nitric oxide in long-term potentiation. *European Journal of Pharmacology*, *199*, 379–381.

Bolhuis, J. J. (1991). Mechanisms of avian imprinting: A review. *Biological Review of the Cambridge Philosophical Society*, *66*, 303–345.

Bonhoeffer, T., Staiger, V., & Aertsen, A. (1989). Synaptic plasticity in rat hippocampal slice cultures: Local "Hebbian" conjunction of pre- and postsynaptic stimulation leads to distributed synaptic enhancement. *Proceedings of the National Academy of Sciences of the U.S.A.*, *86*, 8113–8117.

Bortolotto, Z. A., & Collingridge, G. L. (1993). Characterisation of LTP induced by the activation of glutamate metabotropic receptors in area CA1 of the hippocampus. *Neuropharmacology*, *32*, 1–9.

Butcher, S. P., Hamberger, A., & Morris, R. G. M. (1991). Intracerebral distribution of D,L-2-amino-phosphonopentanoic acid (AP5) and the dissociation of different types of learning. *Experimental Brain Research*, *83*, 521–526.

Cain, D. P., Hargreaves, E. L., Boon, F., & Dennison, Z. (1993). An examination of the relations between hippocampal long-term potentiation, kindling, afterdischarge, and place learning in the watermaze. *Hippocampus*, *3*, 153–163.

Castro, C. A., Silbert, L. H., McNaughton, B. L., & Barnes, C. A. (1989). Recovery of spatial learning following decay of experimental saturation of LTE at perforant path synapses. *Nature (London)*, *342*, 545–548.

Collingridge, G. L. (1992). The mechanism of induction of NMDA receptor-dependent long-term potentiation in the hippocampus. *Experimental Physiology*, *77*, 771–797.

Collingridge, G. L., Kehl, S. J., & McLennan, H. (1983). Excitatory synaptic transmission in the Schaffer collateral-commissural pathway of the rat hippocampus. *Journal of Physiology (London)*, *334*, 33–46.

Davis, S., Butcher, S. P., & Morris, R. G. M. (1992). The NMDA receptor antagonist, D-2-amino-5-phosphonpentanoate (D-AP5) impairs spatial learning and LTP in vivo at intracerebral concentrations comparable to those that block LTP in vitro. *Journal of Neuroscience*, *12*, 21–34.

Daw, N. W., Stein, P. S. G., & Fox, K. (1992). The role of NMDA receptors in information processing. *Annual Review of Neuroscience, 16*, 207–222.

Deupree, D., Turner, D., & Watters, C. (1991). Spatial performance correlates with in vitro potentiation in young and aged Fisher 344 rats. *Brain Research, 554*, 1–9.

Deutsch, J. A. (1993). Spatial learning in mutant mice. *Science, 262*, 760–763.

Dickinson, A., & Mackintosh, N. J. (1980). Classical conditioning in animals. *Annual Review of Psychology, 29*, 587–612.

Doubell, T. P., & Stewart, M. G. (1993). Short-term changes in the numerical density of synapses in the intermediate and medial hyperstriatum ventrale following one-trial passive avoidance training in the chick. *Journal of Neuroscience, 13*, 2230–2236.

Dudek, S. M., & Bear, M. F. (1992). Homosynaptic long-term depression in area CA1 of the hippocampus and the effects of NMDA receptor blockade. *Proceedings of the National Academy of Sciences of the U.S.A., 89*, 4363–4367.

Eichenbaum, H., & Otto, T. (1993). LTP and memory: Can we enhance the connection? *Trends in Neurosciences, 16*, 163–165.

Falls, W. A., Miserendino, M. J. D., & Davis, M. (1992). Extinction of fear-potentiated startle: Blockade by infusion of an NMDA antagonist into the amygdala. *Journal of Neuroscience, 12*, 854–863.

Gally, J. A., Montague, P. R., Reeke, G. N., Jr., & Edelman, G. M. (1990). The NO hypothesis: Possible effects of a short-lived, rapidly diffusible signal in the development and function of the nervous system. *Proceedings of the National Academy of Sciences of the U.S.A., 87*, 3547–3551.

Garthwaite, J., Charles, S. L., & Chess-Williams, R. (1988). Endothelium-derived relaxing factor release on activation of NMDA receptors suggests role as intercellular messenger in the brain *Nature (London), 336*, 385–388.

Gould, S. J., & Lewontin, R. C. (1979). The spandrels of San Marco and the Panglossina paradigm: A critique of the adaptationist programme. *Proceedings of the Royal Society of London, Series, B, 205*, 581–598.

Grant, S. G. N., O'Dell, T. J., Karl, K. A., Stein, P. L., Soriano, P., & Kandel, E. R. (1992). Tyrosine kinases and long-term potentiation: Impaired spatial learning in *fyn* mutant mice. *Science, 258*, 1903–1910.

Green, E. J., McNaughton, B. L., & Barnes, C. A. (1990). Exploration-dependent modulation of evoked responses in fascia-dentata: Dissociation of motor, EEG, and sensory factors and evidence for a synaptic efficacy change. *Journal of Neuroscience, 10*, 1455–1471.

Greenough, W. T., & Bailey, C. H. (1988). The anatomy of memory: Convergence of results across a diversity of tests. *Trends in NeuroSciences, 11*, 142–147.

Haley, J. E., Wilcox, G. L., & Chapman, P. F. (1992). The role of nitric oxide in hippocampal long-term potentiation. *Neuron, 8*, 211–216.

Hebb, D. O. (1949). *The organization of behavior: A neuropsychological theory.* New York: Wiley.

Horn, G. (1985). *Memory, imprinting and the brain.* Oxford: Clarendon Press.

Jeffery, K. J. (1993). Paradoxical enhancement of long-term potentiation in poor-learning rats at low test stimulus intensities. Manuscript submitted for publication.

Jeffery, K. J., & Morris, R. G. M. (1993). Cumulative long-term potentiation in the rat dentate gyrus correlates with, but does not modify, performance in the watermaze. *Hippocampus, 3*, 133–140.

Jessell, T. M., & Kandel, E. R. (1993). Synaptic transmission: A bidirectional and self-modifiable form of cell–cell communication. *Cell (Cambridge, Mass.), 72*, 1–30.

Johnston, D., Williams, S. H., Jaffe, D., & Gray, R. (1992). NMDA-receptor independent long-term potentiation. *Annual Review of Physiology, 54* 489–505.

Kennedy, M. B., Bennett, M. K., Bulleit, R. F., Erondu, N. E., Jennings, V. R., Miller, S. G., Molloy, S. S., Patton, B. L., & Schenker, L. J. (1990). Structure and regulation of Type II calcium/calmodulin-dependent protein kinase in central nervous system neurons. *Cold Spring Harbor Symposia on Quantitative Biology, 55,* 101–110.

Konorski, J. (1948). *Conditioned reflexes and neuron organisation.* Cambridge: Cambridge University Press.

Korol, D. L., Abel, T. W., Church, L. T., Barnes, C. A., & McNaughton, B. L. (1993). Hippocampal synaptic enhancement and spatial learning in the Morris swim task. *Hippocampus, 3,* 127–132.

Larkman, A., Stratford, K., & Jack, J. (1991). Quantal analysis of excitatory synaptic action and depression in hippocampal slices. *Nature (London), 350,* 344–347.

Laroche, S., Doyère, V., Rédini-Del Negro, C., & Burette, F. (in press). Neural mechanisms of associative memory: Role of long-term potentiation. In J. L. McGaugh, N. M. Weinberger, & G. Lynch, (Eds.), *Brain and memory: Modulation and mediation of neuroplasticity* New York: Oxford University Press.

Levy, W. B., Jr., & Steward, O. (1979). Synapses as associative memory elements in the hippocampal formation. *Brain Research, 175,* 233–245.

Lyford, G. L., Gutnikov, S. A., Clark, A. M., & Rawlins, J. N. P. (1993). Determinants of non-spatial working memory deficits in rats given intraventricular infusions of the NMDA antagonist, AP5. *Neuropsychologia, 31,* 1079–1098.

Lynch, G., & Baudry, M. (1984). The biochemistry of memory: A new and testable hypothesis. *Science, 224,* 1057–1063.

Lynch, G., Dunwiddie, T., & Gribkoff, V. (1977). Heterosynaptic depression: A postsynaptic correlate of long-term potentiation. *Nature (London), 266,* 737–739.

Malenka, R. C. (1993). Long-term depression: Not so depressing after all. *Proceedings of the National Academy of Sciences of the U.S.A., 90,* 31321–3123.

McCabe, B. J., & Horn, G. (1991). Synaptic transmission and recognition memory: Time course of changes in N-methyl-D-aspartate receptors after imprinting. *Behavioural Neuroscience, 105,* 289–294.

McClaren, I. P. L., & Dickinson, A. (1990). The conditioning connection. *Philosophical Transactions of the Royal Society of London, Series B, 329,* 179–186.

McNaughton, B. L., Douglas, R. M., & Goddard, G. V. (1978). Synaptic enhancement in fascia dentata: Cooperativity among co-active afferents. *Brain Research, 157,* 277–293.

McNaughton, B. L., & Morris, R. G. M. (1987). Hippocampal synaptic enhancement and information storage within a distributed memory system. *Trends in Neurosciences, 10,* 408–415.

Miserendino, M. J. D., Sananes, C. B., Melia, K. R., & Davis, M. (1990). Blocking of acquisition but not expression of conditioned fear-potentiated startle by NMDA antagonists in the amygdala. *Nature (London), 345,* 716–718.

Moncada, S. (1992). Nitric oxide gas—mediator, modulator, and pathophysiologic entity. *Journal of Laboratory and Clinical Medicine, 120,* 187–191.

Morris, R. G. M. (1989). Synaptic plasticity and learning: Selective impairment of learning and blockade of long-term potentiation in vivo by the N-methyl-D-aspartate receptor antagonist, AP5. *Journal of Neuroscience, 9,* 3040–3057.

Morris, R. G. M. (1990). Towards a representational hypothesis of the role of hippocampal synaptic plasticity in spatial and other forms of learning. *Cold Spring Harbour Symposia on Quantitative Biology, 55,* 161–174.

Morris, R. G. M., Anderson, E., Lynch, G., & Baudry, M. (1986). Selective impairment of learning and blockade of long-term potentiation by an N-methyl-D-aspartate receptor antagonist, AP5. *Nature (London), 319,* 774–776.

Morris, R. G. M., & Collingridge, G. L. (1993). Expanding the potential. *Nature (London) 364,* 104–105.

Morris, R. G. M., Garrud, P., Rawlins, J. N. P., & O'Keefe, J. (1982). Place navigation impaired in rats with hippocampal lesions. *Nature (London), 297,* 681–683.

Morris, R. G. M., Hagan, J. J., & Rawlins, J. N. P. (1986). Allocentric spatial learning by hippocampectomised rats: A further test of the spatial-mapping and working-memory theories of hippocampal function. *Quarterly Journal of Experimental Psychology, 38B,* 365–395.

Morris, R. G. M., Halliwell, R., & Bowery, N. (1989). Synaptic plasticity and learning. II: Do different kinds of plasticity underly different kinds of learning? *Neuropsychologia, 27,* 41–59.

Morris, R. G. M., & Kennedy, M. B. (1992). The Pierian Spring. *Current Biology, 2,* 511–514.

Morris, R. G. M., Schenk, F., Tweedie, F., & Jarrard, L. E. (1990). Ibotenate lesions of the hippocampus and/or subiculum: Dissociating components of allocentric spatial learning. *European Journal of Neuroscience, 2,* 1016–1028.

Morris, R. G. M., & Willshaw, D. J. (1989). Must what goes up come down? *Nature (London), 339,* 175–176.

Moser, E. I., Mathiesen, I., & Andersen, P. (1993). Association between brain temperature and dentate field-potentials in exploring and swimming rats. *Science, 259,* 1324–1326.

Moser, M.-B., Moser, E. I., & Andersen, P. (1993). Temperature independent enhancement of dentate field potentials during exploration of a novel environment. *Society for Neuroscience Abstracts, 17,* 325.7.

Mulkey, R. M., & Malenka, R. C. (1992). Mechanisms underlying induction of homosynaptic depression in area CA1 of the hippocampus. *Neuron, 9,* 967–975.

Mumby, D. G., Weisand, M. P., Barela, P. B., & Sutherland, R. J. (1993). LTP saturation contralateral to a hippocampal lesion impairs place learning in rats. *Society for Neuroscience Abstracts, 17,* 186.2.

O'Dell, T. J., Hawkins, R. D., Kandel, E. R., & Arancio, O. (1991). Tests of the roles of two diffusable substances in long-term potentiation—Evidence for nitric oxide as a possible early retrograde messenger. *Proc. Natl. Acad. Sci., USA, 88,* 11285–11289.

O'Dell, T. J., Kandel, E. R., & Grant, S. G. N. (1991). Long-term potentiation in the hippocampus is blocked by tyrosine kinase inhibitors. *Nature (London), 353,* 558–560.

O'Keefe, J. (1976). Place units in the hippocampus of the freely moving rat. *Experimental Neurology, 51,* 78–109.

Paulsen, O., Li, Y.-G., Hvalby, Ø., Andersen, P., & Bliss, T. V. P. (1993). Failure to induce long-term depression by an anti-correlation procedure in area CA1 of the rat hippocampal slice. *European Journal of Neuroscience, 5,* 1241–1246.

Racine, R. J., Milgram, N. W., & Hafner, S. (1983). Long-term potentiation in the rat limbic forebrain. *Brain Research, 260,* 217–231.

Ramón y Cajál, S. R. (1911). *Histologie du système nerveux de l'homme et des vertébrès* (Vol. II). Paris: Maluine.

Rescorla, R. A., & Wagner, A. R. (1972). A theory of Pavlovian conditioning: Variations in the effectiveness of reinforcement and nonreinforcement. In A. H. Black & W. F. Prokasy (Eds.), *Classical conditioning II: Current research and theory* (pp. 64–99). New York: Appleton-Century-Crofts.

Rolls, E. T. (1990). Theoretical and neurophysiological analysis of the functions of the primate hippocampus in memory. *Cold Spring Harbor Symposia on Quantitative Biology, 55,* 995–1006.

Rosenzweig, M. R., Bennett, E. L., & Diamond, M. C. (1972). Brain changes in response to experience. *Scientific American, 226,*(2), 22–29.

Schoepp, D. D., Bockaert, J., & Sladaczek, F. (1990). Pharmacological and functional characteristics of metabotropic excitatory amino acid receptors. *Trends in Pharmacological Sciences, 11,* 508–515.

Schumann, E. M., & Madison, D. V. (1991). A requirement for the intercellular messenger nitric oxide in long-term potentiation. *Science, 254,* 1503–1506.

Sejnowski, T. J. (1977). Storing covariance with nonlinearly interacting neurons. *Journal of Mathmatical Biology, 4,* 303–321.

Sharp, P. E., McNaughton, B. L., & Barnes, C. A. (1989). Exploration-dependent modulation of evoked response in fascia dentata: Fundamental observations and time-course. *Psychobiology, 17,* 257–269.

Silva, A. J., Paylor, R., Wehner, J. M., & Tonegawa, A. (1992). Impaired spatial learning in α-calcium-calmodulin kinase II mutant mice. *Science, 257,* 206–211.

Silva, A. J., Stevens, C. F., Tonegawa, S., & Wang, Y. (1992). Deficient long-term potentiation in α-calcium-calmodulin kinase II mutant mice. *Science, 257,* 201–206.

Stanton, P., & Sejnowski, T. J. (1989). Associative long-term depression in the hippocampus induced by Hebbian co-variance. *Nature, 339,* 215–218.

Stent, G. S. (1973). A physiological mechanism for Hebb's postulate of learning. *Proceedings of the National Academy of Sciences of the U.S.A., 70,* 997–1001.

Stevens, C. F., & Wang, Y. (1993). Reversal of long-term potentiation by inhibitors of haem-oxygenase. *Nature (London), 364,* 147–149.

Storm, J. (1993). Evidence for "silent" synapses in the hippocampus. *European Neuroscience Association Abstracts, 16,* 960.

Sutherland, R. J., Dringenberg, H. C., & Hoesing, J. M. (1993). Induction of long-term potentiation at perforant path dentate synapses does not affect place learning or memory. *Hippocampus, 3,* 141–148.

Sutherland, R. J., Whishaw, I. Q., & Kolb, B. (1983). A behavioral analysis of spatial localisation following electrolytic, kainate or colchicine induced damage to the hippocampal formation in the rat. *Behavioural Brain Research, 7,* 133–153.

Tonkiss, J., Morris, R. G. M., & Rawlins, J. N. P. (1989). Intraventricular infusion of the NMDA antagonist AP5 impairs DRL performance in the rat. *Experimental Brain Research, 73,* 181–188.

Treves, A., & Rolls, E. T. (1992). Computational constraints suggest the need for two distinct input systems to the Hippocampal CA3 network. *Hippocampus, 2,* 189–199.

Willshaw, D. J., & Dayan, P. (1990). Optimal plasticity from matrix memories: What goes up must come down. *Neural Computation, 2,* 85–91.

Winson, J., & Abzug, C. (1977). Gating of neuronal transmission in the hippocampus: Efficacy of transmission varies with behavioral state. *Science, 196,* 1223–1225.

Zalutsky, R. A., & Nicoll, R. A. (1990). Comparison of two forms of long-term potentiation in single hippocampal neurons. *Science, 248,* 1619–1624.

Zhuo, M., Small, S. A., Kandel, E. R., & Hawkins, R. D. (1993). Nitric oxide and carbon monoxide produce activity-dependent long-term synaptic enhancement in Hippocampus. *Science, 260,* 1946–1950.

CHAPTER 7

Biological Approaches to the Study of Learning

Sara J. Shettleworth

I. INTRODUCTION

Much research on animal learning is a search for general principles that apply across species and situations. The belief that such laws might be found is encouraged by the Darwinian principle of evolutionary continuity between species and by evidence that species from bees to pigeons to monkeys actually do learn some things in the same way (e.g., Bitterman, 1988; Macphail, 1987). However, the study of learning is not only the domain of psychologists in search of general processes of association and reinforcement. Biologists also study learning, but often for different reasons than do their psychologist colleagues.

The body of data and theory about learning that can be roughly characterized as "biological" includes attempts to answer several distinct questions. This chapter is organized around three of them:

1. What and how do animals learn in the wild? Some "naturalistic" learning phenomena have been thought to reflect processes different in kind from simple associative learning. Three of the best studied examples—imprinting, song learning, and observational learning—are discussed here in some detail to illustrate how apparently "special" kinds of learning can be analyzed within the same general framework as associative learning.

2. How are reinforcement and association formation related to natural behavior? Logically, the first question subsumes this one, but it is treated in a separate section because it has generated a large literature of its own.

3. Why (in terms of evolution and present-day function) do animals learn as they do?

Some who have contrasted biologists' and psychologists' approaches to learning (e.g., Kamil, 1988; Rozin & Schull, 1988) have suggested that traditional learning theory should be assimilated to a more biological view of animal behavior. Such a synthetic approach to learning, it has been suggested, would encompass varieties of behavioral plasticity other than classical and instrumental conditioning, recognize specializations of learning in different species and situations, and firmly tie learning to evolutionary theory. There are signs that psychological and biological approaches to learning are becoming more integrated, and some of them are mentioned in reviewing the bodies of research related to each of our three questions. In the concluding section of the chapter I mention some broader implications of biological approaches for issues in psychology.

II. WHAT AND HOW DO ANIMALS LEARN IN THE WILD?

A. Overview

The hallmark of the biological approach to learning is an interest in phenomena demonstrably important in animals' lives: learning where home is, what song to sing, how to find food, how to behave toward members of a social group. Some of the phenomena that ethologists have uncovered while simply trying to understand why particular species behave as they do seem at first glance quite different from associative learning. Others seem to illustrate rather well the detailed workings of conditioning principles. Although biological studies of learning are often more data than theory driven, some detailed models of specific phenomena have been developed. Primary among these are models of song learning and imprinting. Because such models have been developed in a different tradition from associative learning models, they often have a different flavor from psychological models. Nevertheless, a comprehensive analysis of any learning phenomenon must address certain common questions. These have been made explicit for Pavlovian conditioning (Rescorla, 1988; Rescorla & Holland, 1976), but they have more general relevance.

Any learning phenomenon has at least an implicit operational definition: an experience at one time, T1, causes a change in the animal's state (learning or, equivalently, memory), which is reflected in a test at a later time, T2, in which animals that had the experience behave differently from those that did not (Rescorla & Holland, 1976). This broad definition provides the base for a framework within which candidates for distinct kinds of learning can be compared (see Shettleworth, 1993a, for further discussion). This framework consists of three fundamental questions that can be asked about any learning phenomenon (Rescorla, 1988). First, what are the conditions of learning? That is to say, what experiences at T1 are necessary to bring about the behavioral change of interest? Studies of the conditions of learning include experiments on the effects of delay, magnitude, or quality of reinforcement, CS–US interval, stimulus duration, quality, or intensity, developmental stage, and indeed, any attempt to specify quantitatively or qualitatively the "laws of learning."

Rescorla's other two questions presuppose a theoretical distinction between learning and performance that has not always been explicit in biological studies of learning. These are "What are the contents of learning?" and "What are the effects of learning on behavior?" Analyzing the contents of learning means attempting to characterize the hypothetical representational structure or change in state produced by exposure to the conditions of learning. Excitatory and inhibitory associative bonds, response strengths, cognitive maps (see Chapter 8, this volume, by Gallistel), templates (see Section II.C)—all have been described as contents of learning in one case or another. These and other hypothetical ways of storing the effects of experience fall into two classes: simple excitatory or inhibitory links and more complexly structured representations on which behavior may be based in more flexible ways. Whether the latter must ultimately consist of networks of simple associative links is an issue that not need be considered here (cf. Gallistel, 1990, and McLaren, Kaye, & Mackintosh, 1989, for discussions related to animal learning).

Many biological studies of learning involve distinctive effects of learning on behavior. New songs are sung, specific companions or sexual partners are chosen. These sorts of effects have sometimes seemed to set off the learning phenomena studied by biologists from classical and instrumental conditioning, but evidence reviewed later in this chapter indicates that this is probably an artificial distinction.

In the remainder of this section, I briefly review three of the best studied examples of "learning in the wild." I hope to convey some of the flavor of the original work, while using the framework just outlined to reveal some of the theoretical issues that such work raises.

B. Imprinting

1. Filial Imprinting

Filial imprinting refers to the process through which newly hatched precocial birds form attachments to their parents. *Precocial* refers to species like ducks and chickens that are covered with down, have functioning eyes and ears, and can run around soon after hatch. Such birds need a mechanism for keeping the vulnerable young with a parent during the first part of life.

When Konrad Lorenz (1935) described the way in which young geese or ducks form lifelong social preferences after one brief exposure to their mother or another companion, he seemed to be describing an instantaneous, irreversible process that was "different from ordinary learning," by which he meant Pavlovian and operant conditioning (Lorenz, 1970, p. 377). The research stimulated by Lorenz's observations has been reviewed comprehensively elsewhere (Bateson, 1966; Bolhuis, 1991). The story it tells can be summarized by saying that while imprinting does have some unique features, it also has many features in common with other learning phenomena. For instance, rather than occurring immediately at full strength to whatever object the young bird sees first, imprinting depends to some extent on the characteristics of the imprinting object and how long the bird is exposed to it (see Ten Cate, 1989). Similarly, the description of the period in life when imprinting takes place as a "critical period" has been replaced by the term "sensitive period" on the basis of evidence that imprintability waxes and wanes gradually rather than abruptly (Bateson, 1979).

On the surface, the events during imprinting seem quite different from those necessary for associative learning. The young bird is exposed to a single object, the mother or a surrogate like a flashing light, dangling ball, or rotating stuffed hen. It later exhibits a preference for seeing or approaching the object it has seen over another, otherwise equivalent, object. This approach or choice behavior is a symptom of a social attachment that can also be evident in the bird's snuggling up to the imprinting object and twittering, calling shrilly in its absence, and performing an operant to receive brief views of it. In nature, approaching a mother bird might well be instrumentally reinforced, for example, by the warmth under her feathers, but imprinting proceeds well in the laboratory when there is no possibility of an instrumental contingency. It is less obvious, though, whether or not there is some sort of Pavlovian contingency. Considerable discussion has been devoted to this issue in the interest of deciding whether or not imprinting is a distinct form of learning (cf. Bateson, 1990; Bolhuis, de Vos, & Kruijt, 1990; van Kampen, 1993).

The argument that imprinting is a form of Pavlovian conditioning encounters two interrelated problems. One is to identify the unconditioned stimulus (US) with which the presumably neutral, somewhat arbitrary, features of the imprinting object become associated. The other is to show how the putative CS and US can be dissociated from one another to test whether conditions of learning such as contingency between CS and US influence imprinting in the same way as they influence Pavlovian conditioning. The fact that moving objects are more effective in imprinting than are stationary ones has led to suggestions (Hoffman & Ratner, 1973) that motion functions as the US in imprinting. This notion leads to the prediction that if a bird is exposed to a stationary object after seeing it in motion, imprinting should extinguish. Like several related predictions, this one is not supported by the relevant data (Eiserer, 1980). It has also been claimed (Bateson, 1990) that the first imprinting experience is preemptive, in that once a bird is well imprinted to one object, it cannot be imprinted to another. If true, this would make imprinting different from Pavlovian conditioning with, say, food or shock, in which animals can be trained with a succession of different signals for the same US. However, under at least some conditions, imprinting is possible with up to three successive objects, and the "secondary" attachments do not differ qualitatively from the primary one (de Vos & van Kampen, 1993).

More fruitful than to pursue the problem of partitioning the experience necessary for imprinting into CS-like and US-like events is to ask how the animal learns to recognize the imprinting object. This question is distinct from that of why social attachment develops. Bateson (1990; Hollis, Ten Cate, & Bateson, 1991) has proposed a connectionist model of recognition learning in imprinting, similar to other associationist models of perceptual learning (e.g., McLaren et al., 1989). Some researchers (e.g., Ryan & Lea, 1990; van Kampen, 1993) have shown how separable features of imprinting objects can be manipulated to discover how they combine—whether, for example, overshadowing, potentiation, blocking, or configuring occur. In a model of imprinting that embodies a distinction between learning and performance, Bateson (1990; see Figure 1) has pointed out that recognition learning is only part of imprinting, in that birds can learn to recognize many more objects than the one they imprint to. It is as if the stimulus analysis and recognition system can form only a limited number of connections to the "executive system" for filial behavior. Identifying object recognition as a distinct subprocess of imprinting (Hollis et al., 1991) is valuable in allowing one to ask what characteristics this process shares with learning the features of objects in other situations. Exposure to an imprinting object, for example, may lead to perceptual learning that influences the outcome of later discrimination training just as does perceptual learning in other situations (Honey, Horn, & Bateson, 1993).

2. Sexual Imprinting

One way in which Lorenz (1935) claimed that imprinting differed from other forms of learning was that exposure to a particular individual, the mother, early in life influenced what species was chosen for mating later in life. In his account, sexual imprinting was the automatic result of filial imprinting. In the normal course of events in nature, this must be so, even if two processes are involved. Filial imprinting in the first few hours or days

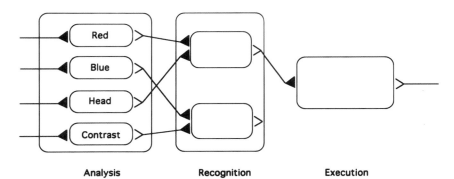

Analysis   Recognition   Execution

FIGURE 1   A depiction of the processes involved in filial imprinting. Stimulus analysis (only a sample of possible stimulus elements is given) and recognition are common to many behavior systems, but connections to the filial behavior system ("execution") are limited and are formed under only some conditions. (Redrawn from Bateson, 1990.)

guarantees that the young bird will stay with its family. Therefore, as long as the sensitive period for acquisition of sexual preferences occurs before the family breaks up, the bird will automatically be exposed to others of its species at the right time. And evidence from at least some species indicates that filial and sexual imprinting do occur at separate times (Bateson, 1979; Vidal, 1980; review in Bolhuis, 1991). A functional reason for this separation may be that while as a youngster the bird needs to learn the features of a particular individual, its mother, as an adolescent it needs to learn the features of its siblings in order to choose a mate optimally different from them (Bateson, 1979). Such selective mate choice (Bateson, 1983) implies an unusual performance rule: choose a mate somewhat different from your learned representation of "sibling."

Early discussions of sexual imprinting include descriptions of birds preferring to mate with the species they were exposed to early in life even after many years of living with other species (Immelmann, 1972). However, for obvious practical reasons, sexual preferences were studied much less than filial ones. Recently, however, work in this area has had a resurgence (see Bolhuis, 1991). Some of it is being done with fast-maturing altricial species like zebra finches, birds quite different from the geese and ducks observed by Lorenz and other early workers. Thus, here, as elsewhere (see especially Section II.C), one must beware of generalizing too much across species. However, the recent work does indicate that sexual preferences may be acquired through a process different from filial imprinting. In zebra finches, social interactions with siblings and during feeding by parents may play an important role. The apparent bias of these birds to imprint on their own species rather than on Bengalese finch foster parents can be explained by differences in the quality and quantity of feeding by zebra finch versus Bengalese finch parents with young zebra finches in the nest (Ten Cate, Los,

& Schilperood, 1984). In this species, at least, it might be more appropriate to refer not to "sexual imprinting" but to "acquisition of social preferences." I use *social* here deliberately, as in most cases it needs to be determined whether the preferences are specifically sexual in nature.

3. Conclusions

The history of research on imprinting is a story of extreme claims for a special form of all-or-none learning in a short, critical period being tempered by data. At the same time as this research was going on, research on learning more generally was developing in ways that made some of the apparently "special" features of filial imprinting easier to accommodate. For example, the fact that it influences a whole system of behavior rather than a single, trained response is more compatible with the contemporary view of conditioning as influencing the organization of behavior systems than with the traditional Pavlovian response rule (see Section III). Animals being imprinted are learning the features of parental or sexual objects, and it is of interest to ask how the rules by which separable features are learned and combined compare to those operating in other situations. Thus, one thing the study of imprinting can contribute to a broader study of behavioral plasticity and its neural basis (cf. Horn, 1990) is a tractable and well-analyzed example of social recognition learning.

C. Song Learning

1. Overview

If imprinting has a competitor in the stakes for "learning with the most remarkable features," it is song learning. Like social preferences, song learning can be the result of experience during a sensitive period early in life with long-lasting effects. Just as with imprinting, early research presented a clear picture of a striking phenomenon, and later research, some of it using other species and some examining more closely the original results, indicated that the first simple sketch was too simple. Also, like imprinting (see Ten Cate, Vos, & Mann, 1993, for further comparisons), song learning is a form of learning defined in terms of its outcome, and thus it does not necessarily reflect the same underlying process in all species that show it. Many species of birds—though far from all (Kroodsma, 1982)—learn their songs, but the details of how they do so vary markedly from one species to another. (For reviews of song learning, see deVoogd, 1994; Marler, 1990; Slater, 1989.)

2. The White-Crowned Sparrow and the Template Model

One of the classic species for studies of song learning is the white-crowned sparrow of North America (*Zonotrichia leucophrys*). Each male white-crowned sparrow has a single, rather simple song, which he shares with his neighbors. Males in different geographic areas have different songs, that is, there are local dialects. The fact that each male has a single, loud, whistled song simplifies the analysis of song in this species, and the existence of dialects suggests that learning is important in its development.

White-crowned sparrows reared in isolation from a few days old to adulthood sing abnormal songs (Marler, 1970). Their songs still undergo some development, however. Each male progresses from a stage of disorganized and variable vocalizations (subsong) to one in which it produces a single stereotyped song (crystallized song), which still has some species-typical characteristics. Deafened birds also fail to develop normal song, but their vocalizations are much more abnormal. The contrast between isolated and early-deafened birds in this and other species indicates that the progression from subsong to crystallized but atypical song shown by isolates depends on auditory feedback from the bird's own vocalizations. White-crowned sparrows given normal early experience and deafened after they have developed full song continue to sing a species-typical song, however, indicating that auditory feedback may not play a role in maintaining the structure of crystallized song (Konishi, 1965; but see Nordeen & Nordeen, 1992).

White-crowned sparrows taken from the nest at a few days of age and reared in isolation acquire normal song if they are played tape-recorded white-crowned sparrow song. They acquire the song they hear whether or not it is from the same dialect area where they were born. They do not, however, acquire the songs of other species from tape recordings, nor do they learn from tapes outside the ages of about 10 to 50 days (Marler, 1970). Since white-crowned sparrows do not start to sing until early in the spring after they hatch, effects of the experience of hearing song in the first two months of life must be retained for many months (Marler & Peters, 1981). Song learning in this species can thus be thought of as consisting of two phases, a sensory phase, in which the bird stores auditory information, and a motor phase, in which it uses this information to shape its own song. The content of learning in the first phase is a representation of song; what is learned in the second phase is a specific motor pattern. Just as with imprinting, the conditions of learning include both the age and species of bird and the amount and type of stimulation to which it is exposed (Hultsch, 1993; Ten Cate, 1989; Ten Cate et al., 1993). For example, white-crowned sparrows learn from live tutors of their own or even another species at an age when tapes are no longer effective (Marler, 1987; Petrinovich, 1988).

Descriptions of song learning in the white-crowned sparrow inspired a simple and compelling model of song development (Konishi, 1965; Marler, 1976; see Figure 2). The bird behaves as if it is born with a simple auditory template or internal representation of the rough characteristics of its species-

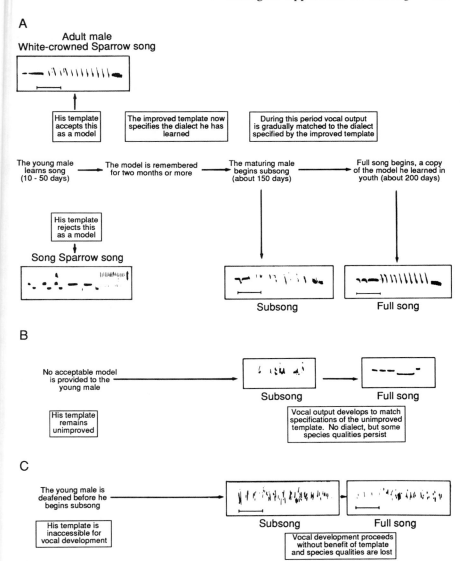

FIGURE 2   A summary of song learning in normally raised (A), socially isolated (B), and deafened (C) male white-crowned sparrows, indicating the ages at which various events occur and the role of the hypothetical template at each stage. Patterns of song experienced and produced are depicted as sonograms, plots of sound frequency against time. (From Marler, 1976.)

typical song. The template selects what songs will be learned during the sensitive period. This learning refines the template, which acts as a standard against which the bird compares its own vocalizations during the motor learning phase. The hypothetical auditory template therefore serves three functions: like a sensory filter, it selects what songs will be memorized; it stores memory between the sensory and motor phases; and it is the standard against which the bird compares its own vocalizations. The abnormal song of birds given appropriate early experience and then deafened demonstrates the necessity of auditory feedback during the motor learning phase. The characteristics of the rough, unmodified template are revealed by the song of birds with no experience of species-typical song and by what songs are selected for learning.

The features of acceptable model songs may vary from species to species, as has been shown in the elegant comparisons of song sparrows (*Melospiza melodia*) and swamp sparrows (*M. georgiana*) by Marler and Peters (1989). Because these two species often breed within earshot of one another, each needs a mechanism for selective learning of its own species song. Selective responsiveness to conspecific song is evident in heart-rate changes of nestlings, and the features that are important for selective learning have been studied with computer-generated song (Figure 3).

3. Complicating the Story: Song Learning in Other Species

This simple story has exceptions in almost every respect (see reviews listed in section II.A.1). For example, some species select what song to learn at least partly on the basis of social or visual characteristics of the singer. Some species, like zebra finches (Slater, Eales, & Clayton, 1988), learn from taped songs only under special conditions (Adret, 1993). Other species, like canaries, learn new songs throughout life, and thus the sensory and motor phases overlap. Some species memorize more song than they eventually sing, suggesting that some, possibly social, selective learning mechanism is operating in adulthood (Marler & Nelson, 1993).

Perhaps the most striking example of the variety in song learning is provided by the brown-headed cowbird (*Molothrus ater*; King & West, 1990). Female cowbirds lay their eggs in other birds' nests and leave the young to be raised by the host. After fledging, young cowbirds join into flocks, where they remain until breeding the next spring. Early exposure to its own species's song is unlikely to influence song development in a bird raised by another species. Indeed, development of species-typical responses in nest parasites seemingly must be an example of a "closed motor program" (Mayr, 1974). On this notion, adult male cowbirds raised in isolation ought to sing normal species-typical song. They do sing a species-typical song, but far from being normal, it is a supernormal releaser of female

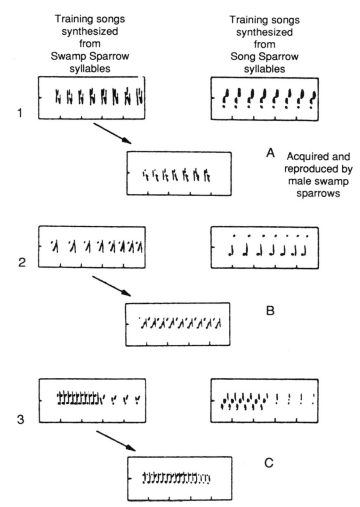

FIGURE 3 Examples of synthetic training songs presented to young male swamp sparrows. The two examples in row 1 have the temporal pattern typical of swamp sparrows; row 3 shows the typical song sparrow temporal pattern. Arrows and central column of panels show which song was imitated. Swamp sparrows selectively learned swamp sparrow syllables regardless of overall song structure and sang them in a swamp sparrow tempo. Song sparrows (data not shown) would accept heterospecific syllables only if they were presented in a song sparrow–like song structure. (From Marler & Peters, 1989.)

copulatory behavior and male aggression. Released into a flock of cowbirds, a male with such a song is the target of aggression from its male companions. It appears that the negative consequences of singing a supernormal song shape the male's song in a socially acceptable direction. At the same time, however, females apparently act in a manner that shapes males' vocal behavior toward the form that they find most stimulating (West & King, 1988).

In other species, as well, social interactions among adults can influence song development, for example, when males on neighboring territories match songs. Marler and Nelson (1993) distinguish this form of song learning, "action-based learning," from the "memory-based learning" characteristic of white-crowned sparrows. Action-based song learning takes place through selective social reinforcement when already-learned vocalizations are produced. In cowbirds, the social effects of vocalizations shape their form in much the same way as auditory feedback seems to selectively reinforce species-typical vocalizations in white-crowned sparrows.

4. Song Perception and Learning: General Implications

Although song learning is a form of plasticity peculiar to only a few of the world's species, the large literature about it is full of information relevant to a general understanding of animal learning and cognition. As will be seen in the next part of this chapter, there are few other examples of animals acquiring specific motor patterns through exposure to the performance of conspecifics. One reason why song learning is apparently more common than other forms of imitation (see Section II.D) may be that in auditory learning the learner can perceive its own output in the same form as that of the model: it hears both of them (cf. Heyes, 1993).

Some species, such as nightingales, zebra finches, and marsh wrens, sing many different songs, and not every individual sings the same ones. In such species, one may ask what rules govern selection of the elements to be learned. For example, are elements that occur together in a model's song sung together in the same sequence by the pupil (Hultsch, 1993; ten Cate & Slater, 1991)? How do frequency and order of exposure influence learning (Kroodsma & Pickert, 1980)? To what extent are the principles found here the same as those applying to the learning of lists by people?

Birds learn others' songs not only to sing them but also for individual recognition (Kroodsma & Byers, 1991; McGregor, 1991). Territorial males recognize their neighbors by song, and attack only those neighbors that are not singing from their accustomed locations (Falls, 1982). In species where each of several neighbors can have a repertoire of different songs, accurate neighbor recognition requires the capacity to memorize many, possibly similar songs. Operant methods can be used to ask whether species differ adaptively in song recognition memory capacity (Stoddard, Beecher, Loesche, & Campbell, 1992).

In some species, parents discriminate their own from alien young on the basis of voice. Such discrimination is more important for colonial species than for solitary nesters. Beecher and his colleagues (see Beecher & Stoddard, 1990) have investigated how the signals produced by the baby birds and the recognition abilities of the parents differ in an adaptive way to produce the expected species differences in response to alien offspring in closely related species of swallows. Recent years have seen an increasing number of interdisciplinary studies in which standard psychological methods have been used to analyze how birds perceive, classify, and remember their songs (for examples, see Cynx, 1990; Dooling, Brown, Park, & Okanoya, 1990; Nelson & Marler, 1990). Such work promises to provide increasing insight into the general issue (Guilford & Dawkins, 1991) of how species-specific fine-tuning and general perceptual mechanisms interact in animal signal systems.

D. Social Learning

1. Issues and Perspectives

Like song learning and imprinting, social learning (which I define as learning from conspecifics) has been the subject of whole books (Zentall & Galef, 1988) and many reviews (e.g., Davis, 1973; Galef, 1976, 1988; Lefebvre & Palameta, 1988; Whiten & Ham, 1992). Like song learning or imprinting, the topic of social learning illustrates some key differences between studying functionally defined and mechanistically defined learning phenomena and thus between biological and psychological approaches.

Why some species of animals live in groups, why groups are the size they are, and how individuals benefit from group living are important questions in behavioral ecology. One potential benefit of group living is the opportunity to learn from conspecifics what foods to eat, where to find them, and what is dangerous. Thus, it is of some theoretical interest to know which species learn socially, what they learn, and whether any have evolved behaviors designed to teach conspecifics (Caro & Hauser, 1992; Lefebvre & Palameta, 1988). From this point of view, it matters little *how* animals learn from others, that is, what is the mechanism of social learning. Even though it may be the viewpoint of experimental psychology that the mechanism of social transmission is of crucial importance, the fact is that behavior can be transmitted from one individual to another in myriads of ways.

2. Social Transmission of Information about Food

The most thorough analysis of social learning in any one species is that of social transmission of food preferences in rats carried out by Galef and his colleagues (Galef, 1990; Galef & Beck, 1991). In the wild, Norway rats (*Rattus norvegicus*) are social omnivores. They solve the problem of selecting what to eat by forming preferences for the flavors of foods being eaten safely by other rats in the colony. This learning begins before weaning, when suckling rats acquire preferences for flavors present in their mother's milk. Once they leave the nest, they prefer to feed where other rats are feeding or have recently fed. The food preferences thus acquired are robust and long lasting, even after the young rats are feeding alone with a choice of nutritious foods.

Food preferences can also be transmitted among adult rats, allowing the rat colony to become an information center in which individuals returning from foraging inform others about safe foods to be found in the neighborhood. Acquiring a preference for foods eaten by others is not merely a matter of becoming familiar with their flavors. To be effective, the odor of food has to be experienced in the context of carbon disulfide, a chemical in rats' breath. Thus, a subject rat will acquire a preference for a food recently eaten by an anesthetized rat with which it can interact, but it will not choose a flavor dusted on the rear end of an awake rat (see Galef, 1990). This learning might appear to be an example of simultaneous Pavlovian conditioning, which endows the flavor with positive hedonic properties (Berridge & Schulkin, 1989). However, the one investigation inspired by this analysis to date failed to find evidence of blocking, latent inhibition, or overshadowing (Galef & Durlach, 1993).

An analysis of how learning about food takes place in a social context has also been carried out using feral and laboratory-housed flocks of pigeons by Lefebvre and his colleagues (Lefebvre & Palameta, 1988). A focus of this research has been how a novel feeding technique might spread through a population. For example, one member of a flock might be taught to obtain food by pecking at a novel type of feeder. Whether and how rapidly this technique spreads through the flock is partly determined by whether or not the food available for the trained "producer" can be "scrounged" by others (scroungers are less likely to learn the technique for themselves; Giraldeau & Templeton, 1991; see also Beauchamp & Kacelnik, 1991) and how closed or open is the population of the flock (Lefebvre & Palameta, 1988).

When social learning is studied in its functional context, the exact mechanisms through which learning takes place are of interest mainly as they illuminate the circumstances in which information can be transmitted. For example, the fact that rats acquire preferences for food odors smelled on other rats' breath ensures that they will eat not just any food they have smelled lying around but food being safely consumed by other rats. The circumstances necessary for learning have not always been so carefully analyzed, however. A good example is provided by the famous case of milk-bottle opening by British birds (see Sherry & Galef, 1990). After birds began piercing the foil tops of milk bottles and drinking the cream in one area of Britain, the habit spread to contiguous areas. This suggests that in some way the bottle-opening individuals were providing the conditions for other individuals to acquire the same behavior, but the necessary condition for learning need not have been for one individual to see another feeding from an opened bottle. For example, birds exploring already-opened bottles might have learned to approach other bottles, and subsequently may have been reinforced for pecking through closed bottle tops on their own. The sight of other birds at bottles might have attracted naïve individuals, again, seeing others opening or feeding from bottles not being necessary. Using North American black-capped chickadees, Sherry and Galef (1990) found no evidence that birds learned from seeing others opening milk bottles. In other cases, as well, behaviors may spread through a population more because one individual in some way creates circumstances in which other individuals may learn than because animals directly copy one another's behavior (Whiten & Ham, 1992). Too few of these alternatives to direct copying have been studied as learning processes of potential interest in their own right.

3. Imitation

Whiten and Ham (1992, p. 275) concluded their review of 100 years of research on imitation by stating ". . . we must admit how little is firmly empirically established about which species can and do imitate and through what mechanisms." Since *imitation* refers to copying the form of a motor act as a result of seeing another perform it, it excludes cases like milk-bottle opening, in which what is copied is (at most) where to direct a species-typical motor act. The latter clearly can be learned, as demonstrated, for example, by McQuoid and Galef (1992) with jungle fowl learning which of two bowls to feed from. This form of learning, the conditions for which could well be those of Pavlovian conditioning, has been termed *releaser-induced recognition learning* by Suboski (1990).

Over the years since Thorndike (1911) sought (and failed to find) evidence for imitation in animals, the most popular way to investigate imitation in the laboratory has been to compare acquisition of a task by subjects that have and have not observed another animal perform it. Such a test fails to separate copying of the motor act per se from copying the location or object toward which behavior should be directed. What is needed, as first pointed out by Dawson and Foss (1965), is a situation in which two distinct motor acts can be directed toward the same object. For example, Dawson and Foss used a food bowl from which budgerigars could remove a lid using their beak or a foot. Then, all animals had a standard test (e.g., presentation of the covered bowl) and performance of groups previously exposed to the two alternative actions was compared. Experiments with this design have obtained convincing evidence of imitation in rats (Heyes, Dawson, & Nokes, 1992), while revealing that in budgerigars the effect reported by Dawson and Foss is not robust (Galef, Manzig, & Field, 1986).

"True imitation" has traditionally been of much interest because humans seem to be able to imitate any action of which they are physically capable. Therefore (somewhat paradoxically, as pointed out by Whiten & Ham, 1992; but see Heyes, 1993), imitation is seen as evidence of great intelligence. Testing for it in other species was therefore a natural part of the anthropocentric program of comparative psychology, which sought evidence of evolutionary continuity between humans and other animals (cf. Shettleworth, 1993b). However, it is somewhat surprising to find that recent research indicates that monkeys do not imitate, even in cases where they have been assumed to (see Chapter 10, this volume, by Heyes). The great apes, however, such as chimpanzees and orangutans, can perform otherwise improbable acts like using hammers or paint brushes appropriately after seeing people perform them (Russon & Galdikas, 1993; review in Whiten & Ham, 1992). Thus, insofar as it is possible to compare the procedures and observations with species as different as laboratory rats and free-ranging orangutans, one may conclude that imitation might not be phylogenetically widely distributed. Moore (1992) has suggested that it originated independently in mammals and birds and that imitation evolved in a few families of birds from vocal mimicry. Other forms of social learning such as the process involved in copying where to feed or what to eat (McQuoid & Galef, 1992; Palameta & Lefebvre, 1985) may be more widespread (cf. Suboski, 1992), though there may be interesting differences between social and nonsocial species (Lefebvre & Palameta, 1988).

4. Conclusions

Learning from conspecifics is likely to be important for social species in their natural environments. Such learning can take place in a variety of ways. Its nature has barely begun to be explored, and that in just a small sample of species. It is worth bearing in mind, however, that in order to have biologically important consequences, the effect of social learning alone need not be large or long lasting. For example, seeing a conspecific feed from a distinctive bowl induces jungle fowl to peck only a few times in a similar empty bowl 48 hours later, but if the bowl holds food in the test so that the birds are reinforced for choosing it, as they might be for choosing a reliable food source in nature, they display a long-lasting, socially induced preference (McQuoid & Galef, 1992). The most informative analyses of social learning in its various forms are likely to be those in which theoretically important tests take into account the species and context in which the learning might be expected to occur naturally.

E. Other Examples of Learning in the Wild and Problems for the Future

The foregoing sections of this chapter, on imprinting, song learning, and social learning, are by no means a comprehensive survey of the sorts of learning that animals demonstrate in the wild. Gallistel's Chapter (8, this volume) on space and time adds two topics in which the biological approach has been central. Heyes's Chapter (10, this volume) on social cognition in primates provides further examples. Other examples, primarily from foraging behavior, are described in the next part of this chapter. The three examples discussed in detail here illustrate the most important features of the biological approach and its relationship to other approaches to learning. Perhaps the most distinctive of these is that information about the natural behavior of particular species, rather than some a priori general theory of learning or cognition, dictates the problems and species studied. This approach is not incompatible with a more general and abstract analysis of learning mechanisms, but such an analysis is not essential to it. However, as I have tried to show (see also Shettleworth, 1993a), any learning phenomenon can be analyzed within a broad general framework consisting of questions about the conditions and contents of learning and how learning affects behavior (Rescorla, 1988). Placed within such a framework and compared to more thoroughly analyzed cases such as Pavlovian and instrumental conditioning, imprinting, song learning, observational learning, and other "naturalistic" learning phenomena present a wealth of problems for future investigation.

III. WHAT DO REINFORCEMENT AND ASSOCIATION FORMATION HAVE TO DO WITH NATURAL BEHAVIOR?

A. The Issues

Although the biological approach has contributed many examples of learning that are not obviously Pavlovian or instrumental conditioning, this does not mean that associative learning is confined to situations contrived in the laboratory. In this section I first briefly review evidence that simple Pavlovian conditioning and instrumental reinforcement are important in the natural lives of animals. I then discuss the consequences of turning this approach around and, rather than asking about the role of conditioning in natural behavior, I ask about the role of natural behavior in conditioning. Addressing this issue has led in recent years to new ways of understanding the performance rules in conditioning, in the behavior system approach to learning derived from ethology.

B. The Role of Conditioning in Natural Behavior

The best worked examples of how simple conditioning mechanisms can play a role in nature come from foraging behavior (for reviews, see Shettleworth, 1988; Shettleworth, Reid, & Plowright, 1993). For instance, bees learn where and when flowers are available and how to extract nectar from them through what appear to be simple associative mechanisms. What they learn about flowers has been studied by manipulating the features of artificial nectar sources in the field (e.g., Gould, 1990) and by training wild bees to visit food sources in the laboratory (e.g., Couvillon, Leiato, & Bitterman, 1991). Bitterman and his colleagues have developed a detailed model of some of the learning involved (Couvillon & Bitterman, 1991). Whether the bees also have any specialized associative mechanisms, for example, for learning some features of flowers after visiting them, is a matter of debate (Couvillon et al., 1991).

Simple associative learning is also important for vertebrate predators learning about bad-tasting prey items. Generalization of learned avoidance from distinctively patterned or colored unpalatable prey to similar-appearing but palatable prey is thought to be responsible for the evolution of Batesian mimicry systems (cf. Brower, 1988). These are associations of unpalatable and palatable species (usually insects) in which the palatable species (mimics) have evolved resemblance to unpalatable species (models) and are thereby protected from predation. To develop mathematical models of the relationships among populations of models and mimics it is necessary to consider population features that influence rates of learning and breadth of generalization such as the relative numbers and palatabilities of models and mimics, whether they are dispersed or clumped, and whether models and mimics are available to predators at the same or different times of year. In principle, models of mimicry systems should incorporate psychologically realistic assumptions about learning and generalization, but they have not always done so. For example, one model (Huheey, 1988) assumes that a single encounter with an unpalatable model confers protection on (i.e., conditions avoidance of) a fixed number of mimics whether that encounter is the predator's first or nth (i.e., there are no savings). There is considerable scope here for more detailed integration of psychological and ecological theory and data.

Where psychological and biological data have been most successfully integrated is in tests of optimal foraging models (Shettleworth, 1988). Models of optimal foraging developed in the 1970s suggested simple laboratory tests that were essentially schedules of reinforcement. For example, the handling time necessary to subdue a victim or extract a nut from its shell was equated to a delay of reinforcement (Fantino & Abarca, 1985); prey density in two patches was equated to concurrent ratio schedules (Krebs, Kacelnik, & Taylor, 1978). Those testing foraging models often needed to devise unconventional schedules to fit the models' assumptions, but their subjects' behavior could often be accounted for by principles derived from more conventional schedules (Fantino & Abarca, 1985; Shettleworth, 1988). At the same time, consideration of the variables that ought to affect foraging decisions in the wild sometimes revealed unanswered (even previously unasked) questions about operant behavior. For instance, some models of foraging predict that choice should be influenced by the time horizon, the future time available for foraging. This amounts to saying that session length, a variable generally disregarded, should be important in some sorts of experiments, and experimental tests of optimality predictions show that it is (e.g., Plowright & Shettleworth, 1991). Foraging models and information about learning can interact in the reverse way, as well, when well-established features of learning such as variability in time estimation are incorporated into optimality models as operating characteristics of the forager, or constraints (Brunner, Kacelnik, & Gibbon, 1992).

Learning has a large role to play in the development of feeding behavior. Many animals must learn what to feed on and how to process it most efficiently. For example, yellow-eyed juncos must learn feeding techniques during their first summer (Sullivan, 1988); poorer learners are less likely to survive. The role of learning processes in the development of feeding in ring doves has been studied in the laboratory to understand how simple associative processes contribute to the squab's transition from accepting crop milk from adults to pecking at grain (Balsam & Silver, 1994). The development of food recognition by jungle fowl chicks depends on simple associative learning to some extent but also on some effects of experience that do not fit readily into traditional molds (Hogan, 1984). The study of development in general provides considerable scope for an analysis of how experience affects behavior modeled on the approach developed by learning theorists (Balsam & Silver, 1994).

Conditioning also plays a role in learning about things other than food. Enemy recognition in several bird species is transmitted by a simple conditioning process: mob whatever you have seen other birds mob (Curio, 1988). The role of conditioning in sexual behavior and how sexual conditioning might differ from conditioning in other systems has been studied extensively by Domjan and his colleagues (Crawford, Holloway, & Domjan, 1993; Domjan & Hollis, 1988). Clearly, learning of some kind is what allows individuals to recognize their own nest, mate, eggs, young, or territorial boundaries, although exactly how they do so has not always been thoroughly analyzed. A question of particular interest here is whether species differ in an adaptive way in recognition learning ability, with those having to learn more, or more difficult discriminations, being better able to do so. In the natural situation, however, recognition may be just part of a whole system in which a need for more discriminative behavior can be met by evolution of more discriminable signals from young, eggs, or mates, or by more accurate perceptual mechanisms, as well as by possibly specialized learning abilities (see Beecher & Stoddard, 1990, for one example, and Section V.B for further discussion).

C. The Role of Natural Behavior in Conditioning

Over a quarter of a century ago, the Brelands' description of misbehavior (Breland & Breland, 1961) and the discovery of autoshaping (Brown & Jenkins, 1968) stimulated attempts to incorporate ethological thinking about the organization of behavior into accounts of what animals do when exposed to conditioning paradigms (i.e., performance rules). Of course some of the most popular experimental arrangements for testing learning take advantage of the subject's natural inclinations (cf. Timberlake, 1990). Rats running down alleys to find food or escape shock and pigeons pecking lighted disks for food are not exactly exhibiting responses unrelated to the motivational conditions. However, the need to take the subject's natural behavior into account seems more compelling when preorganized behavior conflicts with what experience is expected to produce. For example, the Brelands described how a pig being trained to deposit tokens in a bank became slower and slower at performing the reinforced behavior and began instead to dig and root with them. "Misbehavior" of this sort can also be produced in rats in the laboratory (see Timberlake, 1990). When an object reliably signals food or water, classically conditioned behaviors to the object result in "misbehavior" to the extent that they are incompatible with the behavior being instrumentally reinforced.

In ethological terms, a behavior system consists of interrelated perceptual, central, and motor units (Hogan, 1988; Figure 4). This organization is expressed in the way in which particular external stimuli and internal states create predispositions to respond in certain ways. An animal that is hungry and confronted with food simply behaves differently from one encountering a predator or sexual partner. As soon as one looks beyond narrowly defined responses such as bar pressing or salivation, it is not hard to see that such predispositions express themselves in conditioning situations. However, it has proven difficult to formalize and give predictive power to this

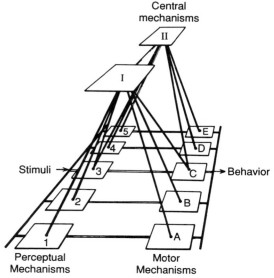

Central
mechanisms

Perceptual
Mechanisms

Motor
Mechanisms

Stimuli →

Behavior

FIGURE 4   Ethological conception of the organization of behavior systems. Stimuli are first analyzed by perceptual mechanisms. These may have direct connections to motor mechanisms, as in reflexes, or may influence behavior via central coordinating mechanisms. A central mechanism specific to each behavior system integrates input from relevant external and internal stimuli (e.g., food and hunger for the hunger system). Some motor patterns (walking, e.g.) are influenced by more than one behavior system. (From Hogan, 1988.)

intuition. Progress has been made for specific species and situations in research programs that combine ethological studies of freely behaving animals with observations of what this same species does during instrumental and/or Pavlovian conditioning (Fanselow & Lester, 1988; Shettleworth & Juergensen, 1980). Timberlake and his colleagues (see Timberlake, 1990) have conducted comparative studies, in which the behavior of different rodents toward signals for food has been predicted from the nature of their predatory behavior.

More general theoretical accounts of how learned behavior is related to pre-existing behavioral organization include Timberlake's (1993) behavior system approach, Suboski's (1990) suggestion that much conditioning or recognition learning consists of the transfer of releasing value from unconditioned to conditioned stimuli, and Holland's (1984) thoughtful discussion of the origins of Pavlovian CRs. Related to these accounts is that of Hollis (1982), who suggests that the form of Pavlovian conditioned responses can be accounted for by the function that conditioning has evolved to serve in particular situations. She has tested hypotheses about the function of conditioning by using signals for sexual and aggressive encounters with blue gouramis as subjects. For example, fish trained with a signal preceding appearance of a rival are more often victorious in signaled aggressive encounters than fish for which the signal was not a Pavlovian CS (Hollis, 1990).

IV. HOW SHOULD ANIMALS LEARN?

A. Optimality Approaches to Learning

In the optimal foraging models mentioned in Section III.B, the question of interest was how to maximize rate of energy intake or some other currency of fitness under given conditions, for example, with patches that deplete in certain ways. Predictions about learning are implicitly involved in some optimal foraging models because the optimal behavior may presuppose learning. For example, optimal choice of prey items may be based on information about the distribution of prey items currently available. However, some models in behavioral ecology make explicit predictions concerning what and when animals should learn.

A conceptually (if not mathematically) simple illustration of the optimality approach is provided by the work of Krebs et al. (1978). They considered what an animal should do when confronted with two nondepleting patches of prey with different densities initially unknown to the animal. Given that it has a certain time available for foraging, how should the animal behave to maximize its food intake? Clearly, an omniscient animal would spend all its time in the better patch, that is, that which yields more prey per unit effort (analogous to the lower of two, concurrent fixed-ratio schedules), but in order to find out which is the better patch a real animal has to first sample both. The optimal (i.e., energy-maximizing) behavior in this situation, according to the model of Krebs et al. (1978), is first to alternate between the two patches (sampling) and then, at a point determined by the relative prey densities and the time available for foraging (the time horizon), to switch abruptly to exclusive use of the one that experience has shown has the higher payoff.

As anyone will know who has trained animals on even a simple discrimination, the predicted all-or-none switch from "sampling" (in this situation, random behavior) to "exploiting" (100 percent choice of one alternative) is not always seen in animals. Nevertheless, Krebs et al.'s model stimulated a number of laboratory tests in which the "patches" were concurrent random-ratio schedules (e.g., Plowright & Shettleworth, 1990; review in Shettleworth, 1988). It was the vanguard of a revival of interest in acquisition processes in operant situations (e.g., Davis, Staddon, Machado, & Palmer, 1993; Mazur, 1992; Shettleworth & Plowright, 1992), usually neglected in the quest for descriptions of behavior in the steady state (see Chapter 4, this volume by Williams). It also introduced an enduring theme for models of optimal learning, the need to strike a balance between sampling the environment to gather information that might be useful in the future and exploiting the currently best resources. Distinguishing sampling from simply making mistakes (i.e., imperfect discrimination) requires an explicit model of the amount and pattern of sampling under specified conditions. One such model is that of Stephens (1987) for an environment consisting of one patch of constant value and a second patch that is sometimes better and sometimes worse than the constant patch. In one test of this model (Shettleworth, Krebs, Stephens, & Gibbon, 1988) the amount of sampling by pigeons changed with the values of the alternatives more or less as predicted by Stephens, but sampling occurred at random rather than at regular intervals. This pattern does not fit Stephens's model, but it is predicted by a general model of choice in time-based schedules of reinforcement (Gibbon, Church, Fairhurst, & Kacelnik, 1988). This model predicts that animals will always choose less rewarded options at least a small part of the time. Thus, it provides a possible mechanism for generating behavior that functions as sampling. (For discussions of sampling in other experimental contexts, see Devenport, 1989.)

A second issue that has been central to many optimality models of learning is how much of the past should be used in evaluating the present value of alternatives. The amount of the past over which experience is averaged is referred to as the *memory window* (Cowie, 1977). Again, the optimal balance has to be struck: the memory window must be long enough to avoid undue influence of random fluctuations in conditions, but short enough for the animal's estimate of environmental quality to be responsive to true change. This issue has been addressed in a number of general models of learning and memory (Mangel, 1990; McNamara & Houston, 1987; Real, 1991), as well as in some specific tests of models based on the linear operator model. In such models, one term summarizes the influence of the past and another the influence of the current trial. The relationship between them corresponds roughly to the memory window. In some experiments animals have behaved as if they average reinforcers over a short memory window, even when this does not lead to optimal behavior (Shettleworth & Plowright, 1992; Todd & Kacelnik, 1993). However, other evidence is consistent with enduring effects of the distant past, in violation of the independence of path assumption of linear models (Davis et al., 1993).

B. When to Learn

It has often been suggested that learning has a cost compared to the alternative of relying on preprogrammed responses (Johnston, 1982). The cost could be in neural or genetic hardware or in less than optimal behavior during the stage while learning is still going on (as in Sullivan, 1988). Thus the question arises: When should animals learn?

Intuitively it seems that learning would not be worthwhile in a completely random environment, while in a completely unvarying one, unlearned behaviors would replace learning (Johnston, 1982). Stephens (1991) has analyzed this idea further and shown that what is needed for learning to evolve is predictability within generations with variability between generations. A simple example might be a species in which adults maintain a fixed territory but the young disperse and acquire their own territories of which they have to learn the characteristics.

As well as models dealing specifically with learning, behavioral ecology has contributed a number of analyses of specific ecological problems that require certain learning abilities. For example, analysis of the circumstances in which individuals will perform acts that benefit others at some cost to themselves (i.e., be altruistic) shows that reciprocal altruism will evolve among unrelated individuals only when they can remember past interactions with specific individuals (Trivers, 1971). However, the example of cleaning symbioses among fish described by Trivers indicates that reciprocal altruists need not be keeping a sophisticated mental balance sheet. Simple conditioning may account for animals directing responses selectively toward past benefactors.

In some puzzling cases animals seem not to learn discriminations of which they ought to be perfectly capable. A number of cases occur in the recognition (or failure thereof) of eggs and offspring by birds that are subject to nest parasitism. Birds parasitized by cuckoos may find themselves feeding one huge baby cuckoo and none of their own young, the latter having been thrown out of the nest by the baby cuckoo. Because the young cuckoo quickly comes to look different from the host's young, it seems surprising that host species have not evolved some sort of discriminative behavior, such as deserting a parasitized nest and starting a new family elsewhere. One reason may be that the possible cost of offspring recognition is too great. If the host relied on learned recognition of its own young to discriminate against parasites, it would run the risk that if its very first brood consisted of only a cuckoo, it would learn to accept cuckoos and reject its own young. The cost of recognition learning is therefore likely to be greater than the cost of raising the occasional parasite in species that do not rear the parasite and host offspring together (Lotem, 1993). This implies that discriminative behavior would be more likely in species in which the parasite leaves the host's own young in the nest, and comparative data suggest that this is indeed the case (Lotem, 1993).

Although the idea that the properties of learning in general might be predicted from first principles of biology by way of an optimality model may be appealing, models for specific situations like Lotem's may be more useful. This is particularly so if the models make predictions about the correlation of specific cognitive abilities with ecological conditions, social structures, or the like, because, as will be discussed further below, such predictions can be subjected to rigorous comparative tests.

V. CONCLUSIONS AND IMPLICATIONS

A. Constraints, Predispositions, and Adaptive Specializations

Twenty-five years ago, the material covered in this chapter would have been presented under the heading "constraints on learning" (Hinde & Stevenson-Hinde, 1973; Shettleworth, 1972). The perhaps unfortunate terms *constraints* (Hinde, 1970) or *biological boundaries* (Seligman & Hager, 1972) refer to the notion that associative learning is totally general across species and situations, save for certain exceptions, or biological constraints. However, as others pointed out, a constraint from one point of view is a predisposition from another (Timberlake, 1990). The Brelands' pigs were predisposed to root with tokens that had been associated with food; they were therefore constrained not to drop the tokens with a short latency, even when dropping tokens in a piggy bank was reinforced.

It is one thing to describe such preorganized behavior or tendency to learn as a predisposition or as prepared (Seligman, 1970) behavior or responding. It is quite another to refer to such predispositions as *adaptive specializations* (Gallistel, 1992; Rozin & Kalat, 1971; Rozin & Schull, 1988; Sherry & Schacter, 1987), for this implies that they have been selected during evolution. When Rozin and Kalat (1971) introduced the term *adaptive specializations* into discussions of learning, it seemed to refer to differences in learning mechanisms. For example, they argued that conditioned taste aversion was different in kind from other forms of learning (but see Domjan, 1983). Song learning and imprinting seemed to obey different principles from classical and instrumental conditioning, principles that seemed suited to doing a specialized job. In their elaboration of the notion of adaptive specialization, Sherry and Schacter (1987) captured this idea with the term *functional incompatibility*. They suggested that different learning mechanisms (or memory systems or cognitive modules) would be expected to evolve when the computational requirements of different tasks requiring learning were incompatible. For example, remembering a single pattern of input experienced during a sensitive period for many months, as in song learning, seems incompatible with remembering the locations of a constantly changing inventory of stored food items, as food-storing birds do (Gallistel, 1990; see Sherry & Schacter, 1987, and Shettleworth, 1990, for further discussion). Thus, within the brain of a food-storing bird there should be separate learning modules for song and stored food, having different operating characteristics and perhaps occupying different neural structures.

Because it seems obvious that different modules are suited to performing different tasks, they are referred to as *adaptive specializations*. However, adaptation is a controversial notion in biology, and there is more than one way to try to decide whether some feature is an evolved adaptation. An adaptation has been selected through evolutionary time because it increased the fitness (i.e., ability to transmit genetic copies of themselves) of individuals possessing it. This is not necessarily the same as serving a function in present-day circumstances (cf. Sherry & Schacter, 1987, on exaptations). One of the most compelling arguments for adaptation is the argument from design: the feature of interest, the intricate structure of the vertebrate eye, for example, seems exquisitely suited to do a particular job, more like what an engineer would have designed than something arisen by chance. However, the adaptationist program has been criticized for too facile use of such verbal arguments (Gould & Lewontin, 1979). It is not always easy to decide what are adaptations and what are incidental by-products of characters that have been selected (Williams, 1966).

One way out of verbal arguments about adaptation is to gather data comparing species that face the putative selection pressure to different degrees (Harvey & Pagel, 1991; see also Domjan & Galef, 1983; Riley & Langley, 1993). For example, a comparative investigation of whether taste-aversion learning is an adaptation to omnivory might involve studying a large number of species with different degrees of dietary specialization and relating degree of specialization to the species' performance on some standard test of conditioned taste aversion (Daly, Rauschenberger, & Behrends, 1982, provide an example with two species). As in other examples of this comparative approach (cf. Harvey & Pagel, 1991), the most convincing evidence consists of differences between closely related species and similarities between distantly related ones. Any thoughtful student of learning will immediately appreciate the daunting nature of such a task (e.g., how should the test of learning be standardized for different species?) and perhaps begin to wonder whether we should expunge *adaptive specialization* from our vocabulary instead. To do that, however, would be to place psychology outside of evolutionary biology, whereas much of what has been presented in this chapter is an argument for a closer integration of psychology with the rest of biology.

Notice that the comparative analysis of adaptation described above can deal comfortably with quantitative variations in a character. No one would quarrel with the notion that quantitative variation in some morphological feature such as beak length or thickness can be adaptive. Similarly with possible adaptive specializations of learning. Learning phenomena can differ from one another in a number of ways, and particular aspects of learning such as the types of events that can be learned about or the intervals over which they can be remembered can differ in an adaptive way (see Shettleworth, 1993a). Thus, adaptive specializations of learning include fine-tuning of particular parameters as much as evolution of wholly different learning mechanisms.

B. Biological Approaches and Comparative Psychology

As the preceding section should make clear, the notion of adaptive specialization implies comparative tests in which variations in the putative adaptation are correlated with ecological variables. Because such comparisons are less confounded by phylogenetic variables the more closely related the species being compared, this approach to comparative psychology contrasts markedly with traditional comparative studies of learning (see Riley & Langley, 1993, and Shettleworth, 1993b for further discussion). Until recently, most comparative studies of learning have compared distantly related species (e.g., goldfish, painted turtles, pigeons, and rats; Bitterman, 1975; Macphail, 1987) with the aim of discovering how phylogenetically widespread are particular learning phenomena such as improvement on successive habit reversals. If the aim is to discover whether there are *any* differences in learning across species, the best way to find out is to start by comparing species that are as different as possible. The contrasting approach sketched in the preceding section provides a rationale for selecting species on the basis of their ecology (see Kamil, 1988, for further discussion). Examples of how this approach has been used include studies of memory in food-storing birds, which have tested the hypothesis that the more a species depends on remembering locations of stored food in the wild, the more capacious and durable should be its spatial memory (Kamil, Balda, & Olson, in press; Krebs, 1990). Another example is the analysis of offspring recognition in swallows (Beecher & Stoddard, 1990).

C. Biological Approaches to Learning and the Rest of Psychology

The days are past when findings from animal learning laboratories were expected to explain all of human behavior (Jenkins, 1979; see Chapter 12, this volume, by Shanks for a discussion of the current relationship between animal and human learning). Nevertheless, many of the themes discussed in this chapter, if not the specific findings, are echoed in current work on human cognition. In some cases, the way in which issues have been approached with animals could serve as a model for other areas, especially because evolutionary concerns can be addressed more directly with, say, animals working for food than with people responding to verbal material. The notion that psychological processes can best be understood as adaptations to particular ecological conditions has begun to be explored for human cognition and social behavior (Barkow, Cosmides, & Tooby, 1992). Sherry and Schacter (1987; see also Gallistel, 1990, 1992) drew attention to the parallels between adaptive specializations of learning in animals and memory systems in people and explored the implication of the idea that what neuropsychologists regard as separate memory systems might be viewed as functionally incompatible modules for different cognitive tasks. Formal optimality arguments for deriving properties of learning and memory have also appeared in the literature on human memory (Anderson, 1991), and the notion that psychologists should be studying cognitive processes demonstrably important in their subjects' lives rather than performance in abstracted laboratory paradigms also has its counterpart in the study of "everyday memory" (Bruce, 1985; Loftus, 1991). Proponents of studying everyday memory must address much the same issue that has been discussed in this chapter, that of whether or not processes studied in "ecologically relevant" situations are really any different from those traditionally studied in the laboratory.

D. Future Directions

Although the findings that fired interest in "biological constraints" in the 1960s have been largely assimilated to evolving and liberalized views of the nature of conditioning (Domjan, 1983), they have not been without lasting effect. Articles on learning pay more attention to what role learning might have in animals' lives, while biologists studying animal behavior have become more knowledgeable about principles of animal learning and cognition. Throughout this chapter have been mentioned numerous examples in which biological and psychological approaches have been integrated, and at least as many possibilities for future integrative research. Such integration will have been achieved when some future edition of this handbook does not need a separate chapter for "biological approaches."

Acknowledgments

Preparation of this chapter was supported by a research grant from the Natural Science and Engineering Research Council of Canada. I thank Rick Westwood for help and Bjorn Forkman, Robert Hampton, Catherine Plowright, and Hendrik van Kampen for comments.

References

Adret, P. (1993). Vocal learning induced with operant techniques: An overview. *Netherlands Journal of Zoology, 43*, 125–142.
Anderson, J. R. (1991). Is human cognition adaptive? *Behavioral and Brain Sciences, 14*, 471–517.

Balsam, P. D., & Silver, R. (1994). Behavioral change as a result of experience: Toward principles of learning and development. In J. A. Hogan & J. J. B. Bolhuis (Eds.), *Causal mechanisms of behavioral development* (pp. 327–357). Cambridge: Cambridge University Press.

Barkow, J. H., Cosmides, L., & Tooby, J. (Eds.). (1992). *The adapted mind.* New York: Oxford University Press.

Bateson, P. P. G. (1966). The characteristics and context of imprinting. *Biological Reviews of the Cambridge Philosophical Society, 41,* 177–220.

Bateson, P. P. G. (1979). How do sensitive periods arise and what are they for? *Animal Behaviour, 27,* 470–486.

Bateson, P. P. G. (1983). Optimal outbreeding. In P. P. G. Bateson (Ed.), *Mate choice* (pp. 257–277). Cambridge: Cambridge University Press.

Bateson, P. P. G. (1990). Is imprinting such a special case? *Philosophical Transactions of the Royal Society of London, Series B, 329,* 125–131.

Beauchamp, G., & Kacelnik, A. (1991). Effects of the knowledge of partners on learning rates in zebra finches *Taeniopygia guttata. Animal Behaviour, 41,* 247–253.

Beecher, M. D., & Stoddard, P. K. (1990). The role of bird song and calls in individual recognition: Contrasting field and laboratory perspectives. In W. C. Stebbins & M. A. Berkley (Eds.), *Comparative perception: Vol. II. Complex signals* (pp. 375–408). New York: Wiley.

Berridge, K. C., & Schulkin, J. (1989). Palatability shift of a salt-associated incentive during sodium depletion. *Quarterly Journal of Experimental Psychology, 41B,* 121–138.

Bitterman, M. E. (1975). The comparative analysis of learning. *Science, 188,* 699–709.

Bitterman, M. E. (1988). Vertebrate-invertebrate comparisons. In H. J. Jerison & I. Jerison (Eds.), *Intelligence and evolutionary biology* (pp. 251–276). Berlin: Springer-Verlag.

Bolhuis, J. J. (1991). Mechanisms of avian imprinting: A review. *Biological Reviews of the Cambridge Philosophical Society, 66,* 303–345.

Bolhuis, J. J., de Vos, G. J., & Kruijt, J. P. (1990). Filial imprinting and associative learning. *Quarterly Journal of Psychology, 42B,* 313–329.

Breland, K., & Breland, M. (1961). The misbehavior of organisms. *American Psychologist, 61,* 681–684.

Brower, L. P. (Ed.). (1988). *Mimicry and the evolutionary process.* Chicago: University of Chicago Press.

Brown, P. L., & Jenkins, H. M. (1968). Auto-shaping of the pigeon's key-peck. *Journal of the Experimental Analysis of Behavior, 11,* 1–8.

Bruce, D. (1985). The how and why of ecological memory. *Journal of Experimental Psychology: General, 114,* 78–90.

Brunner, D., Kacelnik, A., & Gibbon, J. (1992). Optimal foraging and timing processes in the starling, *Sturnus vulgaris:* Effect of inter-capture interval. *Animal Behaviour, 44,* 597–613.

Caro, T. M., & Hauser, M. D. (1992). Is there teaching in nonhuman animals? *Quarterly Review of Biology, 67,* 151–174.

Couvillon, P. A., & Bitterman, M. E. (1991). How honeybees make choices. In L. J. Goodman & R. C. Fischer (Eds.), *The behaviour and physiology of bees* (pp. 116–130). Wallingford, UK: CAB International.

Couvillon, P. A., Leiato, T. G., & Bitterman, M. E. (1991). Learning by honeybees (*Apis mellifera*) on arrival at and departure from a feeding place. *Journal of Comparative Psychology, 105,* 177–184.

Cowie, R. J. (1977). Optimal foraging in great tits (*Parus major*). *Nature (London), 268,* 137–139.

Crawford, L. L., Holloway, K. S., & Domjan, M. (1993). The nature of sexual reinforcement. *Journal of the Experimental Analysis of Behavior, 60,* 55–66.

Curio, E. (1988). Cultural transmission of enemy recognition by birds. In T. R. Zentall & B. G. Galef, Jr. (Eds.), *Social learning: Psychological and biological perspectives* (pp. 75–97). Hillsdale, NJ: Erlbaum.

Cynx, J. (1990). Experimental determination of a unit of song production in the zebra finch *Taeniopygia guttata*. *Journal of Comparative Psychology, 104,* 3–10.

Daly, M., Rauschenberger, J., & Behrends, P. (1982). Food aversion learning in kangaroo rats: A specialist-generalist comparison. *Animal Learning and Behavior, 10,* 314–320.

Davis, D. G. S., Staddon, J. E. R., Machado, A., & Palmer, R. G. (1993). The process of recurrent choice. *Psychological Review, 100,* 320–341.

Davis, J. M. (1973). Imitation: A review and critique. *Perspectives in Ethology, 1,* 43–71.

Dawson, B. V., & Foss, B. M. (1965). Observational learning in budgerigars. *Animal Behaviour, 13,* 470–474.

Devenport, L. (1989). Sampling behavior and contextual change. *Learning and Motivation, 20,* 97–114.

deVoogd, T. J. (1994). The neural basis for the acquisition and production of bird song. In J. A. Hogan & J. J. Bolhuis (Eds.), *Causal mechanisms of behavioral development* (pp. 49–81). Cambridge: Cambridge University Press.

de Vos, G. J., & van Kampen, H. S. (1993). Effects of primary imprinting on the subsequent development of secondary filial attachments in the chick. *Behaviour, 125,* 245–263.

Domjan, M. (1983). Biological constraints on instrumental and classical conditioning: Implications for general process theory. *Psychology of Learning and Motivation, 17,* 215–277.

Domjan, M., & Galef, B. G., Jr. (1983). Biological constraints on instrumental and classical conditioning: Retrospect and prospect. *Animal Learning and Behavior, 11,* 151–161.

Domjan, M., & Hollis, K. L. (1988). Reproductive behavior: A potential model system for adaptive specializations in learning. In R. C. Bolles & M. D. Beecher (Eds.), *Evolution and learning* (pp. 213–237). Hillsdale, NJ: Erlbaum.

Dooling, R. J., Brown, S. D., Park, T. J., & Okanoya, K. (1990). Natural perceptual categories for vocal signals in budgerigars (*Melopsittacus undulatus*). In W. C. Stebbins & M. A. Berkley (Eds.), *Comparative perception: Vol. II. Complex signals* (pp. 345–374). New York: Wiley.

Eiserer, L. A. (1980). Development of filial attachment to static visual features of an imprinting object. *Animal Learning and Behavior, 8,* 159–166.

Falls, J. B. (1982). Individual recognition by sound in birds. In D. E. Kroodsma & E. H. Miller (Eds.), *Acoustic communication in birds* (Vol. 2, pp. 237–278). New York: Academic Press.

Fanselow, M. S., & Lester, L. S. (1988). A functional behavioristic approach to aversively motivated behavior: Predatory imminence as a determinant of the topography of defensive behavior. In R. C. Bolles & M. D. Beecher (Eds.), *Evolution and learning* (pp. 185–212). Hillsdale, NJ: Erlbaum.

Fantino, E., & Arbaca, N. (1985). Choice, optimal foraging, and the delay-reduction hypothesis. *Behavioral and Brain Sciences, 8,* 315–330.

Galef, B. G., Jr. (1976). Social transmission of acquired behavior: A discussion of tradition and social learning in vertebrates. *Advances in the Study of Behavior, 6,* 77–99.

Galef, B. G., Jr. (1988). Imitation in animals: History, definition, and interpretation of data from the psychological laboratory. In T. R. Zentall & B. G. Galef, Jr. (Eds.), *Social learning: Psychological and biological perspectives* (pp. 3–28). Hillsdale, NJ: Erlbaum.

Galef, B. G., Jr. (1990). An adaptationist perspective on social learning, social feeding, and social foraging in norway rats. In D. A. Dewsbury (Ed.), *Contemporary issues in comparative psychology* (pp. 55–79). Sunderland, MA: Sinauer Assoc.

Galef, B. G., Jr. & Beck, M. (1991). Diet selection and poison avoidance by mammals individually and in groups. In E. M. Stricker (Ed.), *Handbook of behavioral neurobiology* (Vol. 10, pp. 329–349). New York: Plenum.

Galef, B. G., Jr., & Durlach, P. J. (1993). Absence of blocking, overshadowing, and latent inhibition in social enhancement of food preferences. *Animal Learning and Behavior, 21,* 214–220.

Galef, B. G., Jr., Manzig, L. A., & Field, R. M. (1986). Imitation learning in budgerigars: Dawson and Foss (1965) revisited. *Behavioural Processes, 13,* 191–202.

Gallistel, C. R. (1990). *The organization of learning.* Cambridge, MA: MIT Press.

Gallistel, C. R. (1992). Classical conditioning as an adaptive specialization: A computational model. *Psychology of Learning and Motivation, 28,* 35–67.

Gibbon, J., Church, R. M., Fairhurst, S., & Kacelnik, A. (1988). Scalar expectancy theory and choice between delayed rewards. *Psychological Review, 95,* 102–114.

Giraldeau, L., & Templeton, J. J. (1991). Food scrounging and diffusion of foraging skills in pigeons, *Columba livia:* The importance of tutor and observer rewards. *Ethology, 89,* 63–72.

Gould, J. L. (1990). Honey bee cognition. *Cognition, 37,* 83–103.

Gould, S. J., & Lewontin, R. C. (1979). The spandrels of San Marco and the Panglossian paradigm: A critique of the adaptationist program. *Proceedings of the Royal Society of London, Series B, 205,* 581–598.

Guilford, T., & Dawkins, M. S. (1991). Receiver psychology and the evolution of animal signals. *Animal Behaviour, 42,* 1–14.

Harvey, P. H., & Pagel, M. D. (1991). *The comparative method in evolutionary biology.* Oxford: Oxford University Press.

Heyes, C. M. (1993). Imitation, culture and cognition. *Animal Behaviour, 46,* 999–1010.

Heyes, C. M., Dawson, G. R., & Nokes, T. (1992). Imitation in rats: Initial responding and transfer evidence. *Quarterly Journal of Experimental Psychology, 45B,* 229–240.

Hinde, R. A. (1970). *Animal behaviour* (2nd ed.). New York: McGraw-Hill.

Hinde, R. A., & Stevenson-Hinde, J. (Eds.). (1973). *Constraints on learning: Limitations and predispositions.* New York: Academic Press.

Hoffman, H. S., & Ratner, A. M. (1973). A reinforcement model of imprinting: Implications for socialization in monkeys and men. *Psychological Review, 80,* 527–544.

Hogan, J. A. (1984). Pecking and feeding in chicks. *Learning and Motivation, 15,* 360–376.

Hogan, J. A. (1988). Cause and function in the development of behavior systems. In E. M. Blass (Ed.), *Handbook of behavioral neurobiology* (Vol. 9, pp. 63–106). New York: Plenum.

Holland, P. C. (1984). Origins of behavior in Pavlovian conditioning. *Psychology of Learning and Motivation, 18,* 129–174.

Hollis, K. L. (1982). Pavlovian conditioning of signal-centred action patterns and autonomic behavior: A biological analysis of function. *Advances in the Study of Behavior, 12,* 1–64.

Hollis, K. L. (1990). The role of Pavlovian conditioning in territorial aggression and reproduction. In D. A. Dewsbury (Ed.), *Contemporary issues in comparative psychology* (pp. 197–219). Sunderland, MA: Sinauer Assoc.

Hollis, K. L., Ten Cate, C., & Bateson, P. P. G. (1991). Stimulus representation: A subprocess of imprinting and conditioning. *Journal of Comparative Psychology, 105,* 307–317.

Honey, R. C., Horn, G., & Bateson, P. (1993). Perceptual learning during filial imprinting: Evidence from transfer of training studies. *Quarterly Journal of Experimental Psychology, 46B,* 253–269.

Horn, G. (1990). Neural bases of recognition memory investigated through an analysis of imprinting. *Philosophical Transactions of the Royal Society of London, Series B, 329,* 133–142.

Huheey, J. E. (1988). Mathematical models of mimicry. In L. P. Brower (Ed.), *Mimicry and the evolutionary process* (pp. 22–41). Chicago: Chicago University Press.

Hultsch, H. (1993). Tracing the memory mechanisms in the song acquisition of nightingales. *Netherlands Journal of Zoology, 43,* 155–171.

Immelmann, K. (1972). Sexual and other long-term aspects of imprinting in birds and other species. *Advances in the Study of Behavior, 4,* 147–174.

Jenkins, H. M. (1979). Animal learning and behavior theory. In E. Hearst (Ed.), *The first century of experimental psychology* (pp. 177–228). Hillsdale, NJ: Erlbaum.

Johnston, T. D. (1982). Selective costs and benefits in the evolution of learning. *Advances in the Study of Behavior, 12,* 65–106.

Kamil, A. C. (1988). A synthetic approach to the study of animal intelligence. *Nebraska Symposium on Motivation, 35,* 257–308.

Kamil, A. C., Balda, R. P., & Olson, D. J. (in press). Performance of four seed-caching species in the radial-arm maze analog. *Journal of Comparative Psychology.*

King, A. P., & West, M. J. (1990). Variation in species-typical behavior: A contemporary issue for comparative psychology. In D. A. Dewsbury (Ed.), *Contemporary issues in comparative psychology* (pp. 321–339). Sunderland, MA: Sinauer Assoc.

Konishi, M. (1965). The role of auditory feedback in the control of vocalization in the white-crowned sparrow. *Zeitschrift für Tierpsychologie, 22,* 770–783.

Krebs, J. R. (1990). Food-storing birds: Adaptive specialization in brain and behaviour? *Philosophical Transactions of the Royal Society of London, Series B, 329,* 153–160.

Krebs, J. R., Kacelnik, A., & Taylor, P. (1978). Tests of optimal sampling by foraging great tits. *Nature (London), 275,* 27–31.

Kroodsma, D. E. (1982). Learning and the ontogeny of sound signals in birds. In D. E. Kroodsma & E. H. Miller (Eds.), *Acoustic communication in birds* (Vol. 2, pp. 1–23). New York: Academic Press.

Kroodsma, D. E., & Byers, B. E. (1991). The function(s) of bird song. *American Zoologist, 31,* 318–328.

Kroodsma, D. E., & Pickert, R. (1980). Environmentally dependent sensitive periods for avian vocal learning. *Nature (London), 288,* 477–479.

Lefebvre, L., & Palameta, B. (1988). Mechanisms, ecology, and population diffusion of socially learned, food-finding behavior in feral pigeons. In T. R. Zentall & B. G. Galef, Jr. (Eds.), *Social learning: Psychological and biological perspectives* (pp. 141–164). Hillsdale, NJ: Erlbaum.

Loftus, E. F. (1991). The glitter of everyday memory . . . and the gold. *American Psychologist, 46,* 16–18.

Lorenz, K. (1935). Der Kumpan in der Umwelt des Vogels. Journal für ornithologie, *83,* 137–213, 289–413. [Reprinted in R. Martin (Ed.), *Studies in animal and human behavior* (Vol. 1, pp. 101–258). London: Methuen, 1970.]

Lorenz, K. (1970). Notes. In R. Martin (Ed.), *Studies in animal and human behavior* (Vol. 1, pp. 371–380). London: Methuen.

Lotem, A. (1993). Learning to recognize nestlings is maladaptive for cuckoo *Cuculus canorus* hosts. *Nature (London), 362,* 743–745.

Macphail, E. M. (1987). The comparative psychology of intelligence. *Behavioral and Brain Sciences, 10,* 645–695.

Mangel, M. (1990). Dynamic information in uncertain and changing worlds. *Journal of Theoretical Biology, 146,* 317–332.

Marler, P. (1970). A comparative approach to vocal learning: Song learning in white-crowned sparrows. *Journal of Comparative and Physiological Psychology, 71,* 1–25.

Marler, P. (1976). Sensory templates in species-specific behavior. In J. Fentress (Ed.), *Simpler networks and behavior* (pp. 314–329). Sunderland, MA: Sinauer Assoc.

Marler, P. (1987). Sensitive periods and the roles of specific and general sensory stimulation in birdsong learning. In P. Marler & J. P. Rauschecker (Eds.), *Imprinting and cortical plasticity* (pp. 99–135). New York: Wiley.

Marler, P. (1990). Song learning: The interface between behaviour and neuroethology. *Philosophical Transactions of the Royal Society of London, Series B, 329,* 109–114.

Marler, P., & Nelson, D. A. (1993). Action-based learning: A new form of developmental plasticity in bird song. *Netherlands Journal of Zoology, 43,* 91–103.

Marler, P., & Peters, S. (1981). Sparrows learn adult song and more from memory. *Science, 213,* 780–782.

Marler, P., & Peters, S. (1989). Species differences in auditory responsiveness in early vocal learning. In R. J. Dooling & S. Hulse (Eds.), *The comparative psychology of audition: Perceiving complex sounds* (pp. 243–273). Hillsdale, NJ: Erlbaum.

Mayr, E. (1974). Behavior programs and evolutionary strategies. *American Scientist, 62,* 650–659.

Mazur, J. E. (1992). Choice behavior in transition: Development of preference with ratio and interval schedules. *Journal of Experimental Psychology: Animal Behavior Processes, 18,* 364–378.

McGregor, P. K. (1991). The singer and the song: On the receiving end of bird song. *Biological Reviews of the Cambridge Philosophical Society, 66,* 57–81.

McLaren, I. P. L., Kaye, H., & Mackintosh, N. J. (1989). An associative theory of the representation of stimuli: Applications to perceptual learning and latent inhibition. In R. G. M. Morris (Ed.), *Parallel distributed processing: Implications for psychology and neurobiology* (pp. 102–130). Oxford: Oxford University Press.

McNamara, J. M., & Houston, A. I. (1987). Memory and the efficient use of information. *Journal of Theoretical Biology, 125,* 385–395.

McQuoid, L. M., & Galef, B. G., Jr. (1992). Social influences on feeding site selection by burmese fowl (*Gallus gallus*). *Journal of Comparative Psychology, 106,* 137–141.

Moore, B. R. (1992). Avian movement imitation and a new form of mimicry: Tracing the evolution of a complex form of learning. *Behaviour, 122,* 213–263.

Nelson, D. A., & Marler, P. (1990). The perception of birdsong and an ecological concept of signal space. In W. C. Stebbins & M. A. Berkley (Eds.), *Comparative perception: Vol. II. Complex signals* (pp. 443–478). New York: Wiley.

Nordeen, K. W., & Nordeen, E. J. (1992). Auditory feedback is necessary for the maintenance of stereotyped song in adult zebra finches. *Behavioral and Neural Biology, 57,* 58–66.

Palameta, B., & Lefebvre, L. (1985). The social transmission of a food-finding technique in pigeons: What is learned? *Animal Behaviour, 33,* 892–896.

Petrinovich, L. (1988). The role of social factors in white-crowned sparrow song development. In T. P. Zentall & B. G. Galef, Jr. (Eds.), *Social learning: Psychological and biological perspectives* (pp. 255–277). Hillsdale, NJ: Erlbaum.

Plowright, C. M. S., & Shettleworth, S. J. (1990). The role of shifting in choice behavior of pigeons on a two-armed bandit. *Behavioural Processes, 21,* 157–178.

Plowright, C. M. S., & Shettleworth, S. J. (1991). Time horizon and choice by pigeons in a prey-selection task. *Animal Learning and Behavior, 19,* 103–112.

Real, L. A. (1991). Animal choice behavior and the evolution of cognitive architecture. *Science, 253,* 980–986.

Rescorla, R. A. (1988). Pavlovian conditioning: It's not what you think it is. *American Psychologist, 43,* 151–160.

Rescorla, R. A., & Holland, P. C. (1976). Some behavioral approaches to the study of learning. In M. R. Rosenzweig & E. L. Bennett (Eds.), *Neural mechanisms of learning and memory* (pp. 165–192). Cambridge, MA: MIT Press.

Riley, D. A., & Langley, C. M. (1993). The logic of species comparisons. *Psychological Science, 4,* 185–189.

Rozin, P., & Kalat, J. W. (1971). Specific hungers and poison avoidance as adaptive specializations of learning. *Psychological Review, 78,* 459–486.

Rozin, P., & Schull, J. (1988). The adaptive-evolutionary point of view in experimental psychology. In R. C. Atkinson, R. J. Herrnstein, G. Lindzey, and R. D. Luce (Eds.), *Stevens's handbook of experimental psychology* (2nd ed., Vol. 1, pp. 503–546). New York: Wiley.

Russon, A. E., & Galdikas, B. M. F. (1993). Imitation in free-ranging rehabilitant orangutans (*Pongo pygmaeus*). *Journal of Comparative Psychology, 107*, 147–161.

Ryan, C. M. E., & Lea, S. E. G. (1990). Pattern recognition, updating, and filial imprinting in the domestic chick (*Gallus gallus*). In M. L. Commons, R. J. Herrnstein, S. M. Kosslyn, & D. B. Mumford (Eds.), *Quantitative analyses of behavior* (Vol. 8, pp. 89–110). Hillsdale, NJ: Erlbaum.

Seligman, M. E. P. (1970). On the generality of the laws of learning. *Psychological Review, 77*, 406–418.

Seligman, M. E. P., & Hager, J. L. (Eds.). (1972). *Biological boundaries of learning*. New York: Appleton-Century Crofts.

Sherry, D. F., & Galef, B. G., Jr. (1990). Social learning without imitation: More about milk bottle opening by birds. *Animal Behaviour, 40*, 987–989.

Sherry, D. F., & Schacter, D. L. (1987). The evolution of multiple memory systems. *Psychological Review, 94*, 439–454.

Shettleworth, S. J. (1972). Constraints on learning. *Advances in the Study of Behavior, 4*, 1–68.

Shettleworth, S. J. (1988). Foraging as operant behavior and operant behavior as foraging: What have we learned? *Psychology of Learning and Motivation, 22*, 1–49.

Shettleworth, S. J. (1990). Spatial memory in food-storing birds. *Philosophical Transactions of the Royal Society of London, Series B, 329*, 143–151.

Shettleworth, S. J. (1993a). Varieties of learning and memory in animals. *Journal of Experimental Psychology: Animal Behavior Processes, 19*, 5–14.

Shettleworth, S. J. (1993b). Where is the comparison in comparative cognition? Alternative research programs. *Psychological Science, 4*, 179–184.

Shettleworth, S. J., & Juergensen, M. R. (1980). Reinforcement and the organization of behavior in golden hamsters: Brain stimulation reinforcement for seven action patterns. *Journal of Experimental Psychology: Animal Behavior Processes, 6*, 352–375.

Shettleworth, S. J., Krebs, J. R., Stephens, D. W., & Gibbon, J. (1988). Tracking a fluctuating environment: A study of sampling. *Animal Behaviour, 36*, 87–105.

Shettleworth, S. J., & Plowright, C. M. S. (1992). How pigeons estimate rates of prey encounter. *Journal of Experimental Psychology: Animal Behavior Processes, 18*, 219–235.

Shettleworth, S. J., Reid, P. J., & Plowright, C. M. S. (1993). The psychology of diet selection. In R. N. Hughes (Ed.), *Diet selection: An interdisciplinary approach to foraging behaviour* (pp. 56–77). Oxford: Blackwell.

Slater, P. J. B. (1989). Bird song learning: Causes and consequences. *Ethology, Ecology and Evolution, 1*, 19–46.

Slater, P. J. B., Eales, L. A., & Clayton, N. S. (1988). Song learning in zebra finches (*Taeniopygia guttata*): Progress and prospects. *Advances in the Study of Behavior, 18*, 1–33.

Stephens, D. W. (1987). On economically tracking a variable environment. *Theoretical Population Biology, 32*, 15–25.

Stephens, D. W. (1991). Change, regularity, and value in the evolution of animal learning. *Behavioral Ecology, 2*, 77–89.

Stoddard, P. K., Beecher, M. D., Loesche, P., & Campbell, S. E. (1992). Memory does not constrain individual recognition in a bird with song repertoires. *Behaviour, 122*, 274–287.

Suboski, M. D. (1990). Releaser-induced recognition learning. *Psychological Review, 97*, 271–284.

Suboski, M. D. (1992). Releaser-induced recognition learning by amphibians and reptiles. *Animal Learning and Behavior, 20*, 63–82.

Sullivan, K. A. (1988). Age-specific profitability and prey choice. *Animal Behaviour, 36*, 613–615.

Ten Cate, C. (1989). Behavioral development: Towards understanding processes. *Perspectives in Ethology, 8*, 243–269.

Ten Cate, C., Los, L., & Schilperood, L. (1984). The influence of differences in social experience on the development of species recognition in zebra finch males. *Animal Behaviour, 32,* 852–860.

Ten Cate, C., & Slater, P. J. B. (1991). Song learning in zebra finches: How are elements from two tutors integrated? *Animal Behaviour, 42,* 150–152.

Ten Cate, C., Vos, D. R., & Mann, N. (1993). Sexual imprinting and song learning: Two of one kind? *Netherlands Journal of Zoology, 43,* 34–45.

Thorndike, E. L. (1911). *Animal intelligence.* Darien, CT: Hafner Publishing. (Facsimile edition published 1970)

Timberlake, W. (1990). Natural learning in laboratory paradigms. In D. A. Dewsbury (Ed.). *Contemporary issues in comparative psychology* (pp. 31–54). Sunderland, MA: Sinauer Assoc.

Timberlake, W. (1993). Behavior systems and reinforcement: An integrative approach. *Journal of the Experimental Analysis of Behavior, 60,* 105–128.

Todd, I. A., & Kacelnik, A. (1993). Psychological mechanisms and the Marginal Value Theorem: Dynamics of scalar memory for travel time. *Animal Behaviour, 46,* 765–775.

Trivers, R. L. (1971). The evolution of reciprocal altruism. *Quarterly Review of Biology, 46,* 35–57.

van Kampen, H. S. (1993). Filial imprinting and associative learning: Similar mechanisms? *Netherlands Journal of Zoology, 43,* 143–154.

Vidal, J. M. (1980). The relations between filial and sexual imprinting in the domestic fowl: Effects of age and social experience. *Animal Behaviour, 28,* 880–891.

West, M. J., & King, A. P. (1988). Female visual displays affect the development of male song in the cowbird. *Nature (London), 334,* 244–246.

Whiten, A., & Ham, R. (1992). On the nature and evolution of imitation in the animal kingdom: Reappraisal of a century of research. *Advances in the Study of Behavior, 21,* 239–283.

Williams, G. C. (1966). *Adaptation and natural selection.* Princeton, NJ: Princeton University Press.

Zentall, T. R., & Galef, B. G., Jr. (Eds.). (1988). *Social learning: Psychological and biological perspectives.* Hillsdale, NJ: Erlbaum.

CHAPTER 8

Space and Time

C. R. Gallistel

On a standard neuropsychological evaluation, an important initial assessment is whether the patient is oriented in space and time. Do they know where they are, the season of the year, and how long it has been since breakfast? Disorientation in space and time indicates severe cognitive dysfunction, because knowledge of one's position in time and space plays a fundamental role in human cognition. There is reason to think that this knowledge plays an equally fundamental role in animal learning and cognition.

Many nonhuman animals, including those routinely used in behavioral research, know where they are and they know what time it is. They also know how long was the interval between the last two occurrences of an event and how long it has been since the last occurrence. In this chapter I sketch our emerging understanding of how animals know where they are, how they know what time it is, and how they know currently elapsing and previously elapsed temporal intervals. I also indicate some of the ways in which this knowledge determines responding in classical and instrumental conditioning experiments.

I. THE COGNITIVE MAP

A. Behavioral Consequences of the Position Sense

To know where you are is to know your position in a frame of reference established by the configuration of contours and surfaces in the world around

Animal Learning and Cognition

you. Usually, the greater part of the configuration is not currently perceptible. The patient who answers that she is in the middle of a rectangular beige room does not get credit for knowing where she is. Any patient who is *compos mentis* will say that she is, for example, in such-and-such a hospital in San Jose, California, U.S.A. The specification of position requires an expanded frame of reference, a frame of reference large enough to establish a unique location in the world. There is abundant evidence that animal behavior is likewise determined by a position sense that routinely includes reference to the animal's position within a framework larger than the framework that can be verified from current sensory input (see Gallistel, 1990, for review). The following three examples illustrate this fundamental point about the position sense.

Collett and Kelber (1988) trained honey bees to visit two small huts in succession, one 70 m from their hive and one 33 m farther. The huts were identical 6-foot cubes. The bees entered through a small doorway, with the door closing behind them. Inside each hut was a square array of two yellow and two blue cylinders. The two arrays were identical. Inside one hut, the small drop of sucrose the bee sought was on a glass slide between the two yellow cylinders; inside the other, it was between the two blue cylinders. The position of the cylinder array within each hut—and with it the position of the sucrose—varied from training trial to training trial, so the bee had to use the cylinders to find the sugar. On test trials, there was no sugar in one hut. A video camera recorded the bee's search pattern. The hut serving as the test hut varied from test to test. During these tests, the same test array of cylinders—an array just like the two identical training arrays—was used in each hut, to remove any possibility that the test array itself differed in some way between huts. Where a bee searched within the test array depended on which hut it was in. When it was in the nearer hut, it searched predominantly on, say, the side of the yellow cylinders; in the farther hut, predominantly on the side of the blue cylinders.

The huts differed only with regard to their position in the world. But when a bee was inside a hut, there was no sensory input indicative of that position. From the inside, the huts were indistinguishable. Nonetheless, the bees searched on one side of the array when in one hut and on the other side when in the other, because they knew where they were in a larger frame of reference, a frame that included at least some of the terrain or landmarks outside the huts. They knew their position on their cognitive map of their world (or, at least, on some patch of their world).

Georgakopoulos and Etienne (1994) trained hamsters to make hoarding runs through a sequence of two or three identical square rooms connected by identical tunnels. The first tunnel led from the nesting box into the sequence of rooms. In each room, the hamster picked up seeds and stored them in its cheek pouches. The seeds were in different fixed locations in each room. The rooms were dark, except for a dim, red diode at the same location in each room. To go to the seed cache in a given room, the hamster had to know which room it was in. The diodes did not provide room illumination, so the hamster could not see the location of the seed cache as it emerged from a tunnel. On test trials, extinguishing the diodes or moving the diodes and the seed caches had no effect on the hamsters' behavior. Whether the diodes or the seeds were there or not, as the hamsters emerged from the tunnel into each room, they went straight toward the training position of the cache in that room. This involved a different turn in each room. Sometimes they went straight through a room without picking up its seeds on their outward journey through the boxes. On their way back, they went straight toward the unharvested seed cache when they emerged from the tunnel at the other end of the room.

For the hamster emerging from a tunnel, as for the bee searching in a hut, there is no sensory stimulus acting on it that informs it which room it is in. The tunnels and the rooms are identical. Nonetheless, the hamster makes one turn when coming into one room and a different turn coming into the other. Also it makes a different turn when coming into the first room from the other direction. It does so, because it, like the bee, knows where it is in a frame of reference larger than the frame that acts on its sensory receptors at the time it makes its turn.

The point of both the preceding experiments was to demonstrate that the animal's response to what it could sense at the moment was based on its knowledge of its position in a larger framework. Investigators stumbled on similar phenomena near the beginning of the century. Carr (1917) and Carr and Watson (1908) did a number of experiments designed to determine what sensory cues a rat relied on in making its way through a complex maze. The maze was located inside a square enclosure of heavy, black curtains, designed to prevent the rat in the maze from perceiving anything outside the enclosure. At the beginning of the daily test run, the rat was carried from its home cage elsewhere in the big lab room, across the room, through the curtains, and placed at the start of the maze.

When the rats had learned to run the maze quickly and accurately, sensory inputs were removed by blinding the rats or deafening or making them anosmic, or all three. None of these deprivations had any large effect on the rat's ability to move rapidly and accurately through the maze, which led Carr and Watson to conclude that the rats relied primarily on kinesthetic and tactile cues. Surprisingly, one manipulation that did severely disrupt the rat's maze behavior was to rotate both the maze and its enclosure 90° with respect to the laboratory room. This puzzled Carr and Watson, because rotating the maze together with the enclosing curtains had no effect on the sensory cues acting on the rat while it was in the maze. Carr and Watson (1908) wrote:

> The alleys may be of the same length and be entered by the same direction of turn, but present possible differences in their stimulating effect because they

extend in different directions. It is difficult to conceive why and how this can be so. . . . The successful functioning of an automatic habit depends upon the rat's orientation in relation to cardinal positions [that is, on the rat's compass heading, the direction in which it points in a coordinate framework anchored to the earth]. Change the [compass] direction of the path and the automatic act is disturbed to some extent. The same act accomplished in two different directions is thus *different in some way to the animal.* (p. 31)

The rat knew its orientation with respect to the world outside the maze (its compass heading) even though there was no sensible cue within the curtains that could indicate that orientation.

Animals know their position and compass heading within a framework they cannot presently sense because they construct an internal representation of that framework—a cognitive map. The cognitive map records (at least some of) the geometric relations between sensible portions of their world. Animals continuously update their representation of their position and heading on this map. The resulting knowledge of their current position and heading in the world probably exerts an influence on almost every move they make.

B. Construction of the Cognitive Map

In recent years, there has been keen interest in how the brain, and particularly the hippocampus, represents space (Eichenbaum, Stewart, & Morris, 1990; McNaughton, Chen, & Markus, 1991; Morris, Schenk, Tweedie, & Jarrard, 1990; Muller, Kubie, Bostock, Taube, & Quirk, 1991; O'Keefe, 1990; Sharp, Kubie, & Muller, 1990; Taube, Muller, & Ranck, 1990; Wilson & McNaughton, 1993). This interest rests primarily on the discovery that neurons in the hippocampus and related structures fire only when the animal is in particular places ("places cells") or when its head is oriented in a particular direction ("direction cells"). The hypothesis that animals represent the spaces they move through has also been strengthened by recent purely behavioral work (Cheng, 1986, 1989; Collett, 1987; Collett, Cartwright, & Smith, 1986; Etienne, Sylvie, Reverdin, & Teroni, 1993; Hermer, 1993; Morris, 1981). Nonetheless, we do not know how cognitive maps are constructed. In fact, there is still some dispute about whether they are constructed, at least in insects (Dyer, 1991; Dyer, Berry & Richard, 1992; Gould, 1986, 1990; Wehner, Bleuler, Nievergelt, & Shah, 1990; Wehner & Wehner, 1990). This dispute, discussed later, has arisen despite the fact that some of the most compelling behavioral evidence for cognitive maps comes from experiments with insects.

By combining recent behavioral work with a consideration of how conventional maps are made, we can make a plausible conjecture about how animal brains construct cognitive maps. Most of the assumptions that lead to this conjecture are widely accepted on the basis of strong experimental evidence. It is only the overall conclusion that is disputed. In what follows, the steps in the construction of a conventional map are used to organize pertinent experimental material on how animals may make their map. Most of the data are from experiments with insects.

1. Establish a Direction of Angular Reference

A surveyor defines compass bearings by reference to the trajectories of celestial bodies. The position of the sun over the horizon varies depending on latitude, season, and time of day, but it is always due south at noon in the northern hemisphere. The azimuth position of the sun at midday—or equivalently, the point on the horizon midway between the point where the sun rises and the point where it sets—defines an unvarying direction. At night, the position of the stars also varies as a function of latitude, season, and time of night. But there is one point in the starry array that does not move, the celestial pole. In the present epoch, the northern celestial pole lies near a prominent star, Polaris. The fixed point or center of rotation of the night sky also defines an unvarying line of angular reference, the same line defined by the midpoint in the sun's daily trajectory.

The advantage of using the sun and stars to define a direction of angular reference on the earth is that they are far enough away to be for practical purposes points at infinity. Therefore, moving around on the earth's surface does not change the direction of these reference points. Also, these directional reference points may be seen from anywhere (weather permitting).

By contrast, the direction defined by the line of sight to a terrestrial landmark changes as the surveyor moves, the landmarks change their appearance radically, and they appear and disappear (Figure 1). The change in the direction of a landmark due to a change in the observer's position is called parallax. It is a valuable cue to the distance of the landmark from the observer, but it is a decided hindrance if one wants to use uncharted terrain to maintain a fixed direction of angular reference. To maintain a fixed direction of angular reference solely by looking at the surrounding terrain, the moving map maker must *carry the parallel.* As he moves, he must determine the line of sight from his current position lying parallel to the line of sight he used for angular reference at his base position. Thus, for example, in moving from the hive to the release site in Figure 1, the animal trying to hold a course by reference to the terrain alone (as all animals, including insects, must do on overcast days) has to keep track of which part of the ever-changing horizon profile lies to, for example, the west (contrast Figures 1B and 1C). Carrying the parallel is difficult and subject to cumulative error, which is why lost hunters wander in circles.

With a celestial directional reference, you do not have to carry the parallel; all terrestrial lines of sight on Polaris or the noonday sun are parallel. The use of a celestial directional reference facilitates the construction of a map. Once the map has been constructed, it is possible to maintain one's heading by reference

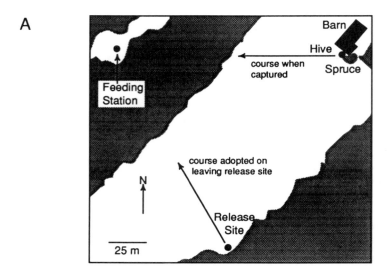

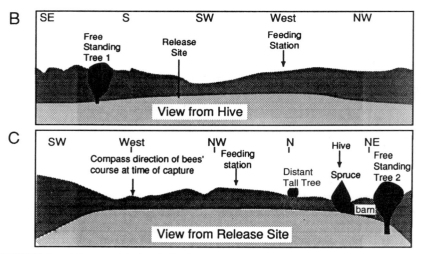

FIGURE 1 (A) Map of the locale for the capture and release experiments of Gould (1986). Gray areas indicates woods. (B) The surrounding terrain as seen from the hive. The barn behind the hive and spruce tree beside it dominated much of the perimeter not shown in this view. (C) The surrounding terrain as seen from the release site shown in (A). The trees behind the release site dominated the part of the perimeter not shown in this view. The free-standing tree in this view is not the same tree as the one in the view from the hive. These views are meant to approximate the retinal snapshots the bee would obtain at each vantage point. Although both snapshots are from vantage points within the same meadow, no two-dimensional feature of the perimeter seen in one view is readily discerned in the other. Even if one could find the same feature in both views, its compass direction would be different in the two views (see, e.g., the mountain perimeter views in Menzel et al., 1990). These parallax changes in the terrain image make it difficult to use the terrain to get compass direction in the absence of a geocentric map of the terrain (compare the view to the west in panels B and C). They also pose a challenge to "retinal snapshot" theories of navigation favored by many

to the charted terrain. The knowledge of where you are on the map facilitates the solving of the terrain recognition problem, because you know from your map (together with your representation of your position on that map) roughly what you should see and roughly where you should see it.

The coding of the spatial relations between landmarks (the map) and of one's own position on the map also makes it possible to use the surrounding terrain to set a compass course (a course with a prespecified geocentric direction). Bees can use the terrain surrounding their hive to set a compass course (Dyer, 1987). Moreover, when a captured bee is released somewhere within its foraging terrain, it can set out again on the compass course it was on when captured, even on days when it cannot see the sun (Wehner & Menzel, 1990). Wehner and Menzel (1990) argue that it is possible to get compass bearings from landmarks without computing one's position on a map of those landmarks. They suggest that doing so is computationally simpler than getting one's bearing in the conventional way (by first determining one's position relative to the landmarks, that is, one's position on a geocentric map of those landmarks). However, an algorithm for getting compass bearings from landmarks without determining one's position on a geocentric map of those landmarks has not been specified. Until a generally workable procedure for doing this has been specified, the fact that bees can get their compass bearings from landmarks surrounding release sites is evidence that they can determine their position on a map of those landmarks.

It may seem recondite to use the trajectories of celestial bodies to define directions of angular reference for terrestrial navigation, but many animals do. Animals that pursue course during the day use the sun as a point of angular reference. Animals that pursue courses at night use the center of rotation of the night sky. (see Gallistel, 1989, 1990, for review).

To use the sun, animals must learn the local and contemporary *solar ephemeris*. The solar ephemeris is the azimuth position of the sun as a function of the time of day. (The azimuth position of the sun is the point on the horizon directly underneath it.) Surprisingly, even insects do learn the solar ephemeris. Moreover, animals then use the reading on their internal circadian clock, together with the azimuth position of the sun (or the resulting sky polarization pattern; Wehner, 1987) to determine compass direction, that is, direction relative to the direction of angular reference. To know how they are oriented in

---

students of insect homing (Collett, 1992; Wehner, 1992). It remains to be demonstrated that the two-dimensional image-matching problem can be solved. This is the problem of matching the retinal image of the terrain seen from a different vantage point to the retinal image made at the base position and stored in memory, without recourse to a three-dimensional reconstruction of the terrain, using information about the compass bearings and distances of the landmarks that correspond to features in the retinal image. (Redrawn with slight alterations from Figures 5, 6, and 7 of Gould, 1990, pp. 92–97, by permission of the author and publisher.)

space (to know their *heading*), animals that rely on the sun must know the time of day because the ephemeris function specifies the sun's direction as a function of the time of day. Thus, orientation in space and time are interrelated functions.

To use the stars at night, animals must learn the celestial pole, the center of rotation of the night sky. The celestial pole varies continuously, on a time scale measured in millennia, due to the precession of the earth's axis. Surprisingly, migratory birds, at least, do learn the location of the celestial pole. They learn it while they nestlings, unable to move out of the nest (Able & Bingham, 1987; Emlen, 1975). The "foresight" implicit in the astronomical self-education of the nestling is presumably mediated by a purpose-specific learning mechanism that has evolved through natural selection. All of the mechanisms by which animals acquire knowledge of their spatial and temporal position seem similarly specialized. They compute specific things from certain classes of inputs using principles peculiar to that particular problem. They are "instincts to learn" (Gould & Marler, 1987). This raises the question whether or not all learning mechanisms, even the general purpose mechanisms that mediate what are thought to be examples of associative learning, might not also turn out to be specialized for a particular task (Gallistel, 1992, 1994; see also Chapter 7, by Shettleworth).

2. Determine the Direction and Distance of Salient Aspects of the Terrain around the Base Position

When a directional reference has been determined, the surveyor begins plotting the positions of salient features of the terrain (landmarks) surrounding his base position. By *plotting,* I mean *coding,* not "putting down on paper." These days, most position information in surveying is entered into a field computer rather than plotted in the literal sense. In an animal, the position codes for salient features of the terrain and the interesting points within that terrain are laid down somehow, somewhere in the nervous system. How is mysterious, because we do not know how the nervous system stores the values of variables (Gallistel, 1990).

To plot the position of a landmark, the surveyor must determine its compass bearing and distance. The *compass bearing* of a landmark is the angle through which the surveyor must turn as he goes from pointing in the reference direction to pointing toward the landmark.

Determining the *distance* to a landmark is more difficult. Nowadays professional surveyors do it with a laser interferometer or a measuring chain. But one can also do it by pacing it off, which is crude, but quick, satisfactory for many purposes, and doable by any animal that can keep track of how far it has moved, which is something most animals can do (see material on *dead reckoning* below). Distance can also be determined by triangulation.

Triangulation is computationally complex, or at least it seems so to many students on first encounter. Nonetheless, it is commonly used, even by insects (see Srinivasan, 1992, for review).

The simplest use of triangulation is to place oneself at some distance from a landmark on the basis of its apparent size. The retinal extent of a landmark's image (its apparent size) is inversely related to your distance from the landmark. Bees learn to search for a feeding site at a certain distance from a single landmark on the basis of its apparent size (Cartwright & Collett, 1983). While this is a primitive form of triangulation, it need not require trigonometric computation. To position oneself a prespecified distance from a landmark of known size does require trigonometric computation, but the bee may not know what the distance is. It may only know that when the retinal image has a certain size, it is at the right distance. On the other hand, the bee may in fact know the distance (from survey flights of the kind described below) and rely on apparent size as a simple indicator of distance.

The locust's use of parallax, on the other hand, clearly requires trigonometric computation. When the locust plans to jump to something, it orients directly toward its target. Then, with almost comic deliberation, it shifts its head and upper body slowly and steadily to one side, stops, shifts slowly and steadily to the other side, and so on. As it shifts its head from side to side, it counterrotates it, so that the head does not change its compass orientation. This behavior is called peering. It was long suspected that its purpose was to estimate the distance of the target by parallax, and this has recently been experimentally proven (Collett & Patterson, 1991; Sobel, 1990). The principle is simple (Figure 2). The distance to the target (landmark) is the long side of a right triangle. The short side or base is the line through which the locust's head moves as it shifts to one side of this line. Kinesthetic and motor signals (the strengths of movement commands) indicate the length (b) of this base. The angle at the end of the base is the change in the visual angle ($\Delta\alpha$) of the target as the head shifts to one side. The long side of the triangle—the distance (D) to the landmark—may be computed from, for example, the relation $D = b/\tan(\Delta\alpha)$. Alternatively, the locust may use kinesthetic and motor signals specifying the speed (s) of its sideways motion, together with visual signals specifying the direction (α) and the angular velocity of the landmark (α'), the rate at which the contours in its image are moving across the retina, to compute the distance to the landmark from the relation:

$$D = s \cdot \sin(\alpha)/\alpha'$$

Similarly, when a gerbil plans to jump across a gap, it bobs its head up and down to determine the distance to be jumped (Ellard, Goodale, & Timney, 1984). The lizards on my terrace spend much of the time between

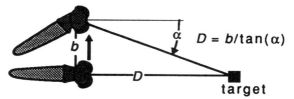

$$D = b/\tan(\alpha)$$

target

FIGURE 2   Estimating distance by parallax in the locust. (Redrawn from Sobel, 1990, by permission of the author and publisher.)

darts bobbing their head up and down. I suspect these bobs serve the same purpose as the gerbil's bobs and the locust's peers. Moving back and forth through a known distance or with a known ground speed to compute the distance of landmarks is a common animal behavior. When a solitary digger wasp leaves a newly dug burrow for the first time, or after any return in which she has had trouble finding it, and similarly, when a yellow jacket or a honey bee approaches or leaves a newly found food source, they engage in a back and forth dance to compute the distance and bearings (compass directions) of the local landmarks (Lehrer, 1993; Zeil, 1993a,b). Bees and wasps buzz around the picnic to get the lay of the land so they can find the food the next time they come.

The best studied base survey behavior is the behavior of the digger wasp emerging from her burrow when it is newly dug or when she has had trouble finding it on her last return (Zeil, 1993a). As the wasp emerges from her burrow, she stops at the exit and looks around by means of several saccadic head movements. Then, she walks out, turns around to face the entrance, lifts off and flies sideways back and forth, in arcs that increase in their height and radius. The arcs are roughly centered on the burrow. Indeed, if a high-contrast annulus around the entrance is moved while the wasp dances back and forth, the center of her dance shifts with it—and so does the center of her search for the entrance when she returns from her foraging flight (with the annulus gone). If there is a prominent landmark near the nest (a cylinder within 10 to 30 cm), the midpoints of the arcs tend to lie near a line from the landmark through the burrow to the arc, so that the wasp sights the landmark across the burrow as she flies back and forth (Figure 3).

As the wasp flies a survey arc, she counterrotates her body axis, holding the nest entrance in a cone of visual angles between 30° and 70° off the midline. She counterrotates at about the same rate no matter how distant the arc is from the burrow. The constant rate of counterrotation holds the burrow within the 30–70° cone of visual angles because the speed at which she flies an arc increases in proportion to her distance from the burrow. In other words, the angular velocity with which she arcs around the burrow is independent of her distance from the burrow.

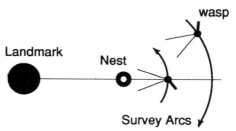

FIGURE 3   Schematic rendering of the arcs in the near-nest survey flight of a digger wasp. The wasp is indicated by a black dot for the head, with a heavy line segment for the body. The hairlines show the cone of visual angles between 30° and 70° off the midline, within which she keeps the nest while arcing back and forth around it. (Based on Zeil, 1993a.)

What has just been described is the survey flight for the immediate vicinity of the nest or food site. Bees perform similar survey flights, both at the hive and at feeding sites (Becker, 1958; Lehrer, 1993; Wehner, 1981). When the zigzag stage is complete, the wasp or bee spirals around the nest or feeding site in loops too big to video tape. This latter stage probably surveys the more distant terrain. It has not been proven that wasps and bees determine landmark distances in the course of these survey flights, but it has been shown that on her return the wasp uses the distance from the landmark to find her burrow (Zeil, 1993b). It has also been shown that bees make use of the true distance of landmarks from feeding sites (Cheng, Collett, Pickhard, & Wehner, 1987), and not, as was earlier thought, simply the apparent size of the landmarks. Also, as already noted, Zeil (1993a) showed that moving a high-contrast annulus that marks the nest hole during the wasp's survey flight moves the position to which she returns later (after the annulus has been removed), implying that she determined the distance of the nest hole from the nearby landmark during her survey flight.

3. Move to a Vantage Point a Known Direction and Distance from the Base Position

Map making requires determining the directions and distances of the landmarks visible from vantage points at known directions and distances from the base position (*known vantage points*). To move to a known vantage point, the surveyor must be able to hold a course (move in a fixed direction) and keep track of the distance moved along it. Bees and ants at least can do this, as can all vertebrates in which the capacity has been experimentally tested. The process is called *dead reckoning* or *path integration*. In dead reckoning, $P(t)$, the position vector at time t after departure from the base position, is computed from the relation:

$$P(t) = \int_0^t \upsilon(\tau)d\tau$$

where $v(\tau)$ is the velocity vector (the direction and speed of movement) at each moment τ between the departure at time 0 and the present time t. In some recent models of this process in the ant (Müller & Wehner, 1988), it is assumed that the integration is done in effect numerically, that is, by summing successive small displacements.

The most direct evidence that bees could keep track of the distance and direction flown between a food source and the hive came from studies of the recruitment dance of foraging honey bees (von Frisch, 1967). The returning honey bee forager does a waggle dance that recruits other foragers to come to the source it has found. The orientation of the dance with respect to gravity tells the recruits the direction of the source relative to the sun. The number of waggles tells them the distance of the food. A more compelling behavioral demonstration that the nervous system of an insect can encode distances and directions could hardly be asked for.

More recently, Wehner and Srinivasan (1981) have produced experimental proof that the foraging desert ant *Cataglyphis bicolor* homes on its nest by dead reckoning. It can carry out the homing run in the correct compass direction and for the correct distance when running over completely unfamiliar terrain to a fictive (nonexistent) nest hole (see also Schmidt, Collett, Dillier, & Wehner, 1992).

The dead reckoning mechanism in the bee can take its directional indication from the sun, but it can also take its directional reference directly from its map of the terrain surrounding the hive (Dyer, 1987; Dyer & Gould, 1981). This latter ability makes it possible for the bee to set and hold a course relative to the terrain itself on days when clouds obscure the sun. As already noted, this ability is evidence that bees have a map of the terrain.

Dead reckoning provides the information about the compass direction and distance of a new vantage point relative to the base position. It gives the vector o, the animal's position relative to the base position (the origin of the coordinate system). Let the vector p be the *egocentric position* of a landmark, that is, the position of a landmark in a coordinate system with the observer at the origin. Then the *geocentric position* of the landmark is $o + p$. The vector summation converts the egocentric (observer-dependent) position code for a landmark—the code presumably generated during survey flights—to a geocentric (observer-independent) code. A geocentric code represents the landmark's position in a coordinate system whose origin is a fixed point on the earth (the base position).

When it is recalled that bees and wasps make survey flights both on leaving their nest and at the sites where they find food, that they demonstrably know the direction and distance of landmarks surrounding their nests and food sites, and that they demonstrably know the direction and distance of food sources from their nests, it does not seem farfetched to imagine that they do the vector summation required to code the positions of the landmarks at the different sites in a common frame of reference. The vector summation, which completes the construction of the map, is a less complex computation than the computations underlying sun-compass navigation, parallax surveys, and dead reckoning, all of which are experimentally established capacities of the insect brain.

C. Using the Map

Putting the position codes for landmarks observed from different vantage points in a common frame of reference to make a geocentric cognitive map makes it possible in principle to set a course from any coded position to any other. The current dispute over whether insects have cognitive maps centers around conflicting experimental evidence about the extent to which bees can do this. Gould (1986) trained bees to forage at a site in a clearing out of sight of the familiar meadow in which their hive was located. He then captured them as they departed the hive bound for that site and released them elsewhere in the meadow (Figure 1A). They departed from the release site in the direction of the feeding site, on a compass course substantially different from the compass course from the hive to the feeding site, which they were about to set out on when captured. They had presumably never before flown the course from the release site to the feeding site, because the release site, although familiar to them from their pre-experimental foraging in the meadow, was far from the intermediate routes they were induced to follow while being trained to forage in the clearing in the woods. As one of several controls, Gould captured other bees returning to the hive from unknown foraging sites at various points of the compass, and released them at the same site. They departed in the direction of the hive rather than in the direction of the feeding site, showing that bees would set different courses from the same release site depending on their destination at the time of capture.

Menzel et al. (1990) and Wehner et al. (1990) failed to replicate Gould's results. However, these experiments seemingly fail to replicate previous and subsequent results by others, so it is hard to know what to make of these failures. For example, when Wehner et al. (1990) displaced bees after they had returned to the hive, they vanished in random directions rather than toward the hive. Wehner et al. interpret this to mean that because these bees had the output of their dead reckoning mechanism set to $<0,0>$ (the base position), they did not know where to go when released away from the hive. Wehner et al. conclude that bees can only follow routes they have learned before. They suggest that "when the bee repeatedly departs from and returns to the hive along the same vector course, landmark-based route information is added to the original vector information" (Wehner et al., 1990, p. 481). (The original vector information is the position of a site relative to the hive given by the dead reckoning mechanism.) They further suggest that "landmarks are used in a route-specific rather than map-specific way. . ." (p. 481).

It is difficult to reconcile this conclusion with many previous findings that honey bees taken from the hive and released at arbitrary sites within their foraging range generally come home fast. For example, Becker (1958), a von Frisch student, moved a hive into territory completely unfamiliar to the bees and allowed experienced foraging bees to make a single wide-ranging but short survey flight lasting only 3–5 min. During these flights, the bees did not gather pollen. Becker captured the bees as they returned from their survey and then released them at various distances south or north of the hive. These bees had never fed at any site in this area, so they had no route-based memories. Their dead-reckoning mechanism could not tell them where they were because they were captured at the hive. All 40 bees released 250 m south or north of the hive came back to the hive—24 of them in less than 5 min, and 14 of those in less than 2 min (Becker, 1958, Table 9, p. 10). In an experiment that makes an instructive contrast, Becker (1958) brought other hives of bees into the same unfamiliar territory and kept them for a day or two with a flight cage in front of the hive. The cage restricted their flights to within 3 m of the hive but gave them a view of the sun and of the terrain to the south of the hive. When these bees—which had never made a survey flight more than 3 m from the hive—were captured and released 200 m north or south from the hive, less than 20% made it back to the hive. The average time for them to get there was about an hour.

Becker did not take departure bearings at the release site and Wehner et al. did not record successful homings or flight times, so the results are not directly comparable. Nonetheless, most of Becker's bees that were allowed a survey flight were clearly on their way toward the hive not long after they were released. They were homing to a hive located in previously unfamiliar terrain after a single 3- to 5-min survey flight. The contrasting results obtained from the bees that were allowed to learn the local solar ephemeris function but not allowed to make a survey flight show that Becker's bees learned something in a 3- to 5-min survey flight that made a big difference in how directly they could return to their hive. What they learned during their survey flight was clearly not a repeatedly followed route to and from a feeding source. One plausible explanation of what they learned is a map of the landmarks surrounding the hive, a map on which they could later compute their position (by "taking a fix" from recognized landmarks) in order to set a course for the hive. R. Wehner and Menzel (personal communications) report that their bees, too, despite departing in random directions, soon returned to the hive. Thus, at some point their bees, though not on a familiar route, nonetheless turned and oriented their flight toward their hive. Before one concludes that bees do not have a map, it would be good to have a clearer understanding of how they manage to set a course for the hive when they realize that they are not getting where they expect to get.

Menzel et al. (1990) captured bees as they left the hive en route to the most recent food site and released them at an earlier food site or at an unfamiliar site around the shoulder of the mountain. No matter where they were released, the bees flew off on the heading they were pursuing when captured. Menzel et al. conclude that "the sun compass [and their heading at the time of capture] dominates far-distance orientation in bees so strongly that even prominent landmarks. . . might be of little importance when the sun is visible" (p. 726). If this were so without qualification, then the Menzel et al. bees should never have come back to the hive after their release. However, they did soon return to the hive (Menzel, personal communication). Menzel et al.'s bees left the release site flying away from the hive into unfamiliar terrain. At some point, the bees must have used landmarks to conclude they were not going where they thought they were. How they then made their way to the hive (or perhaps even to their originally intended destination—see Dyer results below) is an interesting, unsolved problem. It is not clear that it can be solved without recourse to the hypothesis that the bees have a map.

Dyer's (1991) results fail to replicate some of the results of Menzel et al. (1990) and Wehner et al. (1990), and also some of Gould's (1986). Dyer used two feeding/release sites. One (Site B) was down in a quarry over a forested ridge from the hive, while the other (Site A) was across a rising pasture. Bees trained to B (the quarry site), captured while leaving the hive en route to B, and then released at A (the pasture site), headed for B. This replicates Gould and fails to replicate Menzel et al. (1990). On the other hand, when trained to A and captured and released at B (in the quarry), most left B on the same heading they had when captured (replicating Menzel et al., not Gould), but a minority left headed for the hive (not replicating Menzel et al). However, 80% of the bees released down in the quarry arrived at the pasture feeding site (their intended destination when captured) within 20 min. Their mean flight time was 6.8 ± 4.5 min. These arrival-percentage and flight-time data are similar to the flight times and arrival percentages reported by Gould. Menzel et al. do not report where or when their released bees were next sighted, but one would certainly infer that it was not at their intended destination. In any event, Dyer's finding that bees released down in the quarry soon showed up at the feeding site up in the pasture does not support the conclusion reached by Menzel et al. Although the bees released down in the quarry left headed in the wrong direction, away from both the hive and their intended destination, they did not follow an erroneous course oblivious to all landmarks for long. On the contrary, like Gould's bees, most of them got where they were originally headed no matter where they were released, and they got there rather quickly. Again, before concluding that they do not have a map, one would want to understand more clearly how they arrived to soon at their intended destination.

When Dyer first trained bees to the quarry site, then to the pasture site, then captured them while headed to the pasture site, and released them in the quarry, they all left B headed for the hive, as if they could not compute the course from the quarry to the pasture site. While this result fails to replicate Gould, it also fails to replicate Menzel et al.'s main result. In another experiment of a design closely paralleling the design of Menzel et al. (1990), Dyer (1991) began with the feeder down in the quarry, then moved it in stages along the route from there up to the top of the pasture. When the bees had been flying to A (the top of the pasture) for two days, he captured them en route there and released them down in the quarry. These bees made straight for A from B. This is the result that Menzel et al. argued would support a cognitive map conclusion, the result they failed to obtain when they ran this same experiment. The result does not appear to support a route-based navigation account, à le Wehner et al., because the bees never flew the B-to-A route prior to the test. They only flew from the hive to intermediate points on the B-to-A route and back to the hive.

The main point of difference between (at least some versions of) the independent route-based memories and the map hypothesis centers on whether bees can combine the data about distances and directions gained from flying different routes in order to set a course between two points that have not previously been the end points of a single route. Most of Dyer's results, like Gould's, imply that they can. Nonetheless, Dyer argues that when his bees behave like Gould's they do so because they have learned to recognize retinal snapshots of intermediate landmarks—a theory favored by most researchers working on insect navigation (Cartwright and Collett, 1983; Collett, 1992; Collett, Dillmann, Giger, & Wehner, 1992; Dyer, 1991; Wehner, 1992; Wehner et al., 1990; Zeil, 1993a). Before we conclude that this hypothesis is a viable alternative to the map hypothesis, we need to understand better how it answers the following questions.

Due to parallax distortion, the landmarks on the route from the hive to the quarry site will appear different when seen from the top of the pasture. How does the bee recognize familiar terrain seen from novel vantage points? The orientation the bees must adopt with respect to the familiar terrain in order to aim for the quarry when departing the pasture site is different from the orientation they adopt toward those same landmarks when flying from the hive. How does the bee determine what orientation to adopt with respect to the intermediate landmarks, once it has succeeded in recognizing them?

The simulations of Cartwright and Collett (1983), which are often cited as proof of the viability of the retinal snapshot model, fall far short of showing that it can be made to work when confronted with natural terrains.

Moreover, the limited success obtained in these simulations depended fundamentally on the bee's storing the metric relation between the landmarks and the food site. The image-processing algorithm of Cartwright and Collett (1983) used knowledge of both the compass bearings and the distances associated with various parts of the image. The assumption that distances and compass bearings are used in processing the images of landmarks is justified by direct experimental test (Cartwright & Collett, 1983; Cheng et al., 1987). However, the use of such information makes it unclear in what sense landmark recognition and course setting can be said to be based simply on the static retinal image itself, as opposed to a reconstruction of the terrain that generated the moving image (that is, a map). When the course-setting procedure uses knowledge of the distance and compass bearings of segments of the remembered image of the terrain as seen from the destination, and when it relies also on estimates of the observer's current distance and direction from landmarks, it is not clear in what sense it is simply matching snapshots (that is, doing two-dimensional image alignment). When the retinal snapshot model has been elaborated enough to make clear the answers to these questions, there may no longer be a substantive difference between it and models that assume the use of a map.

Clearly, it remains to be determined under what circumstances a bee can set a course to an unseen destination when the bearing of that destination is substantially different from the heading the bee was on when captured. But recent categorical claims to the effect that they cannot do this (Menzel et al., 1990; Wehner et al., 1990) are not readily reconciled with the preponderance of the data from the many capture-and-release experiments done in the last 100 years. Many of these findings, not just the disputed findings of Gould (1986), imply or directly demonstrate that bees can often do this. The ability to do this, at least under favorable circumstances, is the ultimate test of the hypothesis that an animal has a map in the full sense of the term.

II. TIME SENSE

Animals have internal clocks in the form of nerve cells and nerve circuits with activity cycles. These clocks (or endogenous oscillators) differ widely in their *period,* which is the time required to complete one cycle. The periods on these neural clocks range from fractions of a second to a year or years (Aschoff, 1981; Daan & Aschoff, 1981; Farner, 1985). The best studied clock is the daily or circadian clock, which has a period of about 24 hours (Aschoff, 1984). In vertebrates, it is located in a small nucleus in the hypothalamus just above the optic chiasm (Turek, 1985). It gets a synchronizing time signal from the solar light–dark cycle via specialized visual receptors (Foster, 1993). Recently, experimenters have succeeded in taking the clock out of the brain of one animal and installing it in the brain of another

(Ralph, Foster, Davis, & Menaker, 1990), much as one may take the calendar/clock chip out of one computer and install it in another.

From their internal clocks animals get temporal information of two kinds—information about the *phase* of a clock's cycle (where it is in its cycle) at the time they experience something and information about the temporal *intervals* between events or between an event and the present moment (see Gallistel, 1990, for a review).

A. Role of the Time-of-Day Sense in Learned Behavior

One use for the phase information from the circadian clock is in learning the solar ephemeris, the azimuth position of the sun as a function of the time of day. We know that the time-of-day signal for the solar ephemeris comes from the internal clock (rather than from external cues) because phase shifting the internal clock (putting it out of synchrony with the local day–night cycle) shifts compass orientation in both insects (Renner, 1960) and homing pigeons (Emlen, 1975). When the animal's internal clock indicates the wrong time of day, the animal orients as if the sun were at the azimuth position it would be at if the animal's internal clock were correct. A bird with a 3-hour internal clock retardation that tries to fly north at noon, flies east. That is, it keeps the sun, which is in fact in the south, on its right. This would indeed head it north if it were 9:00 in the morning, the time indicated by the bird's circadian clock, at which time, of course, the sun is in the east.

The position of the sun is not the only datum that is recorded as a function of the time of day. And course setting is not the only behavior that depends on the animal's knowledge of the time of day. If bees experience a heightened concentration of sugar in their nectar between certain hours of the day, they are more likely to visit that feeding station during those hours (Wahl, 1932). If they experience an odor or color while harvesting nectar at one time of day and a different odor or color while harvesting it from the same location at another time of day, they prefer the first odor and color during the hours of the day at which they previously experienced them, and they prefer the other odor and color during the hours they previously experienced them (Koltermann, 1971). Similarly, birds learn to choose feeding sites on the basis of the time indicated by their circadian clock (Biebach, Falk, & Krebs, 1989; Biebach, Gordijn, & Krebs, 1989; Saksida & Wilkie, 1994).

If rats are fed at certain times of the day, they become active an hour or two before each feeding (Bolles & Moot, 1973). The onset of the anticipatory activity is triggered by the reading of the internal clock rather than by external indicators of time, because shifting the phase of the internal clock shifts the time of day at which the anticipatory activity begins (Rosenwasser, Pelchat, & Adler, 1984).

If a rat is shocked for stepping through a doorway from a white compartment into a black compartment, and subsequently tested at different times of day, it is most reluctant to venture through at the training time of day (Holloway & Wansley, 1973). A similar effect is seen in active avoidance tasks (Holloway & Sturgis, 1976) and in appetitive tasks (Wansley & Holloway, 1975). Regardless of task, the animal's performance is most indicative of its previous experience if it is tested at the time of day at which it had that experience. When rats are given conflicting experiences in the black compartment—reward on one occasion, punishment on another—the degree of ambivalence in their responding is greatest when both experiences are given at the same time of day on successive days. Their ambivalence is much less if the conflicting experiences are given at different times of day (Hunsicker & Mellgren, 1977). Learned performance depends on the time of day of testing relative to the time of day of training. This, in turn, implies that part of what is learned is the time of day of the experience (Gallistel, 1990). It also implies that an animal's sense of where it is in time has the same pervasive effect on its learned behavior as does its sense of where it is in space.

B. Learning of Temporal Intervals

Animals also learn the intervals between events. The properties of this interval learning and the decision processes that translate remembered and currently elapsing intervals into observed behavior in a variety of timing tasks have been the focus of extensive experimental and theoretical analysis (see Gallistel, 1990; Gibbon, 1991, 1992; Gibbon & Allan, 1984, for reviews; and Church, Miller, Meck, & Gibbon, 1991, and Wilkie, Gibbon, & Church, 1990, 1992; Saksida, Samson, & Lee, 1994 for more recent work). The properties of the timing process described below and the consequences derivable from them in a variety of tasks constitute what Gibbon has termed "scalar expectancy theory."

1. Timing Tasks

One timing task requires a go/no go response based on the duration of a just-experienced stimulus (the "generalization procedure"; Church and Gibbon, 1982). In another, a response is probabilistically rewarded only after a fixed interval has elapsed since the onset of a stimulus (the "peak procedure"; Roberts & Church, 1978). Both the generalization procedure and the peak procedure require the animal to compare the duration of a currently elapsing interval (the interval since the onset of the trial) to the remembered duration of earlier intervals (the intervals at which reward was obtained on earlier trials). Another task requires a choice between two responses based on a comparison of the remembered durations of two stimuli (the "bisection procedure"; Church & Deluty, 1977; Gibbon, 1981). Yet another requires an animal to judge from moment to moment which of two response options is likely to pay off sooner. The expected time to payoff for one option remains constant regardless of how long the trial has lasted, while the expected time to payoff of the other decreases linearly with the elapsed duration of the trial. This is called the "time-left procedure" (Gibbon & Church, 1981). It requires the animal to compare one remembered interval to the difference between a currently elapsing interval and another remembered interval.

2. Properties of the Timing Mechanism

The average remembered duration of an interval is a linear function of the duration of that interval, with an intercept close to zero, which means that for durations longer than several seconds, subjective duration is proportionate to objective duration (Gibbon & Church, 1981). The variability in the remembered durations of objectively constant intervals obeys Weber's Law, that is, the standard deviation of the distribution of remembered intervals is proportionate to their mean value (Gibbon, 1977, 1992). A notable feature of the decision processes that determine how the animal will respond is their reliance on the ratio of the two temporal quantities being compared. (In the time-left procedure, one of the quantities in this ratio is the computed difference of two intervals.) What matters to the animal is the multiplicative factor by which the test interval (the numerator) exceeds or falls short of the comparison interval (the denominator).

A peculiar feature of the mechanism by which experienced intervals are committed to memory is an animal-specific scalar distortion between the durations initially experienced and the values that get written into memory (Gibbon, Church, & Meck, 1984). The value written to memory differs from the immediately experienced subjective duration of the interval by an animal-specific (not task-specific) percentage. Thus, a particular animal may record durations in memory that are, say, 20% longer than its immediate subjective interval measurement. As a consequence, that animal's peak responding—the point where its anticipation of reward is greatest—will occur 20% later than the time when the reward is scheduled to become available. Another animal may systematically shorten the intervals when writing them to memory; its peak anticipation will precede the schedule interval. The factor by which the remembered duration differs from the originally experienced duration may be altered pharmacologically (Meck & Angell, 1992).

C. Role of Interval Learning in Classical and Instrumental Conditioning

The primary variables in classical conditioning experiments are the temporal intervals separating the onsets and offsets of the stimuli (the CSs and USs). In Pavlov's reports, there was evidence that animals learned these intervals and that the intervals they had learned were an important determinant of the conditioned response. If the food was given awhile after the onset of the conditioned stimulus, then the fully conditioned dog only salivated awhile after the onset of the CS, at about the time the food was due to be given. (Note the close parallel between this result and the results of the peak procedure.) The fact that latency of the conditioned response tends to vary in proportion to the CS–US interval has been confirmed by modern work. For example, Kehoe, Graham-Clarke, and Schreurs (1989) trained rabbits in a conditioned eyeblink procedure with two different, randomly interspersed CS–US intervals. The rabbits learned to blink twice in response to the CS, once just before the first possible occurrence of the US and again before the second possible occurrence.

When fear or shock or illness is the conditioned response, there is notably less tendency for the latency of the response to match the CS–US interval (Brandon & Wagner, 1991). Nonetheless, probing the degree of fear by means of a startle stimulus delivered at various times after CS onset shows that the rat learns the CS–US interval even on its first experience of a pairing between a CS and shock (Davis, Schlesinger, & Sorenson, 1989).

It has recently been argued that cerebellar lesions, which permanently abolish the conditioned but not the unconditioned eyeblink, do so because they knock out the recorded value of the CS–US interval (Ivry, 1993), which is required for the genesis of an appropriately timed conditioned response. In what follows, I give arguments for two hypotheses: (1) in both classical and instrumental conditioning, animals learn the interevent intervals; (2) it is the remembered intervals between, say, onset of CS and US, not simple associations between CS and US, that enter into the decision processes that determine whether a conditioned response will occur and when.

1. Backward Conditioning

Only recently has the broader significance of interval learning in conditioning experiments begun to be appreciated (see Hall, Chapter 2, this volume, for the orthodox associative account). Matzel, Held, and Miller (1988) have shown that rats learn the CS–US interval equally well whether the US onset follows CS onset, as in forward conditioning, or precedes it, as in backward conditioning. Backwardly conditioned subjects do not respond to the CS because they in effect know that by the time the CS has come on it is too late. To show this, Matzel and Miller first taught their subjects the 5-s interval between the onset of a clicking stimulus and the onset of a tone. Then they conditioned the tone to shock—forwardly in one group (CS–US interval = 5 s), simultaneously in another (CS–US interval =0), and backwardly in another (CS–US interval = −5 s). When the simultaneously or backwardly conditioned groups were tested for conditioned responding to the tone, they showed little, but when they were tested for conditioned responding to the clicking, they responded as strongly as the forwardly conditioned group. Note that the only connection between the clicking and shock was via the connection established between the tone and shock (second-order conditioning).

2. Blocking and Background Conditioning

Barnet, Grahame, and Miller (1993) have shown that prior experience with the predictive power of one CS blocks conditioning to another "redundant" CS only if both CSs have the same temporal relation to the US. A simultaneously conditioned CS blocks conditioning to an additional simultaneous CS, and a forwardly conditioned CS blocks conditioning to an additional forward CS, but not the other way around. The conditioning process discounts a redundant CS only if it provides the same information about the timing of the US (see Miller & Barnet, 1993, for review).

Blocking may be understood as a manifestation of a more general principle, to wit: in conditioning experiments, the animal learns the effects of the CSs—including the background, that is, the apparatus itself—on the expected interval between reinforcements. It gives a conditioned response to a CS when the expected interreinforcer interval in the presence of the CS differs from the expected interreinforcer interval in its absence (Gallistel, 1990; Gibbon & Balsam, 1981). For example, when the intershock interval is no different in the presence of a tone CS than in its absence (i.e., than the background expectation), the animal will not respond to the tone no matter how many times the tone is paired with the shock (Rescorla, 1968). This explanation of the effects of background conditioning does not require positing a trial clock, nor does it depend on parametric assumptions. Associative models of conditioning are generally not real-time models. They carve time up into discrete bins and look at the probabilities of reinforcement in each bin (rather than rate of reinforcement, as in real-time models). Thus, to deal with the blocking effects of background conditioning Rescorla and Wagner (1972) posited a subjective trial clock, which divided the animal's continuous experience of the between-CS periods into discrete trials. Their explanation depended on this artifice, and on the values assumed for various free parameters, including the assumed duration of the internally timed "trials."

In a simple blocking experiment (Kamin, 1969), the animal's experience with the blocking CS has taught it the interreinforcer interval to be expected when that CS is present. When the blocked CS is subsequently introduced, the interreinforcer intervals the animal observes in the presence of the two CSs combined are not different than those it expects from the initial CS alone, hence it attributes to the second CS no power to influence interreinforcer interval. Blocking is explained in the same way as are the effects of background conditioning.

3. Effects of the Intertrial Interval

In one version of this kind of analysis of conditioning, the animal begins to respond to a CS only when it has statistically persuasive evidence that the CS has an effect on the expected interreinforcer interval (Gallistel, 1990, 1992). Multiplying the time scale of the experiment by some factor (thereby increasing all intervals by the same factor) has no effect on the statistical implications of the observations made after a given number of trials (CS–US pairings). Thus, increasing the CS–US interval from 4 s to 32 s should have no effect on the rate of conditioning, provided that the intertrial interval is also increased 8-fold, and, in fact, it does not, at least in the autoshaping paradigm (Gibbon, Baldock, Locurto, Gold, & Terrace, 1977). Associative explanations of conditioning, which assume a crucial role for the CS–US interval in the association-forming process (Gluck & Thompson, 1987; Hawkins & Kandel, 1984; see Hall, Chapter 2, this volume), have difficulty explaining why conditioning should be invariant under scalar transformation of the time intervals in the conditioning protocol. They also have trouble explaining the obverse of this, the fact that when the CS–US interval is held constant, the intertrial interval has a dramatic effect on the rate of conditioning (Levinthal, Tartell, Margolin, & Fishman, 1985; Schneiderman & Gormezano, 1964).

4. Effects of Partial Reinforcement and Delay of Reinforcement

In timing experiments, if the interval being timed is interrupted for awhile, the animal ignores the interruption. That is, it behaves as if it stopped its timer at the onset of the interruption and restarted it (without rezeroing it) at the end of the interruption (Roberts & Church, 1978). The effects of partial reinforcement on both the rate of conditioning and the rate of extinction may be understood in the light of this result. Partial reinforcement—pairing a US with the CS on only some of the occasions when the CS occurs—is another way of lengthening proportionately the CS–US interval and the interval when the CS is not on (the intertrial interval). Suppose that we begin with a continual reinforcement protocol with a 2-s CS–US interval and a 20-s intertrial interval. If we use instead a partial reinforcement schedule in which the US is paired with the CS on only one-tenth of its occurrences, we lengthen the interreinforcer interval in the presence of the CS from 2 s to 20 s. However, we also increase the number of intertrial intervals per reinforcement by the same factor of 10. Thus, there should be no effect of partial reinforcement on the rate of conditioning if we measure that rate in terms of the number of reinforcements required rather than the number of "trials" (CS occurrences), and, in fact there is not, at least in the autoshaping paradigm (Gibbon, Farrell, Locurto, Duncan, & Terrace, 1980).

In one model that analyzes conditioning in terms of the learned interreinforcer intervals, extinction occurs when an improbably long interval has elapsed since the last reinforcement in the presence of the CS. How long this interval must be to reach any given level of improbability, is, of course, proportionate to the expected interreinforcer interval (that is, the training interval). Thus, partial reinforcement should prolong extinction in proportion to its prolongation of the expected interreinforcer interval, which it does (Gibbon et al., 1980). A partial reinforcement schedule of 1:10 increases the interreinforcer interval in the presence of the CS by a factor of 10 and prolongs trials to extinction by a factor of 10.

In training, the comparison that underlies the emergence of conditioned responding is between the interreinforcer interval in the presence of the CS and the interreinforcer interval in its absence. In extinction, however, the comparison is between the currently elapsed interval of nonreinforcement in the presence of the CS and the previously experienced interreinforcer intervals in the presence of that stimulus. Thus, varying the intertrial interval (while holding the CS–US interval constant), which has a profound effect on the rate of conditioning, should have no effect on the rate of extinction, which it does not (Gibbon et al., 1980).

The analysis of partial reinforcement in terms of its effect on the expected interreinforcer interval in the presence of the CS implies that partial reinforcement and delay of reinforcement should have comparable effects in instrumental conditioning. Mazur (1989, 1991) used a discrete-trials procedure, in which he could titrate delay of reinforcement against probability of reinforcement. Choosing one key, say, the red one, produced a reward with some probability at a fixed delay after the response. Choosing the other key, say, the green one, always produced a reward, but the experimenter increased and decreased the delay of that reward to find the delay that produced indifference between the two keys. He found that "preference for a probabilistic reinforcer is inversely related to the time spent in the presence of stimuli associated with that reinforcer" (Mazur, 1991, p. 71). When allowance is made for the effects of scalar memory variance on the remembered interreinforcer intervals, thinning the reinforcement schedule is equivalent to increasing the delay of reinforcement by the same factor (Gibbon, Church, Fairhurst, & Kacelnik, 1988). Both manipulations increase the expected interreinforcer interval in the presence of some CS by the same factor.

5. Matching Behavior

Matching behavior (see Williams, Chapter 4, this volume) has also been profitably analyzed from an interval-timing perspective (Gibbon et al., 1988). One such analysis (Mark & Gallistel, 1994) has yielded the first "fully dynamic" model of the choice mechanism. Previous accounts have offered at most a partial account of the forces driving the choice behavior to match the relative rate of reward (Staddon, 1988). Most previous accounts of matching have assumed that it was the outcome of a feedback process in which the animal changed its behavioral investment in the options until it got equal payoffs per response (or unit of time) invested in each option (Hernstein & Vaughan, 1980; Lea & Dow, 1984; Staddon, 1988). These accounts did not fully specify the process because they never made fully explicit how the profitability of each side was assessed. All of them, however, took it for granted that the rate- or profitability-determining process averaged over many rewards. This averaging, together with the feedback-governed approach to the matching equilibrium, would make the kinetics of the adjustment to a change in the relative rate of reward sluggish.

However, Mark and Gallistel (1994) have shown that when rats respond for brain stimulation reward on concurrent variable-interval schedules, the individual subject's time-allocation ratio (the ratio of the amounts of time spent on each lever) closely tracks the large, random fluctuations in apparent relative rate (Figure 4). These fluctuations in the apparent relative rate of reinforcement are inherent in estimates of relative rate made from short time samples of two Poisson processes, such as the processes that generate variable-interval schedules. An example of such an estimate of relative rate would be an estimate based on the ratio of the two most recent interreinforcer intervals on each schedule. This estimate fluctuates widely due to large, random variations in both the numerator and the denominator. If matching were based on a feedback process, it would have to average over many past reinforcements (Killeen, 1981; Lea & Dow, 1984); hence, it could not track the rapid, large, random variations in apparent relative rate of reinforcement that are inherent in concurrent variable-interval schedules.

A choice mechanism or decision process that can track the random variations in apparent relative rate of reinforcement is one in which the relative rate of reinforcement is estimated by the ratio of the two most recent interreinforcer intervals and this estimate sets the ratio of stochastic (Poisson) patch-leaving parameters (Mark & Gallistel, 1994). This same model predicts the counterintuitive preferences that subjects exhibit when tested for their preference between two keys associated with the same variable-interval schedules but paired during concurrent variable-interval training with contrasting schedules, one with a relatively richer and one with a

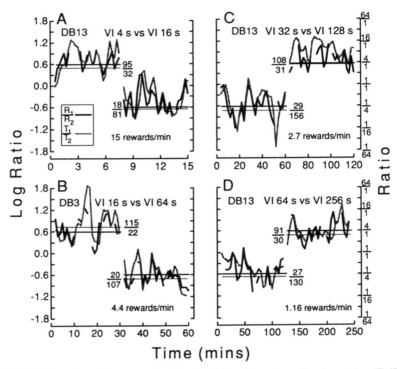

FIGURE 4 Plots of log reward ratios (R_1/R_2) and log time-allocation ratios (T_1/T_2) in successive windows equal to two expected interreward intervals on the leaner schedule, over sessions comprised of two trials, with a 16-fold reversal in the programmed relative rate of reward between the trials. Successive windows overlap by half a window. A gap in the plot line means that the ratio was undefined within that window because no reward occurred on one or both sides. The programmed reward ratios of 4:1 and 1:4 are indicated by the horizontal lines at $+0.6$ and -0.6, respectively. The lighter horizontal lines indicate the actually experienced reward ratio, as calculated by aggregating over the trial. The actually obtained numbers of rewards, which yielded these ratios, are given beside these lighter lines. The actually experienced combined rate of reward across the two trials in a given condition (combined reward density) is given at the lower right of each panel. (A) Programmed reward density = 19.2 rewards/min. (B) Programmed reward density = 4.8 rewards/min. (C) Programmed reward density = 2.4 rewards/min. (The density rate actually experienced, which is greater than the programmed rate, reflects the variability inherent in Poisson schedules.) (D) Programmed reward density = 1.2 rewards/min. At all levels of combined reward density, the rat's time-allocation ratio tracks the large, random variations in the most recently experienced reward ratio, that is, matching occurs on a time scale measured in a few interreward intervals. (Reproduced from Mark & Gallistel, 1994, by permission of the author and publisher.)

relatively leaner schedule (Belke, 1992). The birds exhibit a strong preference for the key paired with the leaner schedule during training (see Williams, Chapter 4, this volume).

D. Role of the Elapsed Interval since Training in Conditioned Responding

The section on the behavioral consequences of the position sense stressed the effect that the animal's knowledge of its position on its cognitive map has on its response to immediate sensory input. Similarly, the section on the role of the time-of-day sense stressed the effect that the animal's knowledge of the time of day has on the use it makes of what it has learned at different times of day. The animal's sense of the interval elapsed since it learned something also has a profound effect on the use it makes of that learning. Again, Pavlov noticed this phenomenon when he described the spontaneous recovery of conditioned responding following extinction. As he realized, and as has been confirmed by contemporary work on extinction (Bouton, 1994), spontaneous recovery of conditioned responding occurs when the spatial or temporal context changes. A change in spatial context is signaled by sensory inputs of various kinds (including the velocity inputs used by the dead-reckoning mechanism to compute position), but it can reasonably be argued that a change in temporal context is not signaled by any stimulus. It is the result of the inherent disposition of animals to use internally generated time signals to monitor the time elapsed since an experience, a disposition demonstrated by the timing experiments reviewed above (see also Wilkie *et al.* 1994).

The phenomenon of spontaneous recovery has been placed in a foraging context by the recent work of J. A. Devenport and Devenport (1993) and L. D. Devenport and Devenport (1994) on what they call the temporal weighting rule. They point out that in the field, food patches vary from visit to visit in how rich or lean they are. If an animal has just sampled a patch, its estimate of the current return to be expected from that patch should be based primarily on this, the most recent sample. But as time elapses, the mean over the last several samples becomes a more reliable indicator of what can be expected on the next visit to that patch. Thus, one would expect to see shifts in choice behavior simply as a function of the interval elapsed since the animal last visited a patch. A patch that was atypically rich or lean on the last visit should be preferred or shunned if the animal chooses soon after that visit, but the atypical visit should be discounted as a function simply of the time elapsed since the visit. After long delays, animals should choose on the basis of patch means, not the most recent observations. L. D. Devenport and J. A. Devenport (1994) have experimentally demonstrated this phenomenon in field experiments with ground squirrels and chipmunks, and also in backyard experiments with dogs (J. A. Devenport & Devenport, 1993).

III. CONCLUSION

In studies of animal learning and cognition rooted in the empiricist/associationist tradition the fundamental role of spatial and temporal knowledge in the determination of behavior has been neglected, because it was unclear what the relevant stimuli were. The contemporary understanding of the mechanisms by which spatial and temporal knowledge are acquired suggests that this knowledge is acquired from sources and by mechanisms not contemplated in the empiricist analysis. There is no stimulus for time. Time is in the mind because clocks are in the brain. And knowledge of position in space is constructed by special-purpose computations such as those that mediate the construction of a geocentric cognitive map and the continuous updating of the representation of the animal's position on that map by dead reckoning. By means of computational mechanisms that process selected inputs in ways designed to extract specific facts about the world, the animal constructs and continuously updates a representation of its position in the space–time manifold. The resulting sense of spatiotemporal position is an important element in every aspect of the animal's behavior.

References

Able, K. P., & Bingham, V. P. (1987). The development of orientation and navigation behavior in birds. *Quarterly Review of Biology, 62,* 1–29.

Aschoff, J. (1981). *Biological rhythms.* New York: Plenum.

Aschoff, J. (1984). Circadian timing. In J. Gibbon & L. Allan (Eds.), *Timing and time perception.* pp. 442–468. New York: New York Academy of Sciences.

Barnet, R. C. Grahame, N. J., & Miller, R. R. (1993). Temporal encoding as a determinant of blocking. *Journal of Experimental Psychology: Animal Behavior Processes, 19,* 327–341.

Becker, L. (1958). Untersuchungen über das Heimfindevermögen der Bienen. *Zeitschrift für Vergleichende Physiologie, 41,* 1–25.

Belke, T. W. (1992). Stimulus preference and the transitivity of preference. *Animal Learning and Behavior, 20,* 401–406.

Biebach, H., Falk, H., & Krebs, J. R. (1989). The effect of constant light and phase shifts on a learned time-place association in garden warblers (*Sylvia borin*): Hourglass or circadian clock? *Journal of Biological Rhythms, 6,* 353–365.

Biebach, H., Gordijn, H., & Krebs, J. R. (1989). Time-place learning by garden warblers, *Sylvia borin. Animal Behaviour, 37,* 353–360.

Bolles, R. C., & Moot, S. A. (1973). The rat's anticipation of two meals a day. *Journal of Comparative and Physiological Psychology, 83,* 510–514.

Bouton, M. E. (1994). Context, ambiguity, and classical conditioning. *Current Directions in Psychological Science, 3,* 49–53.

Brandon, S. E., & Wagner, A. R. (1991). Modulation of a discrete Pavlovian conditioned reflex by a putative emotive Pavlovian conditioned stimulus. *Journal of Experimental Psychology: Animal Behavior Processes, 17,* 299–311.

Carr, H. (1917). Maze studies with the white rat. *Journal of Animal Behavior, 7,* 259–306.

Carr, H., & Watson, J. B. (1908). Orientation of the white rat. *Journal of Comparative Neurology and Psychology, 18,* 27–44.

Cartwright, B. A., & Collett, T. S. (1983). Landmark learning in bees: Experiments and models. *Journal of Comparative Physiology, 151,* 521–543.

Cheng, K. (1986). A purely geometric module in the rat's spatial representation. *Cognition, 23,* 149–178.

Cheng, K. (1989). The vector sum model of pigeon landmark use. *Journal of Experimental Psychology: Animal Behavioral Processes, 15,* 366–375.

Cheng, K., Collett, T. S., Pickhard, A., & Wehner, R. (1987). The use of visual landmarks by honey bees: Bees weight landmarks according to their distance from the goal. *Journal of Comparative Physiology, 161,* 469–475.

Church, R. M., & Deluty, M. Z. (1977). Bisection of temporal intervals. *Journal of Experimental Psychology: Animal Behavior Processes, 3,* 216–228.

Church, R. M., & Gibbon, J. (1982). Temporal generalization. *Journal of Experimental Psychology: Animal Behavior Processes, 8,* 165–186.

Church, R. M., Miller, K. D., Meck, W. H., & Gibbon, J. (1991). Symmetrical and asymmetrical sources of variance in temporal generalization. *Animal Learning and Behavior, 19,* 207–214.

Collett, T. S. (1987). The use of visual landmarks by gerbils: Reaching a goal when landmarks are displaced. *Journal of Comparative Physiology A, 160,* 109–113.

Collett, T. S. (1992). Landmark learning and guidance in insects. *Philosophical Transactions of the Royal Society of London, Series B, 337,* 295–303.

Collett, T. S., Cartwright, B. A., & Smith, B. A. (1986). Landmark learning and visuo-spatial memories in gerbils. *Journal of Comparative Physiology A, 158,* 835–851.

Collett, T. S., Dillmann, E., Giger, A., & Wehner, R. (1992). Visual landmarks and route following in desert ants. *Journal of Comparative Physiology A, 170,* 435–432.

Collett, T. S., & Kelber, A. (1988). The retrieval of visuo-spatial memories by honeybees. *Journal of Comparative Physiology A, 163,* 145–150.

Collett, T. S., & Patterson, C. J. (1991). Relative motion parallax and target localization in the locust, *Shistocerca gragaria. Journal of Comparative Physiology A, 169,* 615–621.

Daan, S., & Aschoff, J. (1981). Short-term rhythms in activity. In J. Aschoff (Ed.), *Biological rhythms* (pp. 491–499). New York: Plenum.

Davis, M., Schlesinger, L. S., & Sorenson, C. A. (1989). Temporal specificity of fear conditioning: Effects of different conditioned stimulus–unconditioned stimulus intervals on the fear-potentiated startle effect. *Journal of Experimental Psychology: Animal Behavior Processes, 15,* 295–310.

Devenport, J. A., & Devenport, L. D. (1993). Time-dependent decisions in dogs. *Journal of Comparative Psychology, 107,* 169–173.

Devenport, L. D., & Devenport, J. A. (1994). Time-dependent averaging of foraging information in least chipmunks and golden-mantle squirrels. *Animal Behavior, 47,* 787–802.

Dyer, F. C. (1987). Memory and sun compensation by honey bees. *Journal of Comparative Physiology, A, 160,* 621–633.

Dyer, F. C. (1991). Bees acquire route-based memories but not cognitive maps in a familiar landscape. *Animal Behaviour, 41,* 239–246.

Dyer, F. C., Berry, N. A., & Richard, A. S. (1993). Honey bee spatial memory: Use of route-based memories after displacement. *Animal Behaviour, 45,* 1028–1030.

Dyer, F. C., & Gould, J. L. (1981). Honey bee orientation: A backup system for cloudy days. *Science, 214,* 1041–1042.

Eichenbaum, H., Stewart, C., & Morris, R. G. (1990). Hippocampal representation in place learning. *Journal of Neuroscience, 10*(11), 3531–3542.

Ellard, C. G., Goodale, M. A., & Timney, B. (1984). Distance estimation in the Mongolian Gerbil: The role of dynamic depth cues. *Behavioural Brain Research, 14,* 29–39.

Emlen, S. T. (1975). Migration: Orientation and navigation. In D. S. Farner & J. R. King (Eds.), *Avian biology* (pp. 129–219). New York: Academic Press.

Etienne, A. S., Sylvie, J. L., Reverdin, R., & Teroni, E. (1993). Learning to recalibrate the role of dead reckoning and visual cues in spatial navigation. *Animal Learning and Behavior, 21* 3(266–280).

Farner, D. S. (1985). Annual rhythms. *Annual Review of Physiology, 47*, 65–82.

Foster, R. G., (1993). Photoreceptors and circadian systems. *Current Directions in Psychological Science, 2*, 34–39.

Gallistel, C. R. (1989). Animal cognition: The representation of space, time and number. *Annual Review of Psychology, 40*, 155–189.

Gallistel, C. R. (1990). *The organization of learning.* Cambridge, MA: Bradford Books/MIT Press.

Gallistel, C. R. (1992). Classical conditioning as an adaptive specialization: A computational model. In D. L. Medin (Ed.), *The psychology of learning and motivation: Advances in research and theory* (pp. 35–67). San Diego: Academic Press.

Gallistel, C. R. (1994). The Replacement of General Purpose Theories with Adaptive Specializations. In M. S. Gazzaniga (Ed.), *The Cognitive Neurosciences* (pp. 1255–1267). Cambridge, MA: MIT Press.

Georgakopoulos, J. & Etienne, A. S. (1994). Identifying location by dead reckoning and external cue. *Behavioural Processes, 31*, 57–74.

Gibbon, J. (1977). Scalar expectancy theory and Weber's Law in animal timing. *Psychological Review, 84*, 279–335.

Gibbon, J. (1981). On the form and location of the psychometric bisection function for time. *Journal of Mathematical Psychology, 24*, 58–87.

Gibbon, J. (1991). Origins of scalar timing. *Animal Learning and Behavior, 22*, 3–38.

Gibbon, J. (1992). Ubiquity of scalar timing with a Poisson clock. *Journal of Mathematical Psychology, 36*, 283–293.

Gibbon, J., & Allan, L. (Eds.). (1984). *Timing and time perception.* New York: New York Academy of Sciences.

Gibbon, J., Baldock, M. D., Locurto, C. M., Gold, L., & Terrace, H. S. (1977). Trial and intertrial durations in autoshaping. *Journal of Experimental Psychology: Animal Behavior Processes, 3*, 264–284.

Gibbon, J., & Balsam, P. (1981). Spreading associations in time. In C. M. Locurto, H. S. Terrace, & J. Gibbon (Eds.), *Autoshaping and conditioning theory* (pp. 219–253). New York: Academic press.

Gibbon, J., & Church, R. M. (1981). Time left: Linear versus logarithmic subjective time. *Journal of Experimental Psychology: Animal Behavior Processes, 7*(2), 87–107.

Gibbon, J., & Church, R. M. (1990). Representation of time. *Cognition, 37*(1–2), 23–54.

Gibbon, J., & Church, R. M. (1992). Comparison of variance and covariance patterns in parallel and serial theories of timing. *Journal of the Experimental Analysis of Behavior, 57*, 393–406.

Gibbon, J., Church, R. M., Fairhurst, S., & Kacelnik, A. (1988). Scalar expectancy theory and choice between delayed rewards. *Psychological Review, 95*, 102–114.

Gibbon, J., Church, R. M., & Meck, W. H. (1984). Scalar timing in memory. In J. Gibbon & L. Allan (Eds.), *Timing and time perception* (pp. 52–77). New York: New York Academy of Sciences.

Gibbon, J., Farrell, L., Locurto, C. M., Duncan, H. J., & Terrace, H. S. (1980). Partial reinforcement in autoshaping with pigeons. *Animal Learning and Behavior, 8*, 45–59.

Gluck, M. A., & Thompson, R. F. (1987). Modeling the neural substrates of associative learning and memory: A computational approach. *Psychological Review, 94*(2), 176–191.

Gould, J. L. (1986). The locale map of honey bees: Do insects have cognitive maps? *Science, 232*, 861–863.

Gould, J. L. (1990). Honey bee cognition. *Cognition, 37*, 83–103.

Gould, J. L., & Marler, P. (1987). Learning by instinct. *Scientific American, 256,* 74–85.

Hawkins, R. D., & Kandel, E. R. (1984). Is there a cell-biological alphabet for simple forms of learning? *Psychological Review, 91,* 375–391.

Hermer, L. (1993). Increases in cognitive flexibility over development and evolution: Candidate mechanisms. *Proceedings of the Cognitive Science Society, 15,* 545–550.

Hernstein, R. J., & Vaughan, W. J. (1980). Melioration and behavioral allocation. In J. E. R. Staddon (Ed.), *Limits to action: The allocation of individual behavior* (pp. 143–176). New York: Academic Press.

Holloway, F. A., & Sturgis, R. D. (1976). Periodic decrements in retrieval of the memory of non reinforcement as reflected in resistance to extinction. *Journal of Experimental Psychology: Animal Behavior Processes, 2,* 335–341.

Holloway, F. A., & Wansley, R. A. (1973). Multiple retention deficits at periodic intervals after active and passive avoidance learning. *Behavioral Biology, 9,* 1–14.

Hunsicker, J. P., & Mellgren, R. L. (1977). Multiple deficits in the retention of an appetitively motivated behavior across a 24-h period in rats. *Animal Learning and Behavior, 5,* 14–26.

Ivry, R. (1993). Cerebellar involvement in the explicit representation of temporal information. In P. Tallal et al. (Eds.), *Temporal processing in the nervous system* (pp. 214–229). New York: New York Academy of Sciences.

Kamin, L. J. (1969). Predictability, surprise, attention, and conditioning. In B. A. Campbell & R. M. Church (Eds.), *Punishment and aversive behavior* (pp. 276–296). New York: Appleton-Century-Crofts.

Kehoe, E. J., Graham-Clarke, P., & Schreurs, B. G. (1989). Temporal patterns of the rabbit's nictitating membrane response to compound and component stimuli under mixed CS-US intervals. *Behavioural Neuroscience, 103,* 283–295.

Killeen, P. R. (1981). Averaging theory. In C. M. Bradshaw, E. Szabadi, & C. F. Lowe (Eds.), *Quantification of steady-state operant behavior.* (pp. 21–34). Amsterdam: Elsevier/North-Holland.

Koltermann, R. (1971). 24-Std-Periodik in der Langzeiterrinerung an Duft- und Farbsignale bei der Honigbiene. *Zeitschrift für Vergleichende Physiologie, 75,* 49–68.

Lea, S. E. G., & Dow, S. M. (1984). The integration of reinforcements over time. In J. Gibbon & L. Allan (Eds.), *Timing and time perception* (pp. 269–277). New York: Annals of the New York Academy of Sciences.

Lehrer, M. (1993). Why do bees turn back and look? *Journal of Comparative Physiology A, 172,* 549–563.

Levinthal, C. F., Tartell, R. H., Margolin, C. M., & Fishman, H. (1985). The CS-US interval (ISI) function in rabbit nictitating membrane response conditioning with very long intertrial intervals. *Animal Learning and Behavior, 13*(3), 228–232.

Mark, T. A., & Gallistel, C. R. (1994). The kinetics of matching. *Journal of Experimental Psychology: Animal Behavior Processes, 20,* 79–95.

Matzel, L. D., Held, F. P., & Miller, R. R. (1988). Information and expression of simultaneous and backward associations: Implications for contiguity theory. *Learning and Motivation, 19,* 317–344.

Mazur, J. E. (1989). Theories of probabilistic reinforcement. *Journal of the Experimental Analysis of Behavior, 51,* 87–99.

Mazur, J. E. (1991). Choice with probabilistic reinforcement: Effects of delay and conditioned reinforcers. *Journal of the Experimental Analysis of Behavior, 55,* 63–77.

McNaughton, B. L., Chen, L. L., & Markus, E. J. (1991). "Dead reckoning," landmark learning, and the sense of direction: A neurophysiological and computational hypothesis. *Journal of Cognitive Neuroscience, 3,* 190–202.

Meck, W. H., & Angell, K. E. (1992). Repeated administration of pyrithiamine leads to a proportinal increase in the remembered duration of events. *Psychobiology, 20,* 39–46.

252   C. R. Gallistel

Menzel, R., Chittka, L., Eichmüller, S., Geiger, K., Peitsch, D., & Knoll, P. (1990). Dominance of celestial cues over landmarks disproves map-like orientation in honey bees. *Zeitschrift der Naturforschung, C: Biosciences 45C*, 723–726.

Miller, R. R., & Barnet, R. C. (1993). The role of time in elementary associations. *Current Directions in Psychological Science, 2*, 106–111.

Morris, R. G. M. (1981). Spatial localization does not require the presence of local cues. *Learning and Motivation, 12*, 239–260.

Morris, R. G. M., Schenk, F., Tweedie, F., & Jarrard, L. E. (1990). Ibotenate lesions of hippocampus and/or subiculum: Dissociating components of allocentric spatial learning. *European Journal of Neuroscience, 2*, 1016–1028.

Müller, M., & Wehner, R. (1988). Path integration in desert ants, *Cataglyphis fortis*. 85, *Proceedings of the National Academy of Sciences*, USA. 5287–5290.

Muller, R. U., Kubie, J. L., Bostock, E. M., Taube, J. S., & Quirk, G. J. (1991). Spatial firing correlates of neurons in the hippocampal formation of freely moving rat. In J. Paillard (Ed.), *Brain and space*, (pp. 296–333). New York: Oxford University Press.

O'Keefe, J. (1990). A computational theory of the hippocampal cognitive map. *Progress in Brain Research, 83*, 301–312.

Ralph, M. R., Foster, R. G., Davis, F. C., & Menaker, D. M. (1990). Transplanted suprachiasmatic nucleus determines circadian period. *Science, 247*, 975–977.

Renner, M. (1960). Contribution of the honey bee to the study of time sense and astronomical orientation. *Cold Spring Harbor Symposia on Quantitative Biology, 25*, 361–367.

Rescorla, R. A. (1968). Probability of shock in the presence and absence of CS in fear conditioning. *Journal of Comparative and Physiological Psychology, 66*(1), 1–5.

Rescorla, R. A., & Wagner, A. R. (1972). A theory of Pavlovian conditioning: Variations in the effectiveness of reinforcement and nonreinforcement. In A. H. Black & W. F. Prokasy (Eds), *Classical conditioning II: Current research and theory* (pp. 64–99). New York: Appleton-Century-Crofts.

Roberts, S., & Church, R. M. (1978). Control of an internal clock. *Journal of Experimental Psychology: Animal Behavior Processes, 4*, 318–337.

Rosenwasser, A. M., Pelchat, R. J., & Adler, N. T. (1984). Memory for feeding time: Possible dependence on coupled circadian oscillators. *Physiology and Behavior, 32*, 25–30.

Saksida, L. M., & Wilkie, D. M. (1994). Time-of-day discrimination by pigeons, *Columbia livida*. *Animal Learning and Behavior, 22*, 143–154.

Schmidt, I., Collett, T. S., Dillier, F.-X., & Wehner, R. (1992). How desert ants cope with enforced detours on their way home. *Journal of Comparative Physiology A, 171*, 285–288.

Schneiderman, N., & Gormezano, I. (1964). Conditioning of the nictitating membrane of the rabbit as a function of CS-US interval. *Journal of Comparative and Physiological Psychology, 57*, 188–195.

Sharp, P. A., Kubie, J. L., & Muller, R. U. (1990). Firing properties of hippocampal neurons in a visually symmetrical environment: Contributions of multiple sensory cues and mnemonic processes. *Journal of Neuroscience, 10*(9), 3093–3105.

Sobel, E. C. (1990). The locust's use of motion parallax to measure distance. *Journal of Comparative Physiology A, 167*, 579–588.

Srinivasan, M. V. (1992). Distance perception in insects. *Current Directions in Psychological Science, 1*, 22–26.

Staddon, J. E. R. (1988). Quasi-dynamic choice models: Melioration and ratio invariance. *Journal of the Experimental Analysis of Behavior, 49*, 303–320.

Taube, J. S., Muller, R. U., & Ranck, J. B. (1990). Head-direction cells recorded from the postsubiculum in freely moving rats. I. Description and quantitative analysis. *Journal of Neuroscience, 10*(2), 420–435.

Turek, F. W. (1985). Circadian neural rhythms in mammals. *Annual Review of Physiology, 47,* 49–64.

von Frisch, K. (1967). *The dance-language and orientation of bees.* Cambridge, MA: Harvard University Press.

Wahl, O. (1932). Neue Untersuchungen über das Zeitgedächtnis der Bienen. *Zeitschrift für Vergleichende Physiologie, 16,* 529–589.

Wansley, R. A., & Holloway, F. A. (1975). Multiple retention deficits following one-trial appetitive training. *Behavioral Biology, 14,* 135–149.

Wehner, R. (1981). Spatial vision in arthropods. In H. Autrum (Ed.), *Comparative physiology and evolution of vision in invertebrates* (pp. 287–617). New York: Springer-Verlag.

Wehner, R. (1987). 'Matched filter'—neural models of the external world. *Journal of Comparative Physiology A, 161,* 511–531.

Wehner, R. (1992). Homing in arthropods. In F. Papi (Ed.), *Animal homing* (pp. 45–144). London: Chapman & Hall.

Wehner, R., Bleuler, S., Nievergelt, C., & Shah, D. (1990). Bees navigate by using vectors and routes rather than maps. *Naturwissenschaften, 77,* 470–482.

Wehner, R., & Menzel, R. (1990). Do insects have cognitive maps? *Annual Review of Neuroscience, 13,* 403–414.

Wehner, R., & Srinivasan, M. V. (1981). Searching behavior of desert ants, genus *Cataglyphis* (*Formicidae,* Hymenoptera). *Journal of Comparative Physiology, 142,* 315–338.

Wehner, R., & Wehner, S. (1990). Insect navigation: Use of maps or Ariadne's thread? *Ethnology, Ecology and Evolution, 2,* 27–48.

Wilkie, D. M., Saksida, L. M., Samson, P., & Lee, A. (1994). Properties of time-place learning by pigeons, *Columbia livida. Behavioral Processes, 31,* 57–74.

Wilson, M. A., & McNaughton, B. L. (1993). Dynamics of the hippocampal code for space. *Science, 261,* 1055–1057.

Zeil, J. (1993a). Orientation flights of solitary wasps (*Cerceris;* Sphecidae; Hymenoptera). I. Description of flight. *Journal of Comparative Physiology A, 172,* 189–205.

Zeil, J. (1993b). Orientation flights of solitary wasps (*Cerceris;* Sphericidae; Hymenoptera). II. Similarities between orientation and return flights and the use of motion parallax. *Journal of Comparative Physiology A, 172,* 207–222.

CHAPTER 9

Animal Memory

The Effects
of Context Change on
Retention Performance

William C. Gordon
Rodney L. Klein

I. INTRODUCTION

In 1969 a symposium held at Dalhousie University brought together several researchers interested in various aspects of memory in nonhuman species. The contributions to this symposium were collected and edited by W. K. Honig and P. H. R. James, and the resulting book, *Animal Memory*, was published in 1971. In many ways the publication of this volume signaled the beginnings of an active and widespread interest in how animals form memories, how they remember and why they forget.

Prior to this landmark book, relatively few researchers interested in animal behavior concerned themselves with memory processes such as information storage, encoding, or retrieval, with issues such as the nature of memory representations, or with the factors that influence retention of learned responses. Instead, the focus for most researchers was how animals learn or how they acquire new, adaptive responses. For the most part, once an acquisition experience had ended, an animal's failure to perform in accord with that acquisition experience was assumed to be the result of either motivational fluctuations or changes in stimulus control.

In the years since 1971, this focus has changed considerably. We now have theoretical accounts of how animals store representations of experience in memory (e.g., McGaugh & Dawson, 1971; McGaugh & Herz, 1972), we

Animal Learning and Cognition

have a body of experimental literature concerning tests of such accounts (e.g., Calhoun, 1971; Gibbs & Mark, 1973), and we have numerous researchers interested in the biological underpinnings for memory storage (e.g., Squire, 1987). Likewise, other researchers in this area have proposed models that emphasize memory encoding processes (e.g., Lewis, 1979) and still others have concerned themselves with models of memory retrieval (e.g., Spear, 1971, 1973). One relatively recent interest has been whether all memory processes in animals can be characterized as automatic or whether animals might also engage in controlled processing (e.g., Grant, 1981; Maki, 1981).

In addition to this emphasis on how animals store, encode, and retrieve information, there are now researchers concerned with the nature of an animal's memory representations (e.g., Grant, 1981; Honig, 1978; Roitblat, 1980; Spear, 1978). Others have focused on such issues as how animals remember spatial locations both in structured laboratory settings (e.g., Olton & Samuelson, 1976) and in more natural environments (e.g., Balda & Turek, 1984; Tomback, 1983). And, there is continuing interest in the causes of forgetting in a variety of experimental paradigms (e.g., Gleitman, 1971; Grant, 1981; Spear, 1978).

Until recently, these and other lines of animal memory research had remained somewhat distinct from traditional studies of animal learning. This, however, is no longer the case as an increasing number of animal learning theorists have begun to incorporate memory concepts into their models of acquisition. It is now quite common, for example, for explanations of animal learning to refer to CS or US representations being acquired at the time of learning (e.g., Rescorla & Durlach, 1981). It is equally common for contemporary learning models to rely heavily on notions of how animals process memories during acquisition (e.g., Wagner, 1981). In effect, over the past ten years, traditional studies of animal learning and the more recent interest in animal memory have begun to merge in interesting and productive ways.

Given the enormous body of animal memory literature that now exists, it would be impossible to present a meaningful overview of this literature in a chapter of this scope. For this reason we have decided to restrict our discussion in the present chapter to a single, fundamental issue—the conditions that influence the retention performance of animals. Specifically, we will focus on how retention performance is affected by changes in an organism's environment between the time of learning and a subsequent retention test.

II. EFFECTS OF CONTEXT CHANGE ON HUMAN VERBAL RETENTION

In studies of human verbal retention it is clear that retention success depends critically on the similarity between learning and testing conditions. For example, subjects who learn verbal materials in a particular external context tend to remember these materials best when testing occurs in that same context. Substantial differences between the learning and testing environments regularly result in significant retention deficits (e.g., Godden & Baddeley, 1975; Greenspoon & Ranyard, 1957; Smith, Glenberg, & Bjork, 1978). Similarly, retention deficits often result from learning–test differences in a subject's internal state. These deficits occur regardless of whether these internal state changes are naturally occurring (e.g., Weingartner, Miller, & Murphy, 1977) or artificially induced (e.g., Eich, Weingartner, Stillman, & Gillin, 1975; Petersen, 1974). Still other studies have shown that poor retention can result from a difference in the specific words accompanying a target word during learning and testing (e.g., Tulving & Osler, 1986) and from presenting a target word in different kinds of verbal contexts during learning and testing (Morris, 1978). In effect, even when verbal target items are well learned, changes in what might be considered irrelevant aspects of the learning situation often produce significant difficulties in the recall or recognition of these items on a later test.

In an effort to explain these effects, Tulving and his colleagues have proposed a theoretical account that has come to be known as the encoding specificity hypothesis (cf. Flexser & Tulving, 1978; Tulving, 1983; Tulving & Thomson, 1973; also see Kintsch, 1974). According to this hypothesis, the poor retention performance that results from various learning–test differences can best be explained as an instance of retrieval failure at the time of testing. This view states that during a learning episode a human does more than simply form memory representations that are direct physical replicas of the target items presented for learning. Instead, a subject's representation of a learning experience depends critically on the nature of the learning context. First, the learning context influences a subject's perception of the target items and, thus, the representation of the target items themselves. Additionally, it is assumed that a subject forms representations of the contextual cues noticed during learning and that these contextual representations become connected to or encoded along with the representations of the target items. Thus, a subject's memory for a learning episode is actually a complex, multidimensional memory comprised of interconnected target item and contextual representations.

Central to this encoding specificity hypothesis is the idea that the subsequent retrieval of some stored target item depends heavily on the similarity between events present at the time of learning and those that occur on a retention test. If a substantial portion of the learning context remains present during testing, then a subject will tend to perceive a target item on the test in the same way the item was perceived during learning and activation of the target item representation will be more likely. In addition, strong similarities between the learning and testing contexts will help to ensure that the contextual representations in the training memory become activated during testing and this also should increase the likelihood that the associated target representations will be retrieved. Thus, according to this hypothesis, differences between a subject's learning and testing contexts should invariably reduce the probability that a target item will be retrieved at the time of a retention test.

In the years since it was first proposed, the encoding specificity hypothesis has gained wide acceptance among researchers interested in human retention performance. However, attempts to apply this hypothesis to the retention performance of animals (see, e.g., Spear, 1973, 1978) have been more problematic. First, it is not at all clear that changes in an animal's learning context invariably produce retention deficits. Second, even when context changes do result in performance deficits, there is substantial disagreement about the mechanisms responsible for such deficits.

In the present chapter we examine the applicability of the encoding specificity hypothesis to the retention performance of nonhuman organisms. We begin by looking at the conditions under which retention performance is affected by context change. Then we assess the various accounts that have been proposed to explain these effects.

III. INFLUENCE OF CONTEXT ON THE RETENTION PERFORMANCE OF ANIMALS

In our discussion of contextual influences on retention performance we focus on three categories of experiments. First, we review the evidence that suggests that retention performance can be reduced by training animals to perform a response in one context and then testing them in a different environment. Second, we review studies in which the test context appears to determine which of several responses are performed when animals have previously acquired multiple responses in different contexts. Finally, we note some specific conditions under which contextual control of retention appears not to occur.

A. Changes in the External Learning Context following Acquisition of a Single Response

There is substantial evidence to indicate that alterations in an animal's external learning environment prior to testing can adversely affect the animal's retention of a previously learned response. In one early study conducted by Steinman (1967), for example, rats were trained to traverse a straight alleyway for a food reward. Half the animals received training in a relatively bright alleyway, while the others were trained in a dark apparatus. Animals tested for retention of the running response one day after the end of acquisition performed well as long as the test occurred in the same alleyway that was used for acquisition. However, rats that were trained in one alleyway and then tested in the alternate apparatus exhibited significant reductions in running speed (also see Perkins & Weyant, 1958). It is notable that similar changes in static apparatus cues have been shown to disrupt retention of an appetitively motivated T-maze discrimination in rats (Chiszar & Spear, 1969) and a learned key-peck response in pigeons (Riccio, Urda, & Thomas, 1966).

Alterations in static background cues also appear to have a deleterious effect on retention of instrumental avoidance responses. In a study conducted by Gordon, McCracken, Dess-Beech, and Mowrer (1981), rats were trained to avoid shock by exiting a white chamber and entering a black compartment within 5 s after the onset of a flashing light. Although all animals were trained to criterion in one of two identically constructed avoidance chambers, half the animals were trained in one distinctive room and the remainder were trained in a different room. The two training rooms differed in terms of size, illumination, odor cues, and background noise. Twenty-four hours after the conclusion of training all rats received an avoidance test either in the original training context or in the alternate context. Those rats that were tested in the same room in which training occurred had a mean avoidance latency of approximately 4 s on the first test trial. However, the mean avoidance latency for rats tested in a different environment was between 12 and 13 s. Clearly, a change in room characteristics between learning and testing can produce substantial and significant retention deficits even when few differences exist between the apparatus used for training and testing.

It is important to note that retention deficits of the kind cited above appear not to be limited to instrumental learning procedures. McAllister and McAllister (1963) have reported that when rats receive CS–shock pairings in one context and then are tested in a different context to determine if they will escape from the previously trained CS, escape latencies are prolonged relative to animals tested in the learning environment. Similarly, rats given CS–shock pairings in one context and then tested for CS-induced suppression of ongoing behavior in another context have been shown to suppress to a lesser degree than animals trained and tested in the same environment (e.g., Balaz, Capra, Hartl, & Miller, 1981). Retention deficits due to external context change have also been found in flavor aversion learning paradigms (e.g., Archer, Sjödén, & Nilsson, 1984), in paradigms involving conditioned approach responses to a food location (e.g., Hall & Honey, 1989), in conditioned "head-jerk" procedures (e.g., Peck & Bouton, 1990), and in autoshaping experiments using pigeons as subjects (Honey, Willis, & Hall, 1990).

One final Pavlovian conditioning procedure that appears to be particularly sensitive to the effects of external context change is the latent inhibition paradigm (Lubow & Moore, 1959). It is now well established that the conditioning of a discrete CS can be retarded significantly if organisms receive substantial exposure to that CS before conditioning trials commence. However, if CS exposures and conditioning trials occur in different contexts, the retardation of conditioning that normally results from these exposures can be dramatically diminished (e.g., Channel & Hall, 1983; Gordon & Weaver, 1989; Hall & Minor, 1984; Lovibond, Preston, & Mackintosh, 1984). This context specificity of CS pre-exposure effects has been demonstrated in a variety of conditioning paradigms.

B. Changes in an Animal's Internal State following Learning of a Single Response

The static background cues present during a learning experience are not restricted to stimuli in a subject's external environment. Other cues that arise from within an organism can also help to comprise the learning context. Based on the evidence reviewed above, one might predict that retention deficits should occur as a result of internal cue changes as well as when external cues are altered.

One method used to test this prediction has been to manipulate an animal's internal state directly by injecting the animal with a particular drug or with a placebo prior to either a learning or a retention test procedure. While not all drugs are effective in producing discriminably different internal cue states, many of those drugs that do produce such an effect have been shown to control an organism's retention of a learned response (Overton, 1972, 1985). In one particularly elegant demonstration of this effect, Bliss (1973) trained monkeys to perform a color discrimination for a food reward after being injected with either sodium pentobarbital or a saline solution. Later these animals were tested for retention following injection with either the same solution given prior to training or the alternate solution. Animals that learned and were tested in the same drug state performed the discrimination well on the test, while animals that learned and were tested in different drug states exhibited substantial retention deficits. Analogous findings were reported using a spatial discrimination task rather than a color discrimination. Such "state-dependent" learning effects have been reported by experimenters utilizing a range of species and learning paradigms as well as a variety of specific drugs (see Overton, 1972).

A second way of assessing the effects of internal cue states on learned behavior is to determine whether changes in naturally occurring states diminish retention performance. One set of data that lends itself to an interpretation in terms of natural state fluctuations has been reported by Holloway and Wansley (1973). These experimenters trained rats to passively avoid a given shock chamber and then tested for retention of the avoidance response after a variety of retention intervals. They found that avoidance performance was excellent if animals were tested either 24, 48, or 72 hours after learning. Performance was diminished, however, after retention intervals that precluded testing at the same time of day as training. Although training and testing at substantially different times of day might be expected to produce learning-test differences in a variety of contextual stimuli, it is well known that many biological functions follow a 24-hour circadian cycle. Thus, it is likely that when animals are tested at the same time of day as when they were trained, they experience a similar internal state during training and testing. Different internal states would be expected when training and testing occur at different times of day. It is likely that in the Holloway and Wansley experiment, the dramatic fluctuations found in retention performance were at least partially due to the circadian fluctuations in the internal state of the organisms (see also Stroebel, 1967).

Aside from the potential effects of circadian fluctuations on retention performance, evidence suggesting that naturally occurring state changes can produce retention deficits can be found in at least two other bodies of literature. First, there is compelling evidence to indicate that the poor retention of avoidance responding normally observed 1–6 hours after training (the "Kamin effect") is the result of changes in an animal's hormonal state triggered by the stress involved in learning (Dunn & Leibmann, 1977; Klein, 1972, 1975; Spear, Klein, & Riley, 1971). Second, there is some indication that different levels of hunger in animals can produce internal states that will influence retention performance. For example, in a study by Yamaguchi (1952) different groups of rats learned to press a lever for a food reward after having been deprived of food for varying periods of time. Yamaguchi found that retention of the lever press response was best when animals were deprived of food for the same period prior to testing as they were prior to learning. Both greater and lesser periods of pretesting deprivation led to reductions in test performance.

C. Retention of Multiple Responses Learned in Different Contexts

All of the studies cited thus far involved training animals to perform a single response in the presence of one set of contextual cues and then testing for retention of that response in either the same or some different context. Although these context-shift experiments do provide evidence suggesting that retention-test context can influence retention performance, some of the strongest support for this conclusion comes from studies in which multiple responses are learned in different contexts. What is most often found in these experiments is that when animals are placed in one of the contexts previously experienced, the response learned in that context tends to occur with a much greater probability than other learned responses. In other words, each context appears to select or retrieve the particular response appropriate to that context.

One of the clearest examples of this finding comes from Pavlovian conditioning experiments that utilize what are called "switching procedures" (see Asratyan, 1965). In one recent autoshaping experiment of this kind reported by Rescorla, Durlach, and Grau (1985), pigeons were exposed to two distinctive contexts during the course of training. In context 1, the birds experienced two key lights (A and B), with A always followed by food reinforcement and B never paired with food. In context 2 these relationships were reversed. That is, B was consistently paired with food and A was always presented alone. Because in this study organisms received equivalent exposure to the two contexts and equivalent reinforcement in the presence of each key light and each context, the question of interest was whether differential responding to the two key lights would develop in a manner that was appropriate to the contingencies existing in each context. The answer to this question was clearly affirmative. Within a relatively few training sessions the placement of the pigeons into context 1 resulted in a high rate of responding to key light A and only minimal responding to key light B. However, when pigeons were placed in context 2, the pecking of key light B was far greater than that to key light A. In effect, each context seemed to signal which key light would be reinforced in its presence and the pigeons learned to respond accordingly.

A second example of this type of effect comes from studies in which animals are trained to emit different and conflicting responses each in a particular context. In one such study (Spear et al., 1980), rats were initially given passive-avoidance training followed 24 hours later by active-avoidance training. Both types of training took place in the same avoidance apparatus, which contained a white start chamber connected to a black compartment. During passive-avoidance training rats learned to remain in the white compartment in order to avoid being shocked in the black chamber. Active-avoidance training involved learning to move to the black chamber in order to avoid a shock in the white start chamber.

Half of the animals in this experiment learned passive avoidance after being injected with sodium pentobarbital, while the remaining animals received an injection of saline solution prior to learning. On day two of the experiment (just prior to active-avoidance acquisition) each rat that had received sodium pentobarbital on day 1 received a saline solution, while the animals that had been trained to passively avoid after saline were injected with sodium pentobarbital. Thus, each rat learned one response after the injection of saline and the other response after receiving pentobarbital.

Twenty-four hours after active-avoidance learning all rats were given a nonshock retention test in the avoidance apparatus to determine whether they would perform in accord with the passive or active avoidance contingencies. However, before being tested, half the animals were injected with the same solution they had received prior to passive-avoidance training and the other half received the solution that had accompanied the active-avoidance sessions. On this test, the rats overwhelmingly performed in accord with the drug state they were in at the time of training. Animals tested in the same drug state that had been present during passive-avoidance training remained in the white compartment of the apparatus for long durations before moving to the black compartment. Animals tested in the active-avoidance drug state quickly escaped to the black compartment on the test. It is notable that these experimenters reported almost identical results from a study in which different rooms were used for each type of training rather than different drug states.

These findings, as well as similar ones (e.g., Gordon, Mowrer, McGinnis, McDermott, 1985; Overton, 1964, 1966; Spear, 1971), make it clear that when organisms have learned conflicting responses in different contexts, the response expressed on a subsequent test is strongly determined by the test context. The greater the match between the test context and the context present during the learning of one of the responses, the greater the likelihood that that response will occur at the time of testing.

Finally, there is a growing body of evidence indicating that when a Pavlovian CR is acquired in one context and extinguished in another, an organism's subsequent performance of that CR depends on whether the CS is presented in the learning or the extinction context. Subsequent CS presentations in the extinction context regularly result in little conditioned responding. However, if an organism is returned to the learning context and the CS is presented, the original CR tends to be "reinstated" (e.g., Baker, Steinwald, & Bouton, 1991; Bouton & Bolles, 1979; Bouton & King, 1983; Bouton & Peck, 1989). Furthermore, this contextual reinstatement following extinction does not appear to be restricted to the extinction of a CS–US association. In a recent master's thesis project conducted in our laboratory, Nelson found that if an organism forms a CS–CS association in one context and this association is then dissolved or "extinguished" in a different context (by single CS presentations), evidence for that CS–CS association may still be found in the original training context. Apparently, whatever an organism learns during an extinction session comes under at least the partial control of the context in which extinction occurs.

D. Limitations on the Contextual Control of Retention Performance

The evidence we have summarized thus far would seem to suggest that learned responses in animals are virtually always context specific. Certainly, these studies indicate that under a variety of conditions responses acquired in one context tend to be expressed with a higher probability when animals are tested in that same or a similar context. However, this is not a conclusion that is universally supported in the literature. Several experimenters have reported results indicating that learned responses can transfer rather easily to new contexts given appropriate parameters or conditions.

One parameter that often influences contextual control of retention performance is the duration of a retention interval. In general, learning–test differences in context tend to produce rather substantial retention deficits if testing occurs shortly after learning. However, as a retention interval is lengthened, the detrimental influence of context change often tends to diminish and it sometimes disappears when retention intervals are relatively long. For example, in the study by McAllister and McAllister (1963) cited earlier, a change in apparatus cues produced a significant deficit in conditioned emotional response (CER) performance when animals were tested 3 min after the conclusion of training. However, in this same experiment, some animals were tested a full 24 hours after training and these rats exhibited virtually no deficit in retention as the result of context change. Similar findings have been reported by Perkins and Weyant (1958) and by Steinman (1967), using appetitively motivated learning tasks. Still, despite this general pattern of effects, several studies have suggested that contextual control over learned responding can be maintained over retention intervals as long as several weeks (e.g., Brooks & Bouton, 1993; Peck & Bouton, 1990; Thomas, Moye, & Kimose, 1984). Thus, while the use of long retention intervals may sometimes diminish the effects of context change, this diminution does not invariably occur.

A second factor that may prove to influence the context specificity of at least a Pavlovian fear CR is degree of learning. For example, Hall and Honey (1990) note that while context change manipulations have been shown to disrupt CER performance in some experiments (e.g., Balaz et al., 1981; Balaz, Capra, Kasprow, & Miller, 1982) several experimenters have reported findings suggesting that context change has little or no effect on retention of a CER (e.g., Bouton & King, 1983; Hall & Honey, 1989; Lovibond et al., 1984). In an attempt to determine why so many experiments using CER procedures have failed to find evidence for context specificity, Hall and Honey (1990) conducted a CER experiment in which they varied the number of CS–shock pairings animals received prior to testing. They found that following a single CS–shock pairing, rats exhibited substantial performance suppression on the retention test if the learning and test contexts were the same. However, when the retention test context was different from that of training, little suppression was evident at the time of testing. This finding contrasted markedly with the suppression performance of animals given multiple CS–shock trials (the procedure used in most studies of this type). Those animals showed clear suppression regardless of the nature of the test context. In effect, this study indicates that a learned CER can be specific to the context in which it was acquired, but that demonstrating this specificity may depend on the use of a minimal number of learning trials.

This finding by Hall and Honey (1990) also bears on a third factor that sometimes seems to affect whether context change reduces performance of a Pavlovian CR. This factor is the nature of the US used in a conditioning procedure. Several experimenters have shown that when an appetitive US is employed in conditioning, the resulting CR undergoes significant disruption when the learning context is changed (e.g., Hall & Honey, 1989; Honey et al., 1990; Peck & Bouton, 1990). However, as we have just noted, this finding is somewhat less common when aversive USs are employed. As Hall and Honey (1990) have pointed out, this difference may be due to the fact that CRs tend to develop differently in appetitive and aversive conditioning paradigms. In appetitive paradigms the CR continues to grow over a number of trials, and once asymptote is reached, the level of responding is maintained. However, when aversive USs are employed, the CR tends to grow rapidly and then diminishes after asymptotic performance is reached.

Based on this pattern of CR development, Hall and Honey speculate that when multiple trials involving an aversive US are used, organisms may acquire some response that actually opposes the excitatory CR. Thus, when a context change occurs, the diminution of the excitatory CR may be offset by a corresponding reduction in the opposing response. For this reason, a change in the CER context following multiple conditioning trials may result in little perceptible change in the excitatory response conditioned, while a context change after only a few conditioning trials leads to a clear reduction in CER performance. Whether this or some other interpretation proves most compelling, it does appear that some types of CRs are more readily disrupted by context change than are others (e.g., Hall & Honey, 1990).

E. Conclusions

The existing literature provides ample evidence to indicate that the retention performance of animals is often influenced by the similarity that exists between learning and test contexts. Specifically, responses acquired in one context tend to be expressed more fully when the retention test context is highly similar to the context of learning, and dramatic differences between learning and test contexts often result in significant retention deficits. Still, it is clear that retention performance can sometimes survive even substantial changes in the learning context. Thus, any comprehensive account of contextual influences on retention must be able to explain both the poor retention that often results from context change and the failure under some conditions to find such an effect. In the following section we examine some of these accounts.

IV. CONTEXT SPECIFICITY EXPLANATIONS

As we have noted earlier, the encoding specificity hypothesis was developed in an effort to explain the poor retention performance exhibited by human subjects who experience context change. Thus, it is hardly surprising that similar hypotheses have been advanced to account for context-specific retention in animals (e.g., Spear, 1971, 1973, 1978). In the present section we assess the applicability of this hypothesis to the animal retention data. First, however, we look at some of the alternative accounts for the findings reviewed above.

A. Reactions to Novel Environments and Altered Conditioned Stimulus Perceptions

In 1984 Lovibond et al. pointed out that some of the performance deficits seen in context-shift experiments may result from rather trivial or theoretically uninteresting factors (also see Spear, 1978). First, they noted that in many context-change experiments animals are trained in one context and then are tested in a context that is not only different from the training environment, but is also a novel environment for the animals. Since it is well known that animals often exhibit freezing or exploratory behaviors in novel contexts, this leaves open the possibility that in some context-shift experiments responses to novelty are occurring during testing and are interfering with an animal's tendency to perform the learned target response.

A second factor that they suggested may be at work in such studies is a decrement in stimulus generalization. If, for example, an animal learns to respond to a CS in one context and then the CS occurs in a radically different test context, the difference between the contexts may alter the animal's ability to fully recognize the training CS, leading to a stimulus generalization decrement. Clearly, training and testing contexts that differ in illumination might be expected to alter the perceived brightness of a visual CS, just as changes in the size or physical construction of an apparatus would be expected to affect the perceived loudness of an auditory stimulus. Obviously, any training–test differences that alter the way a CS or some other critical stimulus is perceived should lead to reductions in learned responding on the basis of a generalization decrement alone.

Given the obvious nature of these potential factors in context-change experiments, it is surprising that relatively few experiments of this type actually include formal control conditions that would allow the impact of these factors to be assessed. However, those experiments that do take such factors into account often find that they contribute only minimally to the retention deficits animals exhibit. For example, in one experiment reported by Lovibond et al. (1984, Experiment 1b), two distinctive contexts were employed. Although the two contexts differed along a number of dimensions (e.g., illumination, olfactory cues), none of these differences was likely to produce differential perceptions of the tone CS utilized in the experiment. All animals in this study were given tone–shock pairings in one context and received equal exposure to the alternate context to ensure that the animals were equally familiar with the two environments prior to testing. During a subsequent conditioned suppression test some animals were tested in the training environment, while the remainder were tested in the alternative context. Those animals trained and tested in the same context exhibited significantly greater suppression than animals trained and tested in different contexts.

Obviously, this result does not imply that novelty responses or changes in CS perception never contribute to the retention deficits found in context-change experiments. As a matter of fact, under some conditions these factors may account for a substantial portion of the performance deficit an animal exhibits in a novel test context. However, these and other similar findings (e.g., Hall & Honey, 1990) indicate that the context specificity of a conditioned response can be clearly demonstrated even when these factors are well controlled. In short, it is not feasible to account for the retention deficits found in context-shift experiments solely on the basis of novelty responses or altered CS perceptions at the time of testing.

B. Loss of Contextual Excitatory Strength

One assumption common to many contemporary conditioning models is that during a conditioning experience contextual stimuli can acquire the capacity to elicit a CR via direct associations with the training US (e.g., Pearce & Hall, 1980; Rescorla & Wagner, 1972). This assumption is supported by the results of several experiments in which direct measures of contextual conditioning have been conducted (e.g., Balsam & Tomie, 1985). One implication of this finding is that when animals are given a retention test involving CS presentations in the training context, the resulting CR is most likely based on the joint excitatory strength held by the CS and the training context. However, when an animal is tested by being given CS presentations outside the training context, the CR should be diminished because of the loss of excitatory strength that results from the absence of the training context. In effect, this position assumes that the poor retention found in context-shift experiments can be explained by the loss of excitatory strength that occurs when excitatory contextual stimuli are eliminated during testing.

Although this hypothesis has great logical appeal, it does not fare well as a general account of context-specificity effects. First, as we have noted, one of the strongest and most consistent examples of context specificity comes from studies of latent inhibition in which the CS pre-exposure and the training phases occur in different contexts. Whatever may be learned during CS pre-exposure that transfers negatively to subsequent conditioning clearly does not involve context–US associations. Thus, in such paradigms a change of context could not be producing a retention deficit through any reduction of excitatory contributions by contextual stimuli. If we assume that this mechanism is responsible for retention deficits in other types of context-shift experiments, then some separate account of the context specificity of latent inhibition would be needed (e.g., Wagner, 1981).

A second problem with this account arises from studies that employ parameters or manipulations specifically designed to minimize context–US associations during conditioning (cf. Miller & Schachtman, 1985). These studies indicate that even when such manipulations are utilized and even when measurable context–US associations have been eliminated prior to testing, animals still perform better in the conditioning context than they do when the test context differs from that of training. In other words, these studies report the existence of substantial context specificity even when the test contexts employed differ little in excitatory value.

Finally, we have referred to a number of studies in which animals receive one CS (CSa) paired with the US in one context and a second CS (CSb) paired with the same US in a different context (e.g., Hall & Honey, 1990; Rescorla et al., 1985). In such studies both contexts should accrue approximately the same excitatory value. However, when animals are tested in these studies they perform well to CSa when it occurs in its own training context, but not when it occurs in the context used for training CSb. Such a finding is difficult to reconcile with the notion that retention deficits in context-change experiments result from reductions in the excitatory value of the test context.

C. Alteration of an Excitatory Configural Cue

A third potential account of context specificity is based on the idea that the effective CS in a conditioning situation is not the discrete CS itself, but is, in actuality, a configural stimulus comprised of the CS and the context in which that CS occurs (see Pearce, Chapter 5, this volume). To the extent that this is true, it follows that if a discrete CS is presented outside the context of training, the contextual components of the configural cue would be missing and a generalization decrement should occur. Such an explanation would seem to account for the context specificity of simple conditioning and latent inhibition, as well as the findings from differential training procedures such as those used in the switching paradigm (see Asratyan, 1965).

Still, the configural account of context specificity is not without difficulties of its own. For example, Hall and Honey (1989) report two context-change experiments that are identical except for the fact that one study utilizes an appetitive US, while in the other an aversive US is used. In the study using an appetitive US, clear performance deficits were found as a result of context change. However, in the second experiment, there was no evidence of such an effect. These two studies used identical CSs and contextual stimuli, therefore, it is unclear why one should expect that a configural cue would form in one of these studies, but not in the other. For the configural cue account to hold, there would need to be some explanation why configuration is more common with an appetitive US than with an aversive US.

A second difficulty with the configural account is made evident in a subsequent series of studies by Hall and Honey (1990). In these experiments Hall and Honey found clear evidence of a context-specificity effect using a CER paradigm when only a single CS–shock pairing preceded testing in a nontrained context. However, control animals that received a single non-reinforced CS presentation in one context and were then tested in a second context exhibited no loss of habituation to that CS as a result of context change. If a single CS presentation is adequate to allow for the formation of a configural cue involving the CS and context, then a change in context should have produced a generalization decrement in habituation as well as in conditioned responding. However, no evidence for such a generalization decrement was obtained, suggesting that the context specificity of the CR was not due to a generalization decrement per se.

D. Encoding Specificity Hypothesis

In the version of the encoding specificity hypothesis proposed by Spear (1973, 1978; see also Estes, 1973; Medin, 1975; Nadel & Willner, 1980), at the time of a learning experience an animal forms representations of the learning context that are integral to the animal's memory for training. It is assumed that these contextual representations are linked in some manner to the animal's representation of the target association. According to Spear, subsequent retrieval of the target association depends to some degree on the activation of the contextual representations that are linked with the target. Thus, if training occurs in one context and a different context is used for testing, performance will be relatively poor because the cues in the test environment will fail to activate the contextual representations in the training memory and the activation of the associated target will be less likely. Conversely, if training and testing occur in the same environment, the contextual representations in the training memory will be activated by the test cues and this will lead to the enhanced activation of the target association linked to these contextual representations. In effect, this view assumes that the poor retention often seen when animals are tested outside the training environment is the result of a failure to retrieve adequately the previously acquired target association.

Clearly, this hypothesis is capable of accounting for the fact that many learned responses appear to be context specific. Additionally, such an account is bolstered by the fact that many instances of forgetting in animals seem to result from inabilities to access information that is clearly stored in memory (e.g., Spear, 1978). The major difficulty with this type of explanation is that it attempts to account for retention performance solely on the basis of the degree of similarity between learning and testing environments. Thus, according to this view, one would always expect retention performance to be optimal when learning and testing contexts are highly similar. Likewise, this hypothesis assumes a monotonic decrease in retention performance as learning and testing conditions become increasingly dissimilar. In effect, while this hypothesis assumes that context change produces poor retention through the mechanism of retrieval failure, it focuses almost entirely on the similarity of learning and testing conditions in predicting retrieval success.

By emphasizing the role of context similarity so strongly, the encoding specificity hypothesis tends to ignore other factors or conditions that might also affect the probability of successful retrieval. For example, one might expect that the retrieval of a target association would vary depending on the distinctiveness of that association in memory, regardless of the context used for testing. Similarly, it would seem reasonable to expect that the impact of context on retention would depend on how strongly linked the contextual and target representations are in a given situation. The encoding specificity hypothesis includes no assumptions concerning the role such factors might play in target retrieval and how such factors might interact with the degree of similarity between learning and test contexts.

Because this hypothesis fails to recognize the myriad factors that can affect retrieval and fails to specify how various retrieval factors might interact to affect retention performance, this approach cannot account for the fact that good retention performance sometimes occurs even when the learning and test contexts differ dramatically. In effect, the predictive power of the encoding specificity hypothesis is limited by its lack of specificity and its restricted scope. Clearly, this hypothesis is singularly successful in providing an adequate explanation for why context change might produce poor retention. However, as it is currently formulated, it cannot stand as a comprehensive account of contextual effects on retention performance.

V. UNDERSTANDING CONTEXTUAL EFFECTS ON RETENTION: CURRENT DIRECTIONS

The strength of the encoding specificity hypothesis is its reliance on retrieval failure as the mechanism underlying the retention deficits that result from context change. However, it is clear that the predictive success of this hypothesis is limited due to its restricted emphasis on the similarity of learning and test contexts as the primary factor governing retrieval. If a retrieval failure explanation of context specificity is to be retained, then, at the very least, we must be able to delineate the conditions under which context change should *not* result in a retrieval failure or a retention deficit. At least three such conditions have been suggested by recent work in this field.

First, Bouton and his colleagues (cf. Bouton & Bolles, 1985; Bouton & King, 1986; Swartzentruber & Bouton, 1992) have noted that a retention test context often has robust effects on performance of a Pavlovian CR, if a CS has previously been associated with different outcomes during training. Specifically, if organisms experience one CS outcome in one context and a different outcome to the same CS in a second context, later responding to that CS is strongly determined by the context of the retention test. Bouton cites as examples the switching paradigm described earlier, the situation in which acquisition and extinction occur in different contexts, and even the latent inhibition paradigm. Similar examples also come from instrumental learning procedures in which conflicting responses are acquired in different environments. Bouton suggests that such differential training leads to much stronger contextual control than that found when organisms learn a single response in one distinctive context.

Based on this conclusion Bouton has proposed that contextual control of responding is heightened when organisms are forced to utilize contextual stimuli to disambiguate a CS or a response requirement. In other words, when a given CS or response can lead to different outcomes and when particular contexts are specifically associated with these various outcomes, the contexts are relied upon to signal or retrieve the appropriate response or CS outcome in a given situation. According to Bouton, in such cases the context may function much like a facilitating or "occasion-setting" stimulus serving to instruct an organism as to which of several contingencies is in effect at a given time (cf. Holland, 1983; Rescorla, 1985).

This notion that contextual stimuli may serve an occasion-setting function under differential training conditions is not inconsistent or at odds with the encoding specificity idea, since Bouton still assumes that contextual stimuli serve to select or retrieve specific representations at the time of a retention test. However, Bouton's proposal suggests that a history of differential training may be necessary in order for contextual stimuli to achieve this capacity. This potential modification of the encoding specificity idea is interesting and it deserves greater experimental attention.

A second idea that could help the encoding specificity hypothesis deal more effectively with the existing data has been suggested by Hall and Honey (1989, 1990). According to these researchers, what appears to be a lack of context specificity in some situations may actually be a lack of understanding about what an organism is learning in a given situation. In other words, an organism may sometimes learn multiple (and often conflicting) responses in a given situation and all of these responses may be specific to the learning context. When this occurs, the effects of a context change may be complex and difficult to predict. In effect, any test of the encoding specificity hypothesis requires a complete understanding of what an organism is learning and what responses might be altered at the time of a context change.

A third possible consideration that might contribute to our understanding of context specificity is suggested by some recent experiments conducted in our own laboratory (see Klein & Gordon, 1993). These experiments were prompted by the growing literature concerning the existence of CS–CS and CS–context associations in Pavlovian conditioning (e.g., Rescorla & Durlach, 1981). In these studies we were concerned with the question of whether individual contextual stimuli become associated with each other during a conditioning episode.

While the answer to this question appears to bear most directly on the issue of associative structure in Pavlovian conditioning, it also has implications for what should occur in a context-shift experiment. For example, in most context-shift experiments, an attempt is made to alter many aspects of a training context prior to testing. However, in virtually all such experiments at least some portion of the training context remains intact as part of the test context. If individual background stimuli do *not* become associated during learning, then even an incomplete context change should still delete many of the contextual cues necessary for retrieval of a learned response. However, if contextual cues do become linked during learning, the presence of even a few of these cues during testing should function to activate an animal's memory of the conditioning context. In effect, the deletion of a portion of the training context should have little detrimental effect on retention, since the presence of some of the training cues should serve to activate the representation of the remaining cues on the test. It follows from this line of reasoning that the context specificity of a learned response should depend on the degree to which training conditions foster within-context associations.

In the first experiment designed to assess the existence of within-context associations, we exposed rats to a black chamber containing an anise odor. During this exposure the rats were given three mild footshocks distributed over a period of approximately 30 s. On subsequent context–preference tests we found that the rats given these exposures avoided not only the original training context containing the "black–anise" compound but also test chambers that either were black or contained an anise odor. In other words, shocking the animals in a "black–anise" context resulted in an aversion to black, to anise, and to the original combination of cues.

In a second experiment of this type we trained animals in the same manner as in Experiment 1, but then gave the rats extensive extinction treatments in a black chamber prior to the context–preference tests. As expected, we found that these extinction sessions had eliminated the aversion to the black chamber. Additionally, however, animals given a preference test between an anise and a neutral odor no longer exhibited an aversion to anise. That is, extinction of one contextual stimulus also served to extinguish the conditioned fear to the other contextual stimulus. Since the fear of anise remained in animals given extinction sessions in a nonblack environment, this finding suggests that during the shock exposures, the animals associated the black and the anise cues so that the subsequent manipulation of one cue correspondingly affected the other.

Having demonstrated that within–context associations may be formed when a US occurs in a given environment, we proceeded to examine other conditions that might result in the formation of such associations. First, we attempted to determine if individual background cues become associated simply as a result of exposing animals to a context containing these cues. In our first experiment we exposed rats to a black chamber containing an anise odor for 30 s. Thirty minutes later the animals were placed in a black chamber without an anise odor and were given three mild footshocks. On a subsequent context–preference test we found that the shock experience had produced an aversion to the black chamber. However, those animals given a preference test between two clear chambers, one containing anise and the other a neutral odor, showed no aversion to anise. In effect, this study provided no evidence to suggest that simple exposure to a "black–anise" environment would result in an association between the black and anise cues.

In subsequent follow–up experiments of this type, we examined the effects of a variety of parametric manipulations. For example, in one study we varied the length of exposure to the "black–anise" environment prior to the shock presentations in the black chamber. In another, we alternated the initial exposures to the "black–anise" environment with exposures to a discriminably different context. In still other studies we varied the number of shocks given in the black chamber and the duration of the shock presentation session in the black chamber. Despite varying these other parameters, we have found no evidence to suggest that simple exposure to an environment fosters associations among the contextual stimuli comprising that environment.

One final experiment in this series is worth noting individually. In this study we exposed rats to the "black-anise" environment just as in our previous experiments, but during this exposure the animals received presentations of a discrete stimulus, a tone. Because presentations of discrete shock stimuli had been successful in promoting a black-anise association in our original experiments, we wanted to determine if such an association depends on the occurrence of any discrete stimulus in a given context. Again, we found no evidence for a black-anise association as the result of tone presentations in the "black-anise" context. Apparently, associations among the individual cues comprising a context depend in some way on the occurrence of a motivationally significant stimulus in that context.

Certainly, these studies represent only a preliminary attempt to examine the existence of associations among individual contextual stimuli. However, the results of these studies are interesting in two regards. First, the studies provide evidence that such associations can be formed. Second, it appears that such associations are fostered by certain conditions and not by others. If we assume that the effects of context change will differ depending on whether within-context associations are formed during original training, then it may be important in predicting the effects of context change to understand the conditions that promote within-context learning.

Currently, we have no completely adequate account for the influence of contextual stimuli on the retention performance of animals. It does not appear possible, for example, to explain retention deficits that result from context change solely on the basis of such factors as the loss of contextual excitatory strength, a change in an animal's perception of key stimuli, or the disruptive effects produced by a novel environment during testing. It also seems unlikely that such retention deficits can be explained adequately by current configural cue accounts. It is possible, of course, that all of these accounts identify factors that contribute to retention deficits following context change. However, none of these explanatory approaches appears capable of standing alone as a general account of context change effects.

The one approach that remains viable, at least to some extent, is the notion that contextual stimuli serve primarily as retrieval cues. Certainly, this approach provides us with a logical and reasonable explanation for how context change might result in retention deficits. Very simply, when the learning context changes, cues necessary for retrieving the memory of a learned response are lost, resulting in a retrieval failure and poor performance. However, this explanatory approach is little more than a restatement of the fact that performance deficits do often result from context change. Furthermore, this hypothesis is silent as to why context change fails to produce performance deficits under certain conditions.

For this approach to become a truly workable and testable explanation, there needs to be some clearer statement as to the rules that govern the effectiveness of contextual stimuli as retrieval cues. For example, such an explanation should include some assumptions as to what learning conditions are likely to foster the use of contextual stimuli as retrieval aids and what conditions are not. Additionally, this form of explanation should include clear performance predictions in cases in which conflicting responses are acquired in the same context. What is necessary in developing these rules and explicit predictions is a better understanding of the associative structure that develops during a learning episode and a more thorough analysis of the multiple responses that may be acquired in a given learning situation.

References

Archer, T., Sjöden, P. O., & Nilsson, L. G. (1984). The importance of contextual elements in taste-aversion learning. *Scandinavian Journal of Psychology, 2,* 251–257.

Asratyan, E. A. (1965). *Compensatory adaptations, reflex activity, and the brain.* Oxford: Pergamon.

Baker, A. G., Steinwald, H., & Bouton, M. E. (1991). Contextual conditioning and reinstatement of extinguished instrumental responding. *Quarterly Journal of Experimental Psychology: Comparative and Physiological Psychology, 43B*(2), 199–218.

Balaz, M. A., Capra, S., Hartl, P., & Miller, R. R. (1981). Contextual potentiation of acquired behavior after devaluing direct context-US associations. *Learning and Motivation, 12,* 383–397.

Balaz, M. A., Capra, S., Kasprow, W., & Miller, R. R. (1982). Latent inhibition of the conditioning context: Further evidence of contextual potentiation of retrieval in the absence of context-US associations. *Animal Learning and Behavior, 10,* 242–248.

Balda, R. P., & Turek, R. J. (1984). The cache-recovery system as an example of memory capabilities in Clark's nutcracker. In H. L. Roitblat, T. G. Bever, & H. S. Terrace (Eds.), *Animal cognition* (pp. 513–532). Hillsdale, NJ: Erlbaum.

Balsam, P. D., & Tomie, A. (Eds.). (1985). *Context and learning.* Hillsdale, NJ: Erlbaum.

Bliss, D. K. (1973). Dissociation learning and state-dependent retention induced by pentobarbital in the rhesus monkey. *Journal of Comparative and Physiological Psychology, 84,* 149–161.

Bouton, M. E., & Bolles, R. C. (1979). Contextual control of conditioned fear. *Learning and Motivation, 10*(4), 445–466.

Bouton, M. E., & Bolles, R. C. (1985). Contexts, event-memories, and extinction. In P. D. Balsam & A. Tomie (Eds.). *Context and learning* (pp. 133–166). Hillsdale, NJ: Erlbaum.

Bouton, M. E., & King, D. A. (1983). Contextual control of the extinction of conditioned fear: Tests for the associative value of the context. *Journal of Experimental Psychology: Animal Behavior Processes, 9,* 248–265.

Bouton, M. E., & King, D. A. (1986). Effect of context on performance to conditioned stimuli with mixed histories of reinforcement and nonreinforcement. *Journal of Experimental Psychology: Animal Behavior Processes, 12,* 4–15.

Bouton, M. E., & Peck, C. A. (1989). Context effects on conditioning, extinction, and reinstatement in an appetitive conditioning preparation. *Animal Learning and Behavior, 17*(2), 188–198.

Brooks, D., & Bouton, M. E. (1993). A retrieval cue for extinction attenuates spontaneous recovery. *Journal of Experimental Psychology: Animal Behavior Processes, 19,* 77–89.

Calhoun, W. H. (1971). Central nervous system stimulants. In E. Furchtgott (Ed.), *Pharmacological and biophysical agents and behavior* (pp. 181–268). New York: Academic Press.

Channell, S., & Hall, G. (1983). Contextual effects in latent inhibition with an appetitive conditioning procedure. *Animal Learning and Behavior, 11*(1), 67–74.

Chiszar, D. A., & Spear, N. E. (1969). Stimulus change, reversal learning, and retention in the rat. *Journal of Comparative and Physiological Psychology, 69*, 190–195.

Dunn, A. J., & Leibmann, S. (1977). The amnestic effect of protein synthesis inhibitors is not due to the inhibition of adrenal corticosteroidogenesis. *Behavioral and Neural Biology, 19*(3), 411–416.

Eich, J., Weingartner, H., Stillman, R., & Gillin, J. (1975). State-dependent accessibility of retrieval cues and retention of a categorized list. *Journal of Verbal Learning and Verbal Behavior, 14*, 408–417.

Estes, W. K. (1973). Memory and conditioning. In F. J. McGuigan & D. B. Lumsden (Eds.), *Contemporary approaches to conditioning and learning* (pp. 265–286). Washington, DC: Winston.

Flexser, A. J., & Tulving, E. (1978). Retrieval independence in recall and recognition. *Psychological Review, 85*, 153–171.

Gibbs, M. E., & Mark, R. F. (1973). *Inhibition of memory formation*. New York: Plenum.

Gleitman, H. (1971). Forgetting of long-term memories in animals. In W. K. Honig & P. H. R. James (Eds.), *Animal memory* (pp. 1–44). New York: Academic Press.

Godden, D. R., & Baddeley, A. D. (1975). Context-dependent memory in two natural environments: On land and under water. *British Journal of Psychology, 66*, 325–331.

Gordon, W. C., McCracken, K. M., Dess-Beech, N., & Mowrer, R. R. (1981). Mechanisms for the cueing phenomenon: The addition of the cueing context to the training memory. *Learning and Motivation, 12*, 196–211.

Gordon, W. C., Mowrer, R. R., McGinnis, C. P., & McDermott, M. J. (1985). Cue-induced memory interference in the rat. *Bulletin of the Psychonomic Society, 23*(3), 233–236.

Gordon, W. C., & Weaver, M. S. (1989). Cue-induced transfer of CS preexposure effects across contexts. *Animal Learning and Behavior, 17*(6), 409–417.

Grant, D. S. (1981). Short-term memory in the pigeon. In N. E. Spear & R. R. Miller (Eds.), *Information processing in animals: Memory mechanisms* (pp. 227–256). Hillsdale, NJ: Erlbaum.

Greenspoon, J., & Ranyard, R. (1957). Stimulus conditions and retroactive inhibition. *Journal of Experimental Psychology, 53*, 55–59.

Hall, G., & Honey, R. C. (1989). Contextual effects in conditioning, latent inhibition, and habituation: Associative and retrieval functions of contextual cues. *Journal of Experimental Psychology: Animal Behavior Processes, 15*(3), 232–241.

Hall, G., & Honey, R. C. (1990). Context-specific conditioning in the conditioned-emotional-response procedure. *Journal of Experimental Psychology: Animal Behavior Processes, 16*(3), 271–278.

Hall, G., & Minor, H. (1984). A search for context-stimulus associations in latent inhibition. *Quarterly Journal of Experimental Psychology: Comparative and Physiological Psychology, 36B*(2), 145–169.

Holland, P. C. (1983). Occasion-setting in Pavlovian feature positive discriminations. In M. L. Commons, R. J. Hernstein, & A. R. Wagner (Eds.), *Quantitative analysis of behavior* (Vol. 4). New York: Ballinger.

Holloway, F. A., & Wansley, R. A. (1973). Multiple retention deficits at periodic intervals after passive avoidance learning. *Science, 80*, 208–210.

Honey, R. C., Willis, A., & Hall, G. (1990). Context specificity in pigeon autoshaping. *Learning and Motivation, 21*(2), 125–136.

Honig, W. K. (1978). Studies of working memory in the pigeon. In S. H. Hulse, H. Fowler, & W. K. Honig (Eds.), *Cognitive processes in animal behavior* (pp. 211–247). Hillsdale, NJ: Erlbaum.

Honig, W. K., & James, P. H. R. (Eds.). (1971). *Animal memory*. New York: Academic Press.

Kintsch, W. (1974). *The representation of meaning in memory*. Hillsdale, NJ: Erlbaum.

Klein, R. L., & Gordon, W. C. (1993). Formation of a within compound association during contextual conditioning. Unpublished manuscript.

Klein, S. B. (1972). Adrenal-pituitary influence in reactivation of avoidance-memory in the rat after immediate intervals. *Journal of Comparative and Physiological Psychology, 79*, 341–359.

Klein, S. B. (1975). ACTH-induced reactivation of prior active avoidance training after intermediate intervals in hypophysectomized, adrenalectomized, and sham-operated rats. *Physiological Psychology, 3*(4), 395–399.

Lewis, D. J. (1979). Psychobiology of active and inactive memory. *Psychological Bulletin, 86*, 1054–1083.

Lovibond, P. F., Preston, G. C., & Mackintosh, N. J. (1984). Context specificity of conditioning, extinction, and latent inhibition. *Journal of Experimental Psychology: Animal Behavior Processes, 10*, 360–375.

Lubow, R. E., & Moore, A. U. (1959). Latent inhibition: The effect of nonreinforced exposure to the conditioned stimulus. *Journal of Comparative and Physiological Psychology, 52*, 415–419.

Maki, W. S. (1981). Directed forgetting in animals. In N. E. Spear & R. R. Miller (Eds.), *Information processing in animals: Memory mechanisms* (pp. 199–225). Hillsdale, NJ: Erlbaum.

McAllister, W. R., & McAllister, D. E. (1963). Increase over time in the stimulus generalization of acquired fear. *Journal of Experimental Psychology, 65*, 576–582.

McGaugh, J. L., & Dawson, R. G. (1971). Modification of memory storage processes. In W. K. Honig & P. H. R. James (Eds.), *Animal memory* (pp. 215–242). New York: Academic Press.

McGaugh, J. L., & Herz, M. J. (1972). *Memory consolidation*. San Francisco: Albion.

Medin, D. L. (1975). A theory of context in discrimination learning. In G. H. Bower (Ed.), *The psychology of learning and motivation* (Vol. 9, pp. 263–314). New York: Academic Press.

Miller, R. R., & Schachtman, T. R. (1985). The several roles of context at the time of retrieval. In P. D. Balsam & A. Tomie (Eds.), *Context and learning* (pp. 167–194). Hillsdale, NJ: Erlbaum.

Morris, C. D. (1978). Acquisition-test interactions between different dimensions of encoding. *Memory & Cognition, 6*(4), 354–363.

Nadel, L., & Willner, J. (1980). Context and conditioning. A place for space. *Journal of Comparative and Physiological Psychology, 8*, 218–228.

Nelson, N. (1993). Extinction of a within-compound association within and outside the learning context. Unpublished M.A. Thesis, University of New Mexico.

Olton, D. S., & Samuelson, R. J. (1976). Remembrance of places passed: Spatial memory in rats. *Journal of Experimental Psychology: Animal Behavior Processes, 2*, 97–116.

Overton, D. A. (1964). State-dependent or "disassociated" learning produced with pentobarbital. *Journal of Comparative and Physiological Psychology, 57*, 3–12.

Overton, D. A. (1966). State-dependent learning produced by depressant and atropine-like drugs. *Psychopharmacologia, 10*, 6–31.

Overton, D. A. (1972). State-dependent learning produced by alcohol and its relevance to alcoholism. In B. Kissin & H. Begleiter (Eds.), *The biology of alcoholism: Vol. II. Physiology and behavior* (pp. 193–214). New York: Plenum.

Overton, D. A. (1985). Contextual stimulus effects of drugs and internal states. In P. D. Balsam & A. Tomie (Eds.), *Context and learning* (pp. 357–384). Hillsdale, NJ: Erlbaum.

Pearce, J. M., & Hall, G. (1980). A model for Pavlovian learning: Variations in the effectiveness of conditioned but not unconditioned stimuli. *Psychological Review, 87*, 532–552.

Peck, C. A., & Bouton, M. E. (1990). Context and performance in aversive-to-appetitive and appetitive-to-adversive transfer. *Learning and Motivation, 21*(1), 1–31.

Perkins, C. C., Jr., & Weyant, R. G. (1958). The interval between training and test trials as determiner of the slope of generalization gradients. *Journal of Comparative and Physiological Psychology, 51,* 596–600.

Petersen, R. (1974). *Isolation of processes involved in state-dependent recall in man.* Paper presented at the meetings of the Federation of American Society for Experimental Biology, Atlantic City, NJ.

Rescorla, R. A. (1985). Conditioned inhibition and facilitation. In R. R. Miller & N. E. Spear (Eds.), *Information processing in animals: Conditioned inhibition* (pp. 299–326). Hillsdale, NJ: Erlbaum.

Rescorla, R. A., & Durlach, P. J. (1981). Within-event learning in Pavlovian conditioning. In N. E. Spear & R. R. Miller (Eds.), *Information processing in animals: Memory mechanisms* (pp. 81–111). Hillsdale, NJ: Erlbaum.

Rescorla, R. A., Durlach, P. J., & Grau, J. W. (1985). Contextual learning in Pavlovian conditioning. In P. D. Balsam & A. Tomie (Eds.), *Context and learning* (pp. 23–56). Hillsdale, NJ: Erlbaum.

Rescorla, R. A., & Wagner, A. R. (1972). A theory of Pavlovian conditioning: Variations in the effectiveness of reinforcement and nonreinforcement. In A. H. Black & W. F. Prokasy (Eds.), *Classical conditioning II: Current research and theory* (pp. 64–99). New York: Appleton-Century-Crofts.

Riccio, D. C., Urda, M., & Thomas, D. R. (1966). Stimulus control in pigeons based on proprioceptive stimuli from the floor inclination. *Science, 153,* 434–436.

Roitblat, H. L. (1980). Codes and coding processes in pigeon short-term memory. *Animal Learning and Behavior, 8,* 341–351.

Smith, S. M., Glenberg, A. M., & Bjork, R. A. (1978). Environmental context and human memory. *Memory & Cognition, 6,* 342–353.

Spear, N. E. (1971). Forgetting as retrieval failure. In W. K. Honig & P. N. R. James (Eds.), *Animal memory* (pp. 45–109). New York: Academic Press.

Spear, N. E. (1973). Retrieval of memory in animals. *Psychological Review, 80,* 163–194.

Spear, N. E. (1978). *The processing of memories: Forgetting and retention.* Hillsdale, NJ: Erlbaum.

Spear, N. E., Klein, S. B., & Riley, E. P. (1971). The Kamin effect as "state-dependent" learning: Memory retrieval failure in the rat. *Journal of Comparative and Physiological Psychology, 74,* 416–425.

Spear, N. E., Smith, G. J., Bryan, R., Gordon, W. C., Timmons, R., & Chiszar, D. (1980). Contextual influences on the interaction between conflicting memories in the rat. *Animal Learning and Behavior, 8,* 273–281.

Squire, L. R. (1987). *Memory and brain.* New York: Oxford University Press.

Steinman, F. (1967). Retention of alley brightness in the rat. *Journal of Comparative and Physiological Psychology, 64,* 105–109.

Stroebel, C. F. (1967). Behavioral aspects of circadian rhythms. In J. Zubin & H. F. Hunt (Eds.), *Comparative psychopathology* (pp. 158–172). New York: Grune & Stratton.

Swartzentruber, D., & Bouton, M. E. (1992). Context sensitivity to the conditioned stimulus. *Animal Learning and Behavior, 20*(2), 97–103.

Thomas, D. R., Moye, T. B., & Kimose, E. (1984). The recency effect in pigeons' long-term memory. *Animal Learning and Behavior, 12*(1), 21–28.

Tomback, D. F. (1983). Nutcrackers and pines: Coevolution or coadaptation? In H. Nitecki (Ed.), *Coevolution.* Chicago: University of Chicago Press.

Tulving, E. (1983). *Elements of episodic memory.* Oxford: Clarendon Press/Oxford University Press.

Tulving, E., & Osler, S. (1968). Effectiveness of retrieval cues in memory for words. *Journal of Experimental Psychology, 77*, 493–601.

Tulving, E., & Thomson, D. M. (1973). Encoding specificity and retrieval processes in episodic memory. *Psychological Review, 80*, 352–373.

Wagner, A. R. (1981). S.O.P.: A model of automatic memory processing in animal behavior. In N. E. Spear & R. R. Miller (Eds.), *Information processing in animals: Memory mechanisms* (pp. 5–47). Hillsdale, NJ: Erlbaum.

Weingartner, H., Miller, H., & Murphy, D. L. (1977). Mood-state-dependent retrieval of verbal associations. *Journal of Abnormal Psychology, 86*(3), 276–284.

Yamaguchi, H. G. (1952). Gradients of drive stimulus (S) intensity generalization. *Journal of Experimental Psychology, 43*, 298–304.

CHAPTER 10

Social Cognition in Primates

C. M. Heyes

I. INTRODUCTION

A. History and Definition

Investigations of complex social behavior in nonhuman primates received new impetus and began to be described as studies of "social cognition" in the late 1970s following publication of the "social function of intellect" or "social intelligence" hypothesis (Humphrey, 1976; see also Chance & Mead, 1953; Jolly, 1966), and the suggestion that chimpanzees may have a "theory of mind" (Premack & Woodruff, 1978). The former proposed that the social, rather than the physical, environment was the principle source of selection pressure for the evolution of primate intelligence, and the latter raised the possibility that individual chimpanzees and other nonhuman primates (henceforward simply "primates") may attribute mental states, such as beliefs and desires, to themselves and to others.

Broad definitions of social cognition portray all recent studies of primate social behavior as investigations of social cognition (Cheney & Seyfarth, 1990b; 1992; de Waal, 1991; Kummer, Dasser, & Hoyningen-Huene, 1990; Quiatt & Reynolds, 1993), but a substantial proportion of these studies are of relatively little interest to psychologists. They either document patterns of spontaneous social behavior without giving any indication of the psychological processes responsible, or are thought to show that perceptual and associative mechanisms well known to be responsible for processing information from the physical or asocial environment are also activated by social stimuli. Studies of this kind will be neglected in the present review in favor of research that provides, or is widely thought to provide, evidence of *distinctively* social cognition, that is, of cognitive processes that operate only or typically on information derived from, or relevant to, other animals. One of the virtues of this restriction is that it means that we do not have to take any stance on the question of whether the processes of associative learning are either "cognitive" or "noncognitive" (Dickinson, 1983; Premack, 1983). It is sufficient to note that, since these processes commonly operate on asocial input, associative learning is not a variety of distinctively social cognition.

B. Overview

The social function of intellect hypothesis is widely cited but its influence on research in social cognition has been indirect. The hypothesis clearly predicts that "there should be a positive correlation across species between 'social complexity' and 'individual intelligence'" (Humphrey, 1976, p. 26), and yet this prediction has not been tested in studies of social cognition (Macphail, 1991). Instead of attempting to correlate these variables across a broad range of primate and nonprimate species, researchers have focused on primates that spend a large proportion of their time engaged in complex social interactions, and have examined, not their general intellectual ability, but their capacity to process social information. Thus, the social function of intellect hypothesis has functioned less as a hypothesis than as a general guide to where in the animal kingdom social cognition is most likely to be found.

The majority of studies reviewed here concern primates but experiments on other animals are mentioned. These are rarely regarded as investigations of social cognition but in many cases the evidence that they provide is as strong as that of primate studies. Consequently, depending on one's assumptions about the evolution of intelligence and the appropriate application of Ockham's razor, research on rodents and birds either indicates that social cognition is relatively widespread among vertebrates, or acts as a reminder that more general processes may be responsible for behavior suggestive of social cognition.

In contrast with the indirect and yet conspicuous influence of the social function of intellect hypothesis, the effect of Premack and Woodruff's (1978) theory of mind hypothesis has been immense but not always obvious. All studies of social cognition in primates have been either explicitly designed to investigate the possibility that they attribute mental states or derive psychological interest from their bearing on this issue. However, the role played by Premack and Woodruff (1978) in stimulating this research is not always apparent because, in addition to "theory of mind," the focus of study is known as "Machiavellian intelligence" (Byrne & Whiten, 1988; Whiten & Byrne, 1988), "metarepresentation" (Whiten & Byrne, 1991), "politics" (de Waal, 1982), "metacognition" (Povinelli, in press), "mind reading" (Krebs & Dawkins, 1984; Whiten, 1991), "perspective-taking" (Povinelli, Nelson & Boysen, 1990), and "mental state attribution" (Cheney & Seyfarth, 1990a; 1990b, 1992). The latter term is used here because it is the most general and soberly descriptive. To say that an animal has a "theory of mind" may imply that the animal infers the presence of particular mental states in others using lawlike generalizations about mental states and behavior (Goldman, 1993). However, this was not the sense in which Premack and Woodruff (1978) used the term, and subsequent research has not addressed the question of how, if at all, animals derive mental state attributions.

Studies of six types of behavior will be reviewed (Section II): imitation, mirror-guided body inspection (or self-recognition), discrimination of social relationships, deception, role taking (or empathy), and perspective taking. The current consensus is that these studies provide convergent evidence that chimpanzees and possibly other apes, but not monkeys, engage in mental state attribution (Byrne, 1993; Cheney & Seyfarth, 1990b; 1992; de Waal, 1991; Gallup, 1982; Jolly, 1991; Povinelli, 1993, in press; Whiten & Byrne, 1991). Research on symbolic communication and teaching has also contributed to this consensus (see Chapter 12 by Savage-Rumbaugh this volume, and Premack, 1991, respectively, for reviews). Within each of the six sections, two questions are addressed: (1) Which primates, if any, exhibit the behavior? and (2) To what extent is the behavior indicative of mental state attribution? Viewed as measures of mental state attribution, the six types of behavior are considered in roughly ascending order of current internal and construct validity (Cook & Campbell, 1979). For example, imitation is considered first because relatively little of the available evidence of behavioral copying is compelling, and behavioral copying only weakly implies a capacity to attribute mental states. Perspective taking is discussed last because there is a procedure that has the potential to detect this behavior fairly reliably, and, under certain conditions, perspective taking would directly imply a capacity for mental state attribution.

II. REVIEW OF EMPIRICAL STUDIES

A. Imitation

Continuity between earlier research in comparative psychology and investigations of social cognition is most apparent in the study of imitation. Comparative psychologists have long regarded motor imitation (the spontaneous reproduction of acts yielding disparate sensory input when observed and executed) as a sign of higher intelligence, and sought evidence that it occurs among nonhuman animals (Thorndike, 1898). However, after nearly 100 years of research, there is still no unequivocal evidence of motor imitation in any primate species (Byrne, 1993; Crawford, 1939; Galef,1988; Tomasello, Davis-Dasilva, Camak, & Bard, 1987; Visalberghi & Fragaszy, 1992).

Under uncontrolled and semicontrolled conditions the occurrence of imitation in monkeys (Beck, 1976; Hauser, 1988; Kawai, 1965; Nishida, 1986; Westergaard, 1988) and chimpanzees (de Waal, 1982; Goodall, 1986; Mignault, 1985; Sumita, Kitahara-Frisch, & Norikoshi, 1985; Terrace, Petitto, Sanders, & Bever, 1979) has been inferred from the performance of a complex, novel, previously observed act by a single animal or a succession of animals within a group. Even if one disregards the potential lack of reliability of these observational or anecdotal data, they are not compelling. In all cases, the observed behavior could have been acquired by a means other than imitation (e.g., trial-and-error learning), and in many cases there is evidence that it was so acquired (Adams-Curtis, 1987; Fragaszy & Visalberghi, 1989; 1990; Galef, 1992; Visalberghi & Trinca, 1989). For example, the habit of potato washing was supposed to have been transmitted through the population of Japanese macaques on Koshima Island through imitation (Kawai, 1965; Nishida, 1986). However, given the order in which members of the troop were observed engaging in this behavior (first a juvenile, Imo, then her playmates, then their mothers), it is equally likely that, rather than copying the actions of potato washers, naïve animals followed or chased them into water while holding a potato. Once in that position, the pursuing animal would only have to drop and then retrieve its potato, now sand free and with a salty taste, to acquire the behavior. Furthermore, the hypothesis that potato washing spread through following rather than imitation is consistent both with the slow rate of transmission on Koshima Island (Galef, 1992), and evidence that isolated monkeys readily learn to wash sandy food when they find it close to water (Visalberghi & Fragaszy, 1992).

A further example concerns a chimpanzee, Nim, that was trained to use American Sign Language (ASL) (Terrace et al., 1979). Analysis of Nim's ASL utterances suggested that, rather than using signs creatively to communicate, he frequently copied sequences of signs that had recently been used by a human trainer. However, there is reason to doubt that this copying behavior is indicative of a capacity beyond that of associative learning. Nim was trained in ASL using a mixture of informal methods including "molding" (physical guidance of the animal's hands by a trainer), and praise for correct or partially correct production of signs. Consequently, his ability to copy signs could have been due, not to spontaneous matching of observed and executed behavior, but to instrumental training in which the trainer's signs acted as discriminative stimuli indicating that matching signs would be rewarded. In this case, the trainer, but not the chimpanzee, would be sensitive to the topographical similarity between the trainer's and the chimpanzee's behavior.

Remarkably few experiments have been conducted on imitation in primates (or in other species), and their results may indicate merely that observation of action can influence the degree to which the observer subsequently attends to certain physical components of a problem situation. This sort of interpretation, in terms of "stimulus enhancement" (Galef, 1988; Heyes, in press; Spence, 1937) is certainly the most natural for early experiments in which monkeys and chimpanzees were presented with a succession of pairs of objects and learned to touch or displace the member of each pair to which they had observed a conspecific making a rewarded response (Crawford, 1939; Warden & Jackson, 1935). Using a similar procedure, Darby and Riopelle (1959) showed that rhesus monkeys can learn to displace the same object as a conspecific when the demonstrator's response revealed food, and to select an alternative object when it did not reveal food. This suggests that, in addition to showing stimulus enhancement, monkeys can learn through observation about relationships among stimuli, but it does not indicate a capacity to imitate, to learn about responses or response-reinforcer relationships by observation.

A concerted attempt to distinguish imitation from stimulus enhancement and other kinds of social learning has been made in only three experiments with primates, all involving chimpanzees (Hayes & Hayes, 1952; Tomasello et al., 1987; Tomasello, Savage-Rumbaugh, & Kruger, in press). In the first (Tomasello et al., 1987), experimental animals that had observed a conspecific demonstrator using three distinctive techniques to rake in food with a T bar were more successful in using the T bar to obtain food than were control animals that had not observed the instrument in use prior to testing. However, the experimental group contacted the T bar more than the controls at pretest, prior to demonstrator observation, and therefore their superior performance may not have reflected social influence of any kind. Further, if demonstrator observation did play a role, it may have been via stimulus enhancement. Unlike the control animals, those in the experimental group saw the demonstrator contacting the T bar, and this experience may have led them to spend more of the test time engaged in trial-and-error learning with the instrument. As Tomasello et al. (1987) pointed out, there was certainly no evidence of imitation in the form of a tendency on the part of experimental animals to apply the same movements to the T bar as had their demonstrator.

In the second experiment (Tomasello et al., in press), "enculturated" chimpanzees (i.e., animals with an extensive training history), relatively

naïve chimpanzees, and young children observed the experimenter manipulating 16 objects in various ways and, after observing each action, were
given access to the same object either immediately or after a 48-hour delay.
When the test was given immediately, and the results for all objects were
combined, the "enculturated" chimpanzees were comparable to the children
in their tendency to act on the same part of the object, and with the same
effect, as the demonstrator. However, for many objects, resemblance between the demonstrator and the observer could have been coincidental or
due to stimulus enhancement rather than imitation. For example, when
presented with a paint brush, the chimpanzees may have squeezed it with
one hand, not because they had observed the trainer executing this or any
other action in relation to the brush, but simply in an effort to grasp a novel
object. Similarly, observation of the trainer turning a spigot to release rope
may have increased the probability that the chimpanzees would touch the
spigot when given the opportunity to do so. Once in contact with the
spigot, they could quickly discover that it can be turned, and that this
operation releases rope. The finding that "enculturated" chimpanzees were
actually superior to children under delayed test conditions casts further
doubt on the view that the chimpanzees were imitating. There can be little
doubt that the children were imitating, and this result suggests that different
processes were responsible for the performance of children and chimpanzees.

The significance of the third experiment on imitation in chimpanzees
(Hayes & Hayes, 1952) is difficult to assess because neither the procedure
nor the results were reported in detail. Hayes and Hayes (1952) gave Viki, a
"home-raised" chimpanzee, a series of 70 "imitation set" tasks. Each task
consisted of the experimenter saying, "Do this," and then performing an
action such as patting his head, clapping his hands, or operating a toy. If
Viki performed a similar action within a few seconds, she was rewarded
with food; otherwise the experimenter repeated the action or helped Viki to
make the response by, for example, manipulating her hands. Viki was said
to have required help in executing each of the first 11 actions, but to have
begun with the 12th item to imitate immediately test actions that were
already part of her repertoire. Further, it was claimed that, beginning with
the 20th task, Viki copied at least 10 completely novel actions.

Taken at face value, the results of this study suggest that Viki did not
show spontaneous imitation, but that she learned to imitate through a procedure in which imitative responding was shaped both manually and by
selective reinforcement, and that she was able to generalize on the basis of
this training. While associative processes may mediate learning to imitate
through shaping (see the discussion above of language-trained apes), learning of this kind would be unlikely to generalize to novel actions. Thus, if
Viki imitated novel actions, even after shaping, there would be reason to

believe that she was capable of some kind of distinctively social cognition. However, this conclusion is not secure because the report on Viki's behavior provided no indication of either the method used to measure the similarity between the experimenter's and the chimpanzee's behavior, or of the degree of similarity observed.

The paucity of evidence of imitation in primates indicates neither that they are unable to imitate nor that such evidence is difficult to obtain for nonhuman animals. Relatively unequivocal evidence of imitation in budgerigars (Galef, Manzig, & Field, 1986) and rats (Heyes & Dawson, 1990; Heyes, Dawson, & Nokes, 1992) has been found by comparing the behavior of subjects that have observed a conspecific acting on a single object in one of two distinctive ways. In addition to providing a methodological lead for primate research, these studies of species that are relatively distantly related to humans are a reminder that, while an imitator may seem to represent the imitated animal's mental state, its point of view, or its beliefs and desires (as suggested by Gallup, 1982; Povinelli, in press; Whiten & Byrne, 1991; Whiten & Ham, 1992), there is no compelling reason to believe that it does so. What is apparently essential for imitation is that the imitating animal represent what the demonstrator did, not what it thought (Heyes, 1993b).

A number of researchers treat imitation as an indicator of mental state attribution, while acknowledging that other processes could also lead to the reproduction of novel, complex acts (Byrne, 1993; Tomasello, Kruger, & Ratner, 1993). However, until mental state attribution can be distinguished empirically from these other processes, imitation must be regarded as a rather poor indicator of mental state attribution in general, and as one that has yielded no evidence of that ability in primates.

B. Mirror-Guided Body Inspection

A series of experiments using a common procedure apparently indicate that chimpanzees and orangutans, but not other primates, are capable of "self-recognition" or mirror-guided body inspection, that is, will use a mirror as a source of information about their own bodies (Cheney & Seyfarth, 1990b; Gallup, 1982; Jolly, 1991; Povinelli, 1987; Whiten & Byrne, 1991). In the standard procedure (Gallup, 1970), an animal with some experience of mirrors is anesthetized and marked on its head with a red, odorless, nonirritant dye; several hours later, the frequency with which the animal touches the marks on its head is measured first in the absence of a mirror and then with a mirror present. Under these circumstances, chimpanzees and orangutans typically touch their head marks more when the mirror is present than when it is absent, while monkeys of various species and gorillas touch their marks with the same low frequency in both conditions (Calhoun &

Thompson, 1988; Gallup, 1970, 1977; Gallup, McClure, Hill, & Bundy, 1971; Ledbetter & Basen, 1982; Platt & Thompson, 1985; Suarez & Gallup, 1981).

There is reason to doubt that: (1) experiments using the marking procedure have demonstrated mirror-guided body inspection; (2) species differences in mark-directed behavior are well established or especially interesting; and (3) mirror-guided body inspection would be sufficient to demonstrate "self-recognition" or possession of a "self-concept." There is an alternative to the standard interpretation of the tendency, shown by chimpanzees and orangutans, to touch their marks more in the presence of the mirror than in its absence: in the mirror-present condition, the animals had had a longer period of time to recover from anesthesia and were therefore more active generally than in the previous, mirror-absent condition. If they were more active generally, they had a higher probability of touching the marked areas of their heads by chance. Thus, chimpanzees and orangutans may touch their marks more when the mirror is present than when it is absent simply because they have had, at the mirror-present stage, a longer period of time to recover from the anesthetic and are therefore more active generally (Heyes, 1994).

According to this hypothesis, which is also consistent with the results of mark tests that vary from the standard procedure (Anderson, 1983; Anderson & Roeder, 1989; Eglash & Snowdon, 1983; Gallup & Suarez, 1991; Lin, Bard, & Anderson, 1992; Robert, 1986; Suarez & Gallup, 1986), species differences arise from the fact that chimpanzees spontaneously touch their faces with a much higher frequency than do either monkeys or gorillas (Dimond & Harries, 1984). However, if the behavior of chimpanzees and orangutans is ascribed to mirror-guided body inspection, rather than an anesthetic artifact, the results of the mark test still do not indicate clearly that other primates lack some cognitive capacity necessary to use a mirror in this way. It has been suggested that monkeys and gorillas may "fail" the mark test because they have an innate tendency to avert their gaze when confronted (Premack, 1983), or because they monitor the state and appearance of their bodies less intensively than do chimpanzees and orangutans (Pearce, 1989). Thus, the results of mark tests leave open the possibilities that all, none, or some primates are capable of mirror-guided body inspection.

Described as "self-recognition," mirror-guided body inspection has been said to imply the possession of a "self-concept" and the potential to imagine oneself as one is viewed by others, that is, to attribute mental states (Gallup, 1982; Povinelli, 1987; Whiten & Byrne, 1991), but this is also doubtful. To use a mirror as a source of information about its body, an animal must be able to distinguish, across a fairly broad range, sensory inputs resulting from the physical state and operations of its own body, from sensory inputs originating elsewhere. If the animal could not do this, if it lacked what might be described loosely as a "body concept," then presumably it could not learn that, when it is standing in front of a mirror, inputs from the mirror correlate with inputs from its body. However, since this "body concept" is equally necessary for mirror-guided body inspection and for collision-free locomotion, it is not clear that the former implies mental state attribution any more than does the latter (Heyes, 1994).

A demonstration that pigeons can be trained to use a mirror to detect paper dots attached to their feathers (Epstein, Lanza, & Skinner, 1981; Gallup, 1983; Premack, 1983) makes it easier to appreciate that mirror-guided body inspection may not imply mental state attribution. However, more direct evidence of a dissociation between the two is provided by studies of autistic children who, although apparently incapable of ascribing beliefs to others, begin to engage in mirror-guided body inspection at the same age as normal children (Ungerer, 1989).

C. Social Relationships

There is a substantial body of evidence suggesting that the social behavior of primates is affected not only by concurrent stimulation and the outcomes of previous, active engagements between the present interactants and third parties. In addition, the behavior of animal A in relation to animal B may be affected by A's prior observations of B in relation to one or a number of other conspecifics, C, D, and so on. Evidence of this kind (Cheney & Seyfarth, 1990b; Hinde, 1983; Smuts, Cheney, Seyfarth, Wrangham, & Struhsaker, 1987) has been derived from observational and experimental studies of chimpanzees (de Waal, 1982; de Waal & Van Roosmalen, 1979), baboons (Bachmann & Kummer, 1980; Kummer, 1968; Smuts, 1985), and various macaques (Anderson & Mason, 1978; Cheney & Seyfarth, 1980, 1986; Dasser, 1988; Datta, 1983; Judge, 1982, 1983; Stammbach, 1988). For example, adult male chimpanzees are more likely to disrupt (through interposition, aggression, or a threat display) social interactions between pairs of high-ranking conspecifics than between pairs of mixed or low-rank conspecifics (de Waal, 1982). Similarly, adult male baboons are. less likely to challenge the resident male of a one-male, multifemale unit if the females groom the resident male with a high frequency (Bachmann & Kummer, 1980), and both baboons and macaques are more likely to respond to an aggressor by attacking a bystander when the bystander is a relative of the aggressor, or when the bystander and aggressor commonly engage in affiliative social interactions (Judge, 1982; Smuts, 1985).

This literature indicates that the social behavior of animals from a broad range of primate species is sensitive to what human observers naturally describe as "social relationships" among conspecifics. It has been said, in addition, to show that primates have knowledge of social relationships (Cheney, Seyfarth, & Smuts, 1986; Cheney & Seyfarth, 1990b; de Waal, 1991; Kummer et al., 1990) and this seems entirely appropriate when the term knowledge is used in a general sense, and social relationships are understood to be observable properties. If, on the other hand, knowledge of social relationships is taken to consist of information about conditions (such as mental states) that are not directly observable, acquired by a means other than associative learning and/or represented in an "abstract" code (Cheney & Seyfarth, 1990b; 1992; Dasser, 1988; de Waal, 1991), then the evidence to date does not support the inference that primates know about social relationships.

Few researchers would contest this conclusion, and therefore consideration of two studies is sufficient to illustrate the plausibility of simple associative accounts of sensitivity to social relationships. In the first (Cheney & Seyfarth, 1980), free-ranging vervet monkeys were played the scream of an absent juvenile from a concealed loudspeaker. The adult female monkeys in the group typically responded to the sound of the juvenile's cry by looking at the juvenile's mother before the mother had responded to the cry herself. In so doing they displayed sensitivity to, or knowledge of, the mother–offspring relationship. But this could clearly have resulted from earlier exposure to a contingency between the cries of a particular juvenile and a vigorous behavioral response from a particular adult female (Cheney et al., 1986).

In the second study (Stammbach, 1988), one subordinate member of each of a number of groups of long-tailed monkeys was trained to obtain preferred food for the group by manipulating three levers. The other monkeys did not acquire the skill themselves, but those that received the most food as a result of the trained animals' activities began to follow them to the lever apparatus, and spent an increasing amount of time sitting beside and grooming the trained animals, even when the apparatus was not in operation. The untrained monkeys may have behaved in this way because they appreciated that the trained individuals had superior knowledge of the workings of the lever apparatus, and wanted to develop friendly relations with them in the hope of gaining more food in the future (Kummer et al., 1990; Stammbach, 1988). However, the results of an experiment with rats show that, rather than attributing superior knowledge, each untrained monkey may have learned an association between the trained animal in their group and receipt of preferred food. In this study (Timberlake & Grant, 1975), rats acquired affiliative social responding to a conspecific that was fastened to a trolley and wheeled into an operant chamber as a signal for the delivery of food.

Experiments by Dasser (1988) used a method that may be useful in future research on primates' knowledge of social relationships. In one of these, a female Java monkey was first rewarded for responding to a photographic slide of a particular mother–daughter duo in her social group, and not rewarded for responding to a simultaneously presented slide of one of five other duos of familiar conspecifics. In each of 14 subsequent transfer trials, she was shown a pair of slides of group members that had not been represented in prior training, and on every trial she chose the slide of a mother–daughter duo rather than the slide of another duo of similar relative size and gender. There are a number of obstacles to the conclusion that the subject in this experiment used an "abstract category analogous to our concept of mother–child affiliation" (Dasser, 1988; p. 229). For example, without detailed analysis of the stimulus materials, it is difficult to exclude the possibility that the monkey was using a relatively simple cue, such as relative posture, to make the discrimination (Chater & Heyes, in press; Premack, 1983). However, with appropriate control over slide content, use of a discrimination learning technique of this kind (see also Demaria & Thierry, 1988) may indicate more clearly what primates know about social relationships, and, consequently, how this information is acquired. As things stand, there is no reason to suggest that primates learn about social relationships in a different way from that by which they learn about other relationships in their environment—through processes of association.

D. Role Taking

In the experiments that gave rise to the suggestion that primates may have a theory of mind (Premack & Woodruff, 1978), a "language-trained" chimpanzee, Sarah, was shown videotapes depicting human actors confronting problems of various kinds, for example, trying to reach inaccessible food, to escape from a locked cage, and to cope with malfunctioning equipment. The final image of each videotape sequence was put on hold, and Sarah was offered a choice of two photographs to place beside the video monitor. Both of these represented the actor in the problem situation, but only one of them showed the actor taking a course of action that would solve the problem. Sarah consistently chose the photographs representing problem solutions, and this was interpreted as evidence that she attributed mental states to the actor (Premack & Woodruff, 1978; see Premack, 1983, 1988, for reservations about this conclusion). It was argued that if Sarah did not ascribe beliefs and desires to the actor, then she would see the video as an undifferentiated sequence of events, rather than a problem. In this case, she would be expected either to respond at random, or to choose from each pair of photographs the one that was more attractive to her, or that bore a greater physical resemblance to the videotape.

Close examination of the videotape experiments (Premack & Premack, 1982; Premack & Woodruff, 1978) suggests that Sarah could, for any given problem, have responded on the basis of familiarity, physical matching, and/or formerly learned associations. For example, when the actor was trying to reach food that was horizontally out of reach, matching could have been responsible for Sarah's success because a horizontal stick was prominent in both the final frame of the videotape and in the photograph depicting a solution. Similarly, when the actor was shivering and looking wryly at a broken heater, Sarah may have selected the photograph of a burning roll of paper, rather than an unlit or spent wick, because she associated the heater with the red-orange color of fire. However, taken together, the results of the videotape experiments are not subject to a single, simple "killjoy" interpretation (Dennett, 1983), and in this respect they are apparently unique within the literature on social cognition in primates.

It is unfortunate that the results of other experiments on role taking (Povinelli, Nelson, & Boysen, 1992; Povinelli, Parks, & Novak, 1992) do not facilitate interpretation of Premack and Woodruff's findings. In one of these (Povinelli, Nelson, & Boysen, 1992), four chimpanzees were initially trained either to choose from an array of containers the one to which an experimenter was pointing (cue-detection task), or to observe food being placed in one of the containers and then to point at the baited receptacle (cue-provision task). Once criterion performance had been achieved on the initial problem, each chimpanzee was confronted with the alternative task, and three of the four animals swiftly attained a high level of accuracy during this second phase of the experiment.

This result was interpreted as evidence of role taking or "cognitive empathy" (Povinelli, Nelson, & Boysen, 1992), but it is subject to another interpretation. The chimpanzees may have quickly achieved a high level of accuracy on the second task, not because the first had allowed them to imagine the situation from another's perspective, but because they had learned most of what they needed to know to solve the second problem during pretraining and outside the experimental situation. The chimpanzees had learned to pull the levers to obtain food during pretraining, and they commonly encountered and exhibited pointing behavior in their day-to-day laboratory lives. When rhesus monkeys, which lacked prior experience of pointing, were switched from cue-provision to cue-detection tasks, or vice versa, they did not immediately succeed on their second problem (Mason & Hollis, 1962; Povinelli, Parks, & Novak, 1992).

If the results of the chimpanzee experiment (Povinelli, Nelson, & Boysen, 1992) had shown that each problem (cue detection and cue provision) was learned faster when it was presented second than when it was presented first, there would be reason to believe that some feature of the first task had facilitated performance in the second. However, even in this case, further experiments, varying the requirements of the first task, would be necessary to find out which feature was enhancing second-task performance, and it is not clear which manipulations, if any, could provide unambiguous evidence that the opportunity for mental state attribution was responsible (Heyes, 1993).

In sum, Sarah, a chimpanzee with an extensive training history, is the only animal that has provided evidence suggestive of mental state attribution in a study of role taking.

E. Deception

When applied to animal behavior, the term deception is often used in a functional sense (Krebs & Dawkins, 1984) to refer to the provision by one animal, through production or suppression of behavior, of a cue that is likely to lead another to make an incorrect or maladaptive response. A mass of observational and anecdotal data leave little doubt that, thus defined, deception occurs in a broad variety of primate and nonprimate species (for recent reviews, see Cheney & Seyfarth, 1991; Krebs & Dawkins, 1984; Mitchell & Thompson, 1986; Whiten & Byrne, 1988). For example, some male scorpion flies adopt the posture and behavior of females, thereby eliciting mating gifts from other males (Thornhill, 1979); chimpanzees occasionally preface aggressive behavior with appeasement gestures (de Waal, in Cheney & Seyfarth, 1991); and various species of birds, vervet monkeys, and chimpanzees sometimes give predator alarm calls in the absence of predators (Cheney & Seyfarth, 1990b; de Waal, 1986; Moller, 1988; Munn, 1986).

While there can be no doubt about the widespread occurrence of functionally deceptive behavior, the research necessary to find out whether it involves mental state attribution has barely begun. There have been many carefully conducted studies of deceptive behavior in nonprimate species but, since this behavior tends to be inflexible and domain specific, they are not thought to indicate intentional deception. Examples include the mating behavior of male scorpion flies (see above), and the practice among male pied flycatchers of tricking females into polygamy by maintaining two, geographically distant territories (Alatalo, Carlson, Lundberg, & Ulstrand, 1981; Alatalo, Lundberg, & Stahlbrandt, 1984). In contrast, it is commonly claimed that primates act with the intention of producing or sustaining a state of ignorance or false belief in another animal, and yet the evidence to date is almost exclusively anecdotal (Cheney & Seyfarth, 1991; Whiten & Byrne, 1988), and the behavior described in each anecdote is subject to at least one alternative interpretation.

In a number of anecdotes, intentional deception is inferred from the fact that one primate has approached another in a friendly way, and then launched an attack. For example, "If Puist [a chimpanzee] is unable to get a hold of her opponent during a fight, we may see her walk slowly up to her and then attack unexpectedly. She may also invite her opponent to reconciliation in the customary way. She holds out her hand and when the other hesitantly puts her hand in Puist's, she suddenly grabs hold of her. This has been seen repeatedly and creates the impression of a deliberate attempt to feign good intentions in order to square accounts" (de Waal, 1982). In cases such as this one, it is undoubtedly natural to assume that the protagonist deliberately deceived its opponent but, as Mackintosh (in press) has pointed out, the behavior is just what one would expect from studies in which animals such as the laboratory rat are confronted with an object that has been associated with both positive and negative consequences. Under these circumstances, the rat will approach the object with increasing hesitation, and dart away again if it gets too close. This suggests that attraction generalizes more widely that aversion, and therefore that Puist may have felt genuinely friendly as she approached the other chimpanzee, a feeling that switched to aggression when she got too close.

Other informal reports of deceptive behavior invite several alternative interpretations: that the behavior occurred: (1) by chance; (2) as a result of associative learning; or (3) as a product of inferences about observable features of the situation rather than mental states (Heyes, 1993; Kummer et al., 1990; Premack, 1988). For example, "One of the female baboons at Gilgil grew particularly fond of meat, although the males do most of the hunting. A male, one who does not willingly share, caught an antelope. The female edged up to him and groomed him until he lolled back under her attentions. She then snatched the antelope carcass and ran" (Jolly, 1985).

The female baboon may have intended to deceive the male about her intentions, but it may also have been no more than a coincidence that she began grooming the male when he was holding the carcass, and made a grab for the carcass when he was lolling back. This could be tested by measuring the frequency with which female baboons groom males who are not in possession of a valuable resource, and thereby assessing the probability that the female in the narrative groomed the male because he was holding a carcass. If the probability turned out to be low, this procedure would allow the "chance" explanation to be discounted.

More intractable problems emerge when one considers a second possibility, that the female's behavior was acquired through associative learning. For example, she may have snatched the carcass when the male was lolling back because in the past similar acts had proved rewarding when executed in relation to supine individuals. That is, the female could have snatched food from conspecifics on many previous occasions, initially without regard to their posture, but if she got away with it when the victim was supine, and not when the victim was upright, she might have acquired an association between snatching food and reward that was activated by the sight of a supine animal. It is not clear how, if at all, observational data could be used to distinguish an associative account of deceptive behavior from an account in terms of mental state attribution. Students of animal behavior used to assume that associative learning occurs gradually, while the effects of inferential learning, or reasoning, suddenly become apparent in behavior (e.g., Kohler, 1925). If this were true, and if observational data could indicate reliably whether learning was gradual or abrupt, then they may be sufficient to distinguish the two accounts. But it would appear not to be true. The many reports of one-trial food aversion learning in rats show that associative learning can be abrupt (e.g., Kaye, Gambini, & Mackintosh, 1988), and evidence that animals acquire beliefs about the relationship between lever pressing and food in the course of instrumental training (Dickinson & Dawson, 1989; Heyes & Dickinson, 1990) imply that the consequences of inferential learning may only gradually become apparent in behavior.

Even if observational studies of deceptive behavior could show that it was acquired through an inferential process, there would remain the possibility that the behavior was based on reasoning about observable features of the situation, rather than mental states. Thus, the female baboon may have inferred from her experience of conspecific behavior that it is relatively safe to snatch food when the other animal is lying back, but she need not have regarded posture as an indicator of mental state. It has been suggested that, if not single anecdotes, collections of such reports, each relating to the behavior of a different individual, could provide clear evidence that reasoning about mental states, rather than observable features of a situation, is responsible for deceptive behavior (Whiten & Byrne, 1988). However, this does not appear to be the case when one considers the hypothetical example of three animals observed, on separate occasions, snatching food that was previously available to a conspecific. The first grooms the conspecific and snatches when it is supine, the second presents and grabs when the male is sexually excited, and the third throws a missile and makes his move when the conspecific is giving chase. We humans might feel inclined to attribute the state of "intending to deceive with intimate behaviour" to all three animals (Whiten & Byrne, 1988), but the potential to attract the same mental state attribution from us might be the sum of what the three have in common regarding mental state attribution. Even if we could be sure that none of them had simply been lucky, and that all of them had acquired the behavior through some inferential process, the possibility would remain that they had learned to snatch from supine, sexually excited, and departing individuals, respectively.

The results of the only experimental investigation of intentional deception in primates (Woodruff & Premack, 1979) are also equivocal. At the beginning of each trial in this study, a chimpanzee was allowed to observe food being placed in one of several inaccessible containers, and then a human trainer, dressed in green ("cooperative" trainer) or white ("competitive" trainer), entered the room and searched one of the containers. The trainer had been instructed to choose the container that the chimpanzee appeared to indicate through pointing, looking, or body orientation. When the cooperative trainer found food, he gave it to the chimpanzee, but the chimpanzee was rewarded on competitive trainer trials only if the trainer chose the incorrect container. After 120 trials, each of the four chimpanzees tested showed a significant tendency to indicate the baited container in the presence of the cooperative trainer, and an empty container in the presence of the competitive trainer. Thus, the chimpanzees' behavior toward the competitive trainer was deceptive, in the functional sense, but the process underlying this behavior is not clear. The animals may have intended to induce in the competitive trainer a false belief about the location of food, or they may have learned, through association or otherwise, that indicating the baited container in the presence of a trainer wearing green led to nonreward (Dennett, 1983; Heyes, 1993).

Recent studies of children underline the difficulty of establishing, using nonverbal tests, that functionally deceptive behavior involves mental state attribution. Chandler, Fritz, and Hala (1989) engaged two- and three-year-old children in a board game in which "treasure" could be hidden in one of several cups with the aid of a puppet that left inky tracks. When they had been familiarized with the situation, the children were encouraged to hide the treasure such that one of the experimenters, who had temporarily left the room, would not find it. Children as young as two and a half years of age used a sponge to wipe away telltale tracks, and/or added false trails to empty cups, and this was taken to indicate that children of this age are capable of attributing ignorance or a false belief to another person, and of acting to encourage such a belief. However subsequent studies (Sodian, Taylor, Harris, & Perner, 1991) showed that children under four years of age who erase tracks and leave misleading trails do not answer questions in a manner that is consistent with this interpretation. When asked where the dupe would believe the treasure to be hidden, they indicated the cup that actually contained the treasure, not the one to which a false trail had been laid. Sodian et al. (1991) concluded that children become capable of intentional, rather than functional, deception at around the age of four years.

F. Perspective Taking

It is a fundamental tenet of the human theory of mind that, under many circumstances, "seeing is believing." When an individual has had visual access to a state of affairs, X, they are likely to know about X, but when they have not, they are likely to be ignorant with respect to X. Consequently, if a nonhuman animal were spontaneously to behave in a different way toward individuals when they have and have not had visual access to an event, and if this behavior were akin to what a human would do when they took another to be either knowledgeable or ignorant with respect to that event, there would be a strong *prima facie* case for mental state attribution by the animal.

Several experiments on "perspective taking" in primates (Cheney & Seyfarth, 1990a; Povinelli et al., 1990; Povinelli, Parks, & Novak, 1991; Premack, 1988; see also Menzel, 1971) have been based on this kind of reasoning. Like studies of deception, they seek evidence that primates attribute beliefs by examining whether the social behavior of protagonists is attuned to the degree to which their interactants have had perceptual access to critical events. However, in research on perspective taking, the focal social behavior is not functionally deceptive, and, to date, vision is the only perceptual modality that has been given explicit consideration.

Two studies of perspective taking in monkeys (Cheney & Seyfarth, 1990a; Povinelli et al., 1991), and several involving chimpanzees (Premack, 1988), reported failure to find evidence that the animals behaved in a different way toward interactants that had, and had not, had visual access to critical events. In the remaining study (Povinelli et al., 1990), chimpanzees were tested in a two-stage procedure. At the beginning of each trial in the first, discrimination training, stage, a chimpanzee was in a room with two trainers. One trainer, designated the "Guesser," left the room, and the other, the "Knower," baited one of four containers. The containers were screened such that the chimpanzee could see who had done the baiting, but not where the food had been placed. After baiting, the Guesser returned to the room, the screen was removed, and each trainer pointed directly at a container. The Knower pointed at the baited container, and the Guesser at one of the other three, chosen at random. The chimpanzee was allowed to search one container, and to keep the food if it was found.

Two of the four animals tested in this way quickly acquired a tendency to select the container indicated by the Knower more often than that indicated by the Guesser, and the second stage of the procedure was designed to find out whether this discrimination was based on the trainers' visual access to the baiting operation. In each trial of this transfer stage, baiting was done by a third trainer in the presence of both the Knower and the Guesser, but during baiting the Guesser had a paper bag over his head. For each chimpanzee, mean choice accuracy in the final 50 trials of Stage 1 was comparable with that in the 30 trials of Stage 2, and this transfer performance was taken to indicate that the chimpanzees were "modelling the visual perspectives of others" (Povinelli et al., 1990).

This experiment provides some of the least ambiguous evidence to date of mental state attribution in any nonhuman animal, and it does so using a transfer procedure with considerable potential (Heyes, 1993). However, two features of the experiment cast doubt on the conclusion that the subjects' behavior was attuned to the trainers' visual access to the baiting procedure. First, since transfer was measured in a less than sensitive way, there may have been an undetected decrement in performance at the beginning of Stage 2 as the subjects learned to base their performance on a new set of cues. Second, it is not clear whether the chimpanzees were accustomed to dealing with "bagged" humans. In the unlikely event that they were, smooth transfer performance might have been due to prior learning that such people provide poor cues. On the other hand, the bagged trainer might have been a novel and rather alarming stimulus from which the chimpanzees averted their gaze, and this may have been responsible for their tendency to continue in Stage 2 to select the container indicated by the Knower.

Even if the experiment by Povinelli et al. (1990) is assumed to demonstrate discrimination on the basis of visual access, studies by Premack (1988) cast doubt on the conclusion that chimpanzees would make this discrimination because they attribute knowledge and ignorance to the trainers. In an experiment similar to that of Povinelli et al. (1990), Premack (1988) allowed chimpanzees to observe that a container had been baited in view of one trainer (the Knower) and not of another (the Guesser), and then offered the animals a choice of two strings to pull, one attached to each trainer. On having his string pulled, the Knower would step forward and tap the baited container, while the Guesser would tap an empty container, and once this had occurred, the chimpanzees were allowed to search one of the containers for food. Two of the four chimpanzees in this experiment learned to pull the string attached to the Knower, and to select for inspection the container indicated by that trainer. This suggests that these two animals spontaneously discriminated between the trainers on the basis of their visual access to the baiting procedure, and it is tempting to infer that they did this because they attributed knowledge of the food's location to the Knower. However, this is unlikely given the outcome of subsequent trials in which the procedure was identical except that the chimpanzees could see the location of food for themselves. Under these conditions, the animals continued to pull the string attached to the Knower before reaching out to claim their prize. This suggests that the chimpanzees selected the trainer that had visual access to baiting, not because they regarded him as a source of knowledge, but because reward had been contingent upon this action in the past.

Rhesus monkeys and human children have been tested using a procedure similar to the one applied by Povinelli et al. (1990) to chimpanzees (Povinelli & deBlois, 1992; Povinelli et al., 1991), but the results do not clarify the significance of the chimpanzees' transfer performance. The monkeys did not learn to choose the Knower during the first stage of the procedure, and consequently could not be given the transfer test (Povinelli et al., 1991). Four-year-old children were more likely than three-year-olds to search the container indicated by the Knower, and independent research has shown that understanding of the relationship between perception and knowledge is usually acquired between the ages of three and four (e.g., Wimmer, Hogrefe, & Perner, 1988). However, among the children tested by Povinelli and deBlois (1992), those that consistently chose the Knower were no more likely than the others to answer correctly questions about how they, and the trainers, knew about the contents of the container.

Through their research on perspective taking, Povinelli and his associates are attempting to develop an equivalent for primates of the standard "false belief" tasks given to young children. In one of these, children witness the transfer of an object from one container to another, and are asked where an individual who was not present during the transfer will look for the object. In another, they observe the contents of a familiar candy box being replaced with, for example, pencils, and are asked what a person who did not witness the replacement will think is inside the box. Below the age of three and a half years, most children fail these tests, saying that the person who did not have visual access to the critical events will, nonetheless, search for the object at the correct location, or think that the candy box contains pencils. However, most four-year-olds say that the stooge will look for the object in the place where he or she last saw it, or think that the box has its usual contents, and they are consequently judged capable of attributing false beliefs (Perner, Leekam, & Wimmer, 1987). There can be little doubt that investigators of social cognition in primates need measures like these, which have relatively high reliability and construct validity. However, it remains to be seen whether a nonverbal test of mental state attribution, such as Povinelli's, can provide such a measure.

III. CONCLUSION

The foregoing review suggests that there is currently no compelling evidence of mental state attribution in nonhuman primates. The research reviewed in Sections II.A and II.B has not provided unequivocal evidence of imitation and mirror-guided body inspection in primates, and these behaviors are, in any event, unlikely to be indicative of mental state attribution. There can be little doubt that the members of many primate and nonprimate species exhibit sensitivity to social relationships and behavior that functions to deceive other animals (Sections II.C and II.E), but in every documented case the behavior could be based on one or a number of psychological processes other than mental state attribution. Finally, research on role taking and perspective taking (Sections II.D and II.F) has provided data suggestive of mental state attribution in a chimpanzee (Premack & Woodruff, 1978), and a transfer test procedure with considerable potential (Povinelli et al., 1990), but no thoroughly compelling evidence of mental state attribution.

In view of the low construct validity of imitation and mirror-guided body inspection, and the anecdotal character of work on social relationships and deception, it is apparent that little research to date has even had the potential to provide strong evidence of mental state attribution. Consequently, the current lack of such evidence indicates not that primates are unable to attribute mental states, but that inadequate empirical methodology has been used in addressing the question. Children's capacity to attribute mental states has been investigated experimentally with considerable success, so why has research on social cognition in primates relied so heavily on observational data?

Perspective taking is likely to be the most profitable focus for future research because it has relatively high construct validity and there is an experimental paradigm, involving conditional discrimination training followed by transfer tests (Povinelli et al, 1990), available for its investigation. In using this paradigm, it would be advisable to apply two or more transfer tests for each trained discrimination and, in each case, to measure transfer performance with as much sensitivity as possible (Heyes, 1993). This would typically involve confining attention to the first transfer trial, and/or withholding differential reinforcement during the transfer test phase.

The hypothesis that primates attribute mental states is intriguing and important; it has implications with respect to the evolution of intelligence, the epistemic status of "folk psychology," and animal welfare issues. Consequently, the hypothesis warrants thorough investigation, and it is hoped that a skeptical approach, of the kind adopted in this review, will stimulate the experimental work necessary for genuine evaluation.

Acknowledgments

I am grateful to N. J. Mackintosh, H. C. Plotkin, and D. Shanks for their comments on an earlier draft of this chapter.

References

Adams-Curtis, L. E. (1987). Social context of manipulative behavior in *Cebus apella*. *American Journal of Primatology, 12,* 325.

Alatalo, R. V., Carlson, A., Lundberg, A., & Ulstrand, S. (1981). The conflict between male polygamy and female monogamy: The case of the pied flycatcher. *American Naturalist, 117,* 738–753.

Alatalo, R. V., Lundberg, A., & Stahlbrandt, K. (1984). Female mate choice in the pied flycatcher. *Behavioural Ecology and Sociobiology, 14,* 253–261.

Anderson, C. O., & Mason, W. A. (1978). Competitive social strategies in groups of deprived and experienced rhesus monkeys. *Developmental Psychobiology, 11,* 289–299.

Anderson, J. R. (1983). Responses to mirror image stimulation and assessment of self-recognition in mirror- and peer-reared stumptail macaques. *Quarterly Journal of Experimental Psychology, 35B,* 201–212.

Anderson, J. R., & Roeder, J. J. (1989). Responses of capuchin monkeys to different conditions of mirror-image stimulation. *Primates, 30,* 581–587.

Bachmann, C., & Kummer, H. (1980). Male assessment of female choice in hamadryas baboons. *Behavioural Ecology and Sociobiology, 6,* 315–321.

Beck, B. B. (1976). Tool use by captive pigtailed monkeys. *Primates, 17,* 301–310.

Byrne, R. W. (1993), The evolution of intelligence. In P. J. B. Slater & T. R. Halliday (Eds.). *Behaviour and evolution.* Cambridge: Cambridge University Press.

Byrne, R. W., & Whiten, A. (Eds.), (1988). *Machiavellian intelligence: Social expertise and the evolution of intellect in monkeys, apes and humans.* Oxford: Oxford University Press.

Calhoun, S., & Thompson, R. L. (1988). Long-term retention of self-recognition by chimpanzees. *American Journal of Primatology, 15,* 361–365.

Chance, M. R. A., & Mead, A. P. (1953). Social behaviour and primate evolution. *Symposia of the Society for Experimental Biology, 7.* 395–439.

Chandler, M. J., Fritz, A. S., & Hala, S. (1989). Small-scale deceit: Deception as a marker of 2-, 3-, and 4-year-olds' theories of mind. *Child Development, 60,* 1263–1277.

Chater, N., & Heyes, C. M. (in press). Animal concepts: Content and discontent. *Mind and Language.*

Cheney, D. L., & Seyfarth, R. M. (1980). Vocal recognition in free-ranging vervet monkeys. *Animal Behaviour, 288,* 362–367.

Cheney, D. L. & Seyfarth, R. M. (1986). The recognition of social allicances among vervet monkeys. *Animal Behaviour, 34,* 1722–1731.

Cheney, D., & Seyfarth, R. (1990a). Attending to behaviour versus attending to knowledge: Examining monkeys' attribution of mental states. *Animal Behaviour, 40,* 742–753.

Cheney, D. L., & Seyfarth, R. M. (1990b). *How monkeys see the world.* Chicago: University of Chicago Press.

Cheney, D. L., & Seyfarth, R. M. (1991). Truth and deception in animal communication. In C. A. Ristau (Ed.), *Cognitive ethology: The minds of other animals,* (pp. 127–151). Hillsdale, NJ: Erlbaum.

Cheney, D. L., & Seyfarth, R. M. (1992). Precis of 'How monkeys see the world.' *Behavioral and Brain Sciences, 15,* 135–182.

Cheney, D. L., Seyfarth, R. M., & Smuts, B. (1986). Social relationships and social cognition in nonhuman primates. *Science, 234,* 1361–1366.

Cook, T. D., & Campbell, D. T. (1979). *Quasi-experimentation: Design and analysis issues for field settings.* Boston: Houghton Mifflin.

Crawford, M. P. (1939). The social psychology of vertebrates. *Psychological Bulletin, 36,* 407–446.

Darby, C. L. & Riopelle, A. J. (1959). Observational learning in the rhesus monkey. *Journal of Comparative and Physiological Psychology, 52,* 94–98.

Dasser, V. (1988). A social concept in lava monkeys. *Animal Behaviour, 36,* 225–230.

Datta, S. B. (1983). Patterns of agnositic interference. In R. A. Hinde (Ed.), *Primate social relationships: An integrated approach.* Oxford: Blackwell.

Demaria, C. & Thierry, B. (1988). Responses to animal stimulus photographs in stumptailed macaques. *Primates, 29,* 237–244.

Dennett, D. C. (1983). Intentional systems in cognitive ethology: The "Panglossian paradigm" defended. *Behavioral and Brain Sciences, 6,* 343–390.

de Waal, F. (1982). *Chimpanzee politics.* London: Jonathan Cape.

de Waal, F. (1986). Deception in the natural communication of chimpanzees. In R. W. Mitchell & N. S. Thompson (Eds.), *Deception: Perspectives on human and nonhuman deceit* (pp. 221–224). New York: State University of New York Press.

de Waal, F. (1991). Complementary methods and convergent evidence in the study of primate social cognition. *Behaviour, 118,* 297–320.

de Waal, F., & Van Roosemalen, A. (1979). Reconciliation and consolation among chimpanzees. *Behavioural Ecology and Sociobiology, 5,* 55–66.

Dickinson, A. (1983). *Contemporary animal learning theory.* Cambridge: Cambridge University Press.

Dickinson, A., & Dawson, G. R. (1989). Incentive learning and the motivational control of instrumental performance. *Quarterly Journal of Experimental Psychology, 41B.* 99–112.

Dimond, S., & Harries, R. (1984). Face touching in monkeys, apes and man: Evolutionary origins and cerebral asymmetry. *Neuropsychologia, 22,* 227–233.

Eglash, A. R., & Snowdon, C. T. (1983). Mirror-image responses in pygmy marmosets. *American Journal of Primatology, 5,* 211–219.

Epstein, R., Lanza, R. P., & Skinner, B. F. (1981). "Self-awareness" in the pigeon. *Science, 212,* 695–696.

Fragaszy, D. M., & Visalberghi, E. (1989). Social influences on the acquisition and use of tools in tufted capuchin monkeys *(Cebus apella). Journal of Comparative Psychology, 103,* 159–170.

Fragaszy, D. M., & Visalberghi, R. (1990). Social processes affecting the appearance of innovative behaviors in capuchin monkeys. *Folia Primatologica, 3,* 54.

Galef, B. G. (1988). Imitation in animals: History, definition, and interpretation of data from the psychological laboratory. In T. R. Zentall & B. G. Galef, Jr. (Eds.). *Social learning: Psychological and biological perspectives* (pp. 3–28). Hillsdale, NJ: Erlbaum.

Galef, B. G. (1992). The question of animal culture. *Human Nature, 3,* 157–178.

Galef, B. G., Manzig, L. A., & Field, R. M. (1986). Imitation learning in budgerigars: Dawson and Foss (1965) revisited. *Behavioural Processes, 13,* 191–202.

Gallup, G. G. (1970). Chimpanzees: Self-recognition. *Science, 167,* 86–87.

Gallup, G. G. (1977). Self-recognition in primates. *American Psychologist, 32,* 329–338.

Gallup, G. G. (1982). Self-awareness and the emergence of mind in primates. *American Journal of Primatology, 2,* 237–248.

Gallup, G. G. (1983). Toward a comparative psychology of mind. R. E. Mellgren (Ed.), *Animal cognition and behavior* (pp. 473–510). Amsterdam: North-Holland Publ.

Gallup, G. G. McClure, M. K., Hill, S. D., & Bundy, R. A. (1971). Capacity for self-recognition in differentially reared chimpanzees. *Psychological Record, 21,* 69–74.

Gallup, G. G., & Suarez, S. D. (1991). Social responding to mirrors in rhesus monkeys: Effects of temporary mirror removal. *Journal of Comparative Psychology, 105,* 376–379.

Goldman, A. (1993). The psychology of folk psychology. *Behavioral and Brain Sciences, 16,* 15–28.

Goodall, J. (1986). *The chimpanzees of Gombe:Patterns of behavior.* Cambridge, MA: Belknap Press.

Hauser, M. D. (1988). Invention and social transmission: New data from wild vervet monkeys. In R. Byrne & A. Whiten (Eds.), *Machiavellian intelligence: Social expertise and the evolution of intellect in monkeys, apes and humans,* (pp. 327–343). Oxford: Oxford University Press.

Hayes, K. J., & Hayes, C. (1952). Imitation in a home-raised chimpanzee. *Journal of Comparative and Physiological Psychology, 45,* 450–459.

Heyes, C. M. (1993a). Anecdotes, training, trapping, and triangulating: Do animals attribute mental states? *Animal Behaviour, 46,* 177–188.

Heyes, C. M. (1993b). Imitation, culture and cognition. *Animal Behaviour, 46,* 999–1010.

Heyes, C. M. (1994). Reflections on self-recognition in primates. *Animal Behaviour.*

Heyes, C. M. (in press). Social learning in animals: Categories and mechanisms. *Biological Reviews.*

Heyes, C. M., & Dawson, G. R. (1990). A demonstration of observational learning using a bidirectional control. *Quarterly Journal of Experimental Psychology, 42B,* 59–71.

Heyes, C. M., Dawson, G. R., & Nokes, T. (1992). Imitation in rats: Initial responding and transfer evidence. *Quarterly Journal of Experimental Psychology, 45B,* 81–92.

Heyes, C. M., & Dickinson, A. (1990). The intentionality of animal action. *Mind and Language,* 5, 87–104.

Hinde, R. A. (Ed.), (1983). *Primate social relationships: An integrated approach.* Oxford: Blackwell.

Humphrey, N. K. (1976). The social function of intellect. In P. P. G. Bateson & R. A. Hinde (Eds.), *Growing points in ethology* (pp. 303–317). Cambridge: Cambridge University Press. [Reprinted in Byrne and Whiten (1988)]

Jolly, A. (1966). Lemur social behaviour and primate intelligence. *Science, 153,* 501–506.

Jolly, A. (1985). *The evolution of primate behaviour* (2nd ed.). New York: Macmillan.

Jolly, A. (1991). Conscious chimpanzees? A review of recent literature. In C. A. Ristau (Ed.), *Cognitive ethology: The minds of other animals* (pp. 231–252). Hillsdale, NJ: Erlbaum.

Judge, P. (1982). Redirection of aggression based on kinship in a captive group of pigtail macaques. *International Journal of Primates, 3,* 301.

Judge, P. (1983). Reconciliation based on kinship in a captive group of pigtail macaques. *American Journal of Primatology, 4,* 346.

Kawai, M. (1965). Newly acquired pre-cultural behavior of the natural troop of Japanese monkeys on Koshima Inlet. *Primates, 6,* 1–30.

Kaye, H., Gambini, B., & Mackintosh, N. J. (1988). A dissociation between one-trial over-shadowing and the effect of a distractor on habituation. *Quarterly Journal of Experimental Psychology, 40B,* 31–47.

Kohler, W. (1925). *The mentality of apes.* London: Routledge & Kegan Paul.

Krebs, J. R., & Dawkins, R. (1984). Animal signals: Mind reading and manipulation. In J. R. Krebs & N. B. Davies (Eds.), *Behavioural Ecology* (pp. 380–402). Oxford: Blackwell.

Kummer, H. (1968). *Social organization of hamadryas baboons.* Basel: Karger.

Kummer, H., Dasser, V., & Hoyningen-Huene, P. (1990). Exploring primate social cognition: some critical remarks. *Behaviour, 112,* 84–98.

Ledbetter, D. H., & Basen, J. A. (1982). Failure to demonstrate self-recognition in gorillas. *American Journal of Primatology, 2,* 307–310.

Lin, A. C., Bard, K. A., & Anderson, J. R. (1992). Development of self-recognition in chimpanzees. *Journal of Comparative Psychology, 106,* 120–127.

Mackintosh, N. J. (in press). The evolution of intelligence. In J. Khalfa (Ed.). *Intelligence.* Cambridge: Cambridge University Press.

Macphail, E. (1991). Review of Whiten & Byrne (1988). *Quarterly Journal of Experimental Psychology, 43B,* 105–107.

Mason, W. A., & Hollis, J. H. (1962). Communication between young rhesus monkeys. *Animal Behaviour, 10,* 211–221.

Menzel, E. W. (1971). Communication about the environment in a group of young chimpanzees. *Folia Primatologica, 15,* 220–232.

Mignault, C. (1985). Transition between sensorimotor and symbolic activities in nursery-reared chimpanzees *(Pan troglodytes). Journal of Human Evolution, 14,* 747–758.

Mitchell, R. W., & Thompson, N. S. (Eds.), (1986). *Deception: Perspectives on human and nonhuman deceit.* New York: State University of New York Press.

Moller, A. P. (1988). False alarm calls as a means of resource usurpation in the great tit *Parus major. Ethology, 79,* 25–30.

Munn, C. A. (1986). Birds that cry 'wolf'. *Nature (London), 319,* 143–145.

Nishida, T. (1987). Local traditions and cultural transmission. In B. B. Smuts, D. L. Cheney, R. M. Seyfarth, R. W. Wrangham, & T. T. Struhsaker (Eds.), *Primate societies* (pp. 462–474). Chicago: University of Chicago Press.

Pearce, J. (1989). *An introduction to animal cognition.* Hillsdale, NJ: Erlbaum.

Perner, J., Leekam, S. & Wimmer, H. (1987). Three-year-olds' difficulty in understanding false belief: Cognitive limitation, lack of knowledge, or pragmatic misunderstanding? *British Journal of Developmental Psychology, 5,* 125–137.

Platt, M. M., & Thompson, R. L. (1985). Mirror responses in Japanese macaque troop. *Primates, 26,* 300–314.

Povinelli, D. J. (1987). Monkeys, apes, mirrors and minds: The evolution of self-awareness in primates. *Human Evolution, 2,* 493–509.

Povinelli, D. J. (1993). Reconstructing the evolution of mind. *American Psychologist, 48,* 493–509.

Povinelli, D. J. (in press). Panmorphism. In R. Mitchell & N. Thompson (Eds.), *Anthropomorphism, anecdotes and animals.* Lincoln: University of Nebraska Press.

Povinelli, D. J., & deBlois, S. (1992). Young children's understanding of knowledge formation in themselves and others. *Journal of Comparative Psychology, 106,* 228–238.

Povinelli, D. J., Nelson, K. E., & Boysen, S. T. (1990). Inferences about guessing and knowing by chimpanzees. *Journal of Comparative Psychology, 104,* 203–210.

Povinelli, D. J., Nelson, K. E., & Boysen, S. T. (1992). Comprehension of role reversal in chimpanzees: Evidence of empathy? *Animal Behaviour, 43,* 633–640.

Povinelli, D. J., Parks, K. A., & Novak, M. A. (1991). Do rhesus monkeys attribute knowledge and ignorance to others? *Journal of Comparative Psychology, 105,* 318–325.

Povinelli, D. J., Parks, K. A., & Novak, M. A. (1992). Role reversal by rhesus monkeys, but no evidence of empathy. *Animal Behaviour, 43,* 269–281.

Premack, D. (1983). Animal cognition. *Annual Review of Psychology, 34,* 351–362.

Premack, D. (1988). 'Does the chimpanzee have a theory of mind?' revisited. In R. W. Byrne & A. Whiten (Eds.). *Machiavellian intelligence: Social expertise and the evolution of intellect in monkeys, apes and humans.* Oxford: Oxford University Press.

Premack, D. (1991). The aesthetic basis of pedagogy. In R. R. Hoffman & D. S. Palermo (Eds.). *Cognition and the symbolic processes.* (pp. 303–325). Hillsdale, NJ: Erlbaum.

Premack, D., & Premack, A. J. (1982). *The mind of an ape.* New York: Norton.

Premack, D., & Woodruff, G. (1978). Does the chimpanzee have a theory of mind? *Behavioral and Brain Sciences, 4,* 515–526.

Quiatt, D., & Reynolds, V. (1993). *Primate behaviour.* Cambridge: Cambridge University Press.

Robert, S. (1986). Ontogeny of mirror behavior in two species of great apes. *American Journal of Primatology, 10,* 109–117.

Smuts, B. B. (1985). *Sex and friendship in baboons.* New York: Aldine.

Smuts, D. L., Cheney, R. M., Seyfarth, R. W., Wrangham, R. W., & Struhsaker, T. T. (Eds.), (1987). *Primate societies.* Chicago: University of Chicago Press.

Sodian, B., Taylor, C., Harris, P. L., & Perner, J. (1991). Early deception and the child's theory of mind: False trails and genuine markers. *Child Development, 62,* 468–483.

Spence, K. W. (1937). Experimental studies of learning and higher mental processes in infra-human primates. *Psychological Bulletin, 34,* 806–850.

Stammbach, E. (1988). Group responses to specially skilled individuals in a *Macaca fascicularis* group. *Behaviour, 107,* 241–266.

Suarez, S. D., & Gallup, G. G. (1981). Self-recognition in chimpanzees and orangutans, but not gorillas. *Journal of Human Evolution, 10,* 175–188.

Suarez, S. D., & Gallup, G. G. (1986). Social responding to mirrors in rhesus macaques: Effects of changing mirror location. *American Journal of Primatology, 11,* 239–244.

Sumita, K., Kitahara-Frisch, J., & Norikoshi, K. (1985). The acquisition of stone-tool use in captive chimpanzees. *Primates, 26,* 168–181.

Terrace, H. S., Petitto, L. A., Sanders, R. J., & Bever, T. G. (1979). Can an ape create a sentence? *Science, 206,* 891–902.

Thorndike, E. L. (1898). Animal intelligence. *Psychological Review Monographs, 2,* (8).

Thornhill, R. (1979). Adaptive female-mimicking behaviour in a scorpion fly. *Science, 205,* 412–414.

Timberlake, W., & Grant, D. L. (1975). Autoshaping in rats to the presentation of another rat predicting food. *Science, 190,* 690–692.

Tomasello, M., Davis-Dasilva, M., Camak, L. & Bard, K. (1987). Observational learning of tool-use by young chimpanzees. *Human Evolution, 2,* 175–183.

Tomasello, M., Kruger, A. C., & Ratner, H. H. (1993). Cultural learning. *Behavioral and Brain Sciences, 16,* 507–545.

Tomasello, M., Savage-Rumbaugh, S., & Kruger, A. C. (in press). Imitative learning of actions on objects by children, chimpanzees, and enculturated chimpanzees. *Child Development.*

Ungerer, J. A. (1989). The early development of autistic children. In G. Dawson (Ed.), *Autism: Nature, diagnosis and treatment* (pp. 75–91). New York: Guilford Press.

Visalberghi, E., & Fragaszy, D. M. (1992). Do monkeys ape? In S. T. Parker & K. R. Gibson (Eds.). *"Language" and intelligence in monkeys and apes: Comparative developmental perspectives* (pp. 247–273). Cambridge: Cambridge University Press.

Visalberghi, E., & Trinca, L. (1989). Tool use in capuchin monkeys, or distinguish between performing and understanding. *Primates, 30,* 511–521.

Warden, C. J., & Jackson, T. A. (1935). Imitative behaviour in the rhesus monkey. *Journal of Genetic Psychology, 46,* 103–125.

Westergaard, G. C. (1988). Lion-tailed macaques *(Macaca silenus)* manufacture and use tools. *Journal of Comparative Psychology, 102,* 152–159.

Whiten, A. (Ed.). (1991). *Natural theories of mind.* Oxford: Basil Blackwell.

Whiten, A. & Byrne, R. W. (1988). Tactical deception in primates. *Behavioral and Brain Sciences, 11,* 233–273.

Whiten, A. & Byrne, R. W. (1991). The emergence of metarepresentation in human ontogeny and primate phylogeny. A. Whiten (Ed.). *Natural theories of mind,* (pp. 267–281). Oxford: Basil Blackwell.

Whiten, A. & Ham, R. (1992). On the nature and evolution of imitation in the animal kingdom: reappraisal of a century of research. P. J. B. Slater, J. S. Rosenblatt, C. Beer & M. Milinski (Eds.), *Advances in the Study of Behavior,* (vol. 21). San Diego: Academic Press.

Wimmer, H., Hogrefe, G. J. & Parner, J. (1988). Children's understanding of informational access as a source of knowledge. *Child Development, 59,* 386–396.

Woodruff, G. & Premack, D. (1979). Intentional communication in the chimpanzee: the development of deception. *Cognition, 7,* 333–362.

CHAPTER 11

Language in Comparative Perspective

Duane M. Rumbaugh
E. Sue Savage-Rumbaugh

An unequivocal demonstration that at least one other species besides man is capable of language would add to the series of man's great reconceptions of himself. . . .

(Ploog & Melnechuk, 1971, p. 607)

I. VIEWS OF LANGUAGE

Traditionally, language has been viewed as either a unique endowment of our species, *or* as the most complex system of behaviors that may be acquired by general learning processes that are *not* specific to our species. Those of the *first* persuasion, and notably Chomsky (1965, 1988), have posited genetic mutations for its presence in humans and have emphasized speech production and "universals" of grammar that transcend specific language systems. Those of the *second* persuasion view language as the culmination of processes that have their roots in evolution of animal life and have emphasized brain complexity and the attendant enhancement of learning and cognition (Bates, Thal, & Marchman, 1991; Rumbaugh, Hopkins, Washburn, & Savage-Rumbaugh, 1991; Savage-Rumbaugh et al., 1993). Although its uniqueness and efficiency as a medium for discourse is fully acknowledged by such advocates, speech is not held to be the sine qua non language. Rather, it is argued that language should be viewed as an emergent behavioral and cognitive system that will benefit from study within a comparative–developmental framework. Accordingly, allowance was and is made for the acquisition of language in some form even by nonhumans. Advocates of this second perspective also have brought attention to *comprehension* (Savage-Rumbaugh et al., 1993) as the foundation of language, from

Animal Learning and Cognition

which productive language emerges, and have argued that media other than speech can serve language and its several functions.

Many readers will recall the names of Noam Chomsky (1965, 1988) and B. F. Skinner (1957) as early advocates of nativistic and empirical perspectives of language origins, respectively. The polarity established by their arguments has fostered substantial bodies of research and new perspectives, which can be found in their contemporary forms in a recent volume by Krasnegor, Rumbaugh, Schiefelbusch, and Studdert-Kennedy (1991). Whatever the perspective, it is helpful to recall that there is no nativist who would argue that learning has no place in the acquisition of language and, similarly, that there is no empiricist so radical as to deny the biological and evolutionary prerequisites of language.

The complexity of language is surely the ultimate orchestration of psychological competencies, and its understanding is to be achieved only through studies of interactions between heredity and environment. And to the degree that a rapprochement between the extreme positions of nativism and empiricism eventually will be forged, it surely will entail an understanding of the interaction between early experience and its impact on brain architectonics and networks.

A. Animals and Language

The last third of the twentieth century has seen a great increase in language research with animals (for recent summaries, see Parker & Gibson, 1990; Roitblat, Herman, & Nachtigall, 1993; Savage-Rumbaugh & Lewin, 1994). The reports include dolphins carrying out novel requests (Herman, 1987; Herman, Pack, & Morrel-Samuels, 1993), sea lions learning the referents for new manual signs apparently on the basis of exclusion (e.g., the object is novel, hence the hand-sign given by the experimenter must be its "name"; Schusterman, Gisiner, Grimm, & Hanggi, 1993), a parrot seemingly able to answer questions regarding multiple characteristics of sets of objects (Pepperberg, 1993), and great apes (*Gorilla, Pan,* and *Pongo*) learning various kinds of symbol sets (e.g., variants of manually produced sign language, plastic tokens, and geometric symbols) for communication with humans and problem solving (Fouts & Fouts, 1989; Gardner & Gardner, 1969; Gardner, Gardner, & Van Cantfort, 1989; Miles, 1990; Patterson & Linden, 1981; Premack, 1971; Premack & Premack, 1983; Rumbaugh, 1977).

Perhaps the most significant of recent reports are those of chimpanzees and bonobos (both of the genus *Pan*) coming to comprehend human speech—not only single words, but sentences as well, notably by the bonobo—at a level that compares with that of a 2 1/2-year-old child (Savage-Rumbaugh et al., 1993). Of particular significance, as detailed later in this chapter, is the fact that these apes acquired their language competencies *not* by specific teaching or training, but rather by being reared from birth in a language-structured environment. This fact serves to validate the argument that language is not uniquely human, and revitalize questions of (1) the evolution and relationship of apes to the early humans and (2) the role of early environment and rearing in language acquisition.

B. Language and Its Definition

Like many other terms in psychology, language cannot easily be defined to everyone's satisfaction, but that problem neither denies the existence of language nor precludes research on it. Whatever controversies might remain on the matter of definition, one may still search for the biobehavioral requisites to language and relationships between various language functions.

For the time being, an approximation or a *working* definition must suffice:

> Language is a neurobehavioral, multidimensional system that provides for the construction and use of symbols in a manner that enables the conveyance and receipt of information and novel ideas between individuals. The meanings of symbols in this system are basically defined and modulated through social interactions.

Language here will be viewed as a constructed system—one that is "open" to the addition of new symbols, new functions and uses, and to the sculpting of meaning through novel social uses of symbols. Language also will be viewed not as a unitary phenomenon, but as a "braided" continuum—where each strand is a process that, through the course of language acquisition, interacts with others.

Perhaps it was the salience of speech that led to the quite incorrect conclusion historically that language *is* speech. Although without question speech affords a highly efficient system that, together with hearing, allows for rapid expression of language, speech can do no more than to disturb the molecules and to establish waves of sounds. But those sounds are without meaning except for that given them by the listener. Were it not thus, there would be no such thing as a "foreign" language.

II. EVOLUTION AND LANGUAGE

The general course of evolution is for the selection of attributes that serve adaptation and reproductive success. Complex brains presumably were selected for their enhanced information processing, learning, and memory that enabled individuals to retain important lessons learned from lifelong cumulative experiences. Nevertheless, selection for them was gradual, and it was likely the selection for specific neural structures laid down in brain evolution that enabled the gradual emergence of language. Large and complex brains made a difference to the interests of survival and reproduction.

When compared with other primates, the great apes (*Pan, Gorilla,* and *Pongo*) have brains that are disproportionately large relative to the sizes of their bodies. The human brain is even larger. Notwithstanding the proportional differences and the fact that the human brain is about three times larger than the ape brain, the basic architecture of the brain remains the same (Jerison, 1985).

Of the over 200 species that comprise the order Primates (Napier & Napier, 1967), the genus most closely related to humans is *Pan. Pan* includes two species—*troglodytes* and *paniscus,* herein called chimpanzee and bonobo, respectively. Human and chimpanzee share 99% of their genetic material (Andrews & Martin, 1987; Sibley & Ahlquist, 1987) and divergence of their evolutionary lineages occurred perhaps as recently as a few million years ago (Sarich, 1983). Human and chimpanzee appear to be more closely related than are chimpanzee and gorilla! Accordingly, it would seem particularly possible, though not inevitable, that human and chimpanzee, and to a lesser degree the other great apes, might share psychological competencies (Darwin, 1859; Rumbaugh & Pate, 1984) as well as the more obvious morphological attributes. It is this probability that directed early language research with animals primarily to the chimpanzee.

A. Constraints on Speech by Apes

Before it became clear that neural and anatomical limitations of the apes' vocal tracts precluded the production of sounds necessary for human speech (Lieberman, 1968; Hopkins & Savage-Rumbaugh, 1991), several attempts were made to cultivate their speech. The first serious effort was by Furness (1916) with an orangutan (*Pongo pygmaeus*), who mastered only the crudest approximations of four words. The best known systematic effort was by Keith and Cathy Hayes (Hayes, 1951; Hayes & Hayes, 1950; Hayes & Nissen, 1971) with Vicki, a chimpanzee (*P. troglodytes*). As with Furness, the success of their student was limited to four words—mama, papa, cup, and up. Although the Hayeses' research is frequently concluded to have been a failure, it is only so in the sense that Vicki failed to acquire facile speech and thereby to define speech as a ready and familiar medium for continued research with apes. By other standards, the Hayeses' effort was a grand success in that it provided a wealth of information about the cognition, perceptions, and affect of a chimpanzee through the course of its early development. Their reports assured researchers for the next half century that *if* apes had language skills, they were not to be accessed by teaching them to speak! And it is to their credit that just because they "failed" to instate language in Vicki, they gave no one reason to believe that they were arguing for the null hypothesis—namely, that because their chimpanzee did not master language, *no* chimpanzee could ever do so.

Intervening between the speech studies of Furness and the Hayes was a study by Kellogg and Kellogg (1933) that consisted of the short-term co-rearing of a chimpanzee, Gua, with their son, Donald. Gua failed to develop speech spontaneously, though in other ways development approximated Donald's. Another important effort by Yerkes and Learned (1925) is notable for the speculation that some sort of manual sign language might be appropriate for studies of language with the chimpanzee.

B. Nonspeech Methods

The mid-1960s saw the beginning of two important chimpanzee projects, each with a unique *non*speech approach. Project Washoe, already referenced, was initiated by Beatrix and Allen Gardner of the University of Nevada, and Project Sarah was started by David Premack, then at the University of California at Santa Barbara. The Gardner's (1969) goal was "two-way communication" with Washoe (a *P. troglodytes*). Given the early view expressed by Yerkes and by the chimpanzee Vicki's use of her hands as she attempted to "talk," the Gardners thought that Washoe might learn American Sign Language (AMESLAN; Bellugi, Bihrle, & Corina, 1991), and thereby acquire a *natural* language. A natural language was used in the hope that whatever Washoe learned would be language and that her achievements would not be lost to quarrels over definitions.

1. Project Washoe

Washoe's contributions were several: (1) her mastery of a large corpus of manual signs, some of her own innovation and generalized use; (2) her performance in blind tests that documented her accurate associations of specific signs with their respective exemplars; and (3) her solitary use of signs, fortuitously captured on film as she looked at magazines and played. The Gardners and others have provided carefully prepared reports that detail Washoe's significant accomplishments (Gardner et al., 1989).

2. Project Sarah

In contrast to the Gardners' choice of manual signs, David Premack (1971) used plastic tokens of various shapes and colors to function as words with his chimpanzee, Sarah. Thus, his was a synthetic, rather than natural, language system. Premack's tactics and goals were quite different from those of the Gardners. He was more concerned with questions regarding how a token might function as a word and the conditions for the generalized or extended use of words. His tactics made extensive use of match-to-sample methods, which, on one hand, simplified language operations, but, on the other hand, served to suggest that Sarah did not have language so much as good "problem-solving skills" (Harlow, 1977; Terrace, 1979a). Project Sarah's contributions included: (1) successful implementation of a synthetic language, comprised of plastic tokens; (2) analyses of word function and use through tests of "analogical" reasoning, that included "same" versus "different" judgments (Oden, Thompson, & Premack, 1990); (3) studies of transitive inference or ordered relationships between items; (4) heuristic experiments that suggested preferential use of persons in problem-solving paradigms; and (5) suggestions of attribution of states of knowledge to others and thereupon the concept of "theory of mind" (see Chapter 10, Heyes, this volume; Premack, 1986; Premack & Premack, 1983). As discussed in greater detail later, in the late 1970s Premack concluded that Sarah's language was quite different from human language and that she was not capable of sentences.

3. Project Lana

The LANA Project was begun in the 1970s by Rumbaugh and his associates of Georgia State University, the University of Georgia, and the Yerkes Primate Center of Emory University. Unique to their effort (Rumbaugh, 1977) was the invention of a computer-monitored keyboard, each key of which had a distinctive geometric pattern called a *lexigram*. Each lexigram was to serve as a word for Lana, their chimpanzee (*Pan troglodytes*) student (see Figure 1). The priority goal of the LANA Project was *not* to determine the chimpanzee's language competence, but rather to determine whether a computer-controlled, language-training system might be perfected to advance research where learning and language abilities were limited, either because of genetics (e.g., in apes, whose brains are one-third the size of our brain) or brain damage (e.g., in children with mental retardation). Additionally, the Project's goals were to objectify the data base obtained by automatically recording it by a computer—a PDP-8 with 8K(!) of memory.

Initial training of Lana entailed teaching her, through use of standard instrumental conditioning principles, a variety of "stock" sentences that would be accepted by the computer's program (von Glasersfeld, 1977) for the activation of appropriate devices that would produce for her a sip of various drinks, foods, a movie, a slide show, the assistance of the person, and so on. Thompson and Church (1980) analyzed a large corpus of Lana's sequences, 1276 of which were produced by Lana when an experimenter was present. They argued that 776 of the specifically trained sequences (such as ⟨object⟩ *name-of this*) could be accounted for by a software-based model, which they devised. Of the 500 remaining sequences, they concluded that

FIGURE 1   Lana, a chimpanzee, at her lexigram-embossed keyboard.

all but 28 probably represented "extensions" rather than departures from prototypic stock sentences and that those extensions were likely produced by Lana when she had not obtained an incentive through use of a stock sentence. Their interpretation was in keeping with ours: Lana elaborated upon her skills whenever the most basic ones proved unsuccessful.

Later in Lana's training, however, and with a much larger corpus, Pate and Rumbaugh (1983) found the Thompson and Church model inadequate. Lana's innovative use of lexigram sequences would now require 69 different sentences, some of which had 135 different alternatives, rather than the six defined in the Thompson and Church model.

The conception of the computer-controlled keyboard proved to be one of those "very good" ideas. Its development succeeded beyond expectations and has served not only our language research programs for the next quarter century, but others as well (studies with chimpanzees by Matsuzawa, 1985a, 1985b and with dolphins by Gory & Xitco, 1993). The LANA system also served as a prototype for portable communication keyboards that are manufactured and now in increasing use by special populations, such as children whose language is deficient due to mental retardation. It has been through use of these devices that children with retardation have become able to learn large numbers of lexigrams and how to use them in real-world situations, including work (Sevcik, Romski, & Wilkinson, 1991).

Project LANA continued until the late 1970s, whenceforward Lana became a mother, and a successful one at that, and then a student in counting research in which she still serves, as she continues, from time to time, to "multiply." The Project put in place a number of important developments and findings: (1) computerized keyboards facilitated and objectified research on apes' language skills; (2) Lana readily learned dozens of lexigrams and how to use them in accordance with the rules of grammar imposed by computer software; (3) Lana innovatively used lexigrams to solve novel problems of communication (Rumbaugh & Gill, 1976); (4) Lana was able to name objects that she could feel with her hand, but could not see, and could more accurately declare whether an object that she could only feel, but not see, was identical to another object that she could see, if the objects had lexigrams that served as their names (Rumbaugh & Gill, 1977; Rumbaugh, Savage-Rumbaugh, & Scanlon, 1982); (5) Lana named hundreds of colored patches and documented that, in the main, her color vision was similar to our own (Essock, 1977; Matsuzawa, 1985a); (6) Lana learned how to name (i.e., label) and to give the color appropriate to 36 objects, composed of duplicates of six objects (bowl, box, can, shoe, cup, and ball) in six different colors (black, white, purple, red, green, and orange); (7) Lana innovatively used the "period" key to erase her errors in sentence construction and also to erase *invalid* sentence stems given to her by experimenters, rather than work on them to no avail; (8) Lana creatively asked for various otherwise "nameless" items, for example, the fruit orange, by asking for it as the "apple which-is orange (color)," a cucumber as the "banana which-is green," an orange-colored Fanta drink as the "coke which-is orange," an overly ripe banana as the "banana which-is black" and (9) on occasion Lana asked for the name of items, then used those names forthwith to request them. The LANA Project also heightened our sensitivity to the need for subsequent research to explore the nature of words. What was a word, and what were the boundaries of its meaning? Thus, at its conclusion, the LANA Project gave rise to continued research led by Savage-Rumbaugh in Project Sherman and Austin and several other projects with apes, discussed later.

4. Project Nim and Sequelae

The 1970s also saw Project Nim, with another chimpanzee, launched by Herbert Terrace (1979b) at Columbia University and Lynn Miles's (1990) Project Chantek (Chantek, an orangutan of the genus *Pongo*) at the University of Tennessee, Chattanooga. Also, the Fouts (Deborah, and Roger, who was the Gardners' first Ph.D. graduate; see Fouts & Fouts, 1989) initiated their own project by continuing research with Washoe and other chimpanzees, first at the University of Oklahoma and then at Central Washington College. Their research focuses on spontaneous sign acquisition and use among members of a social group.

All of these projects used manual signing as the language medium. Primarily through use of molding of the ape's hand (Fouts, 1972), the apes first learned *how,* then *when* to make a given manual sign in association with each of a variety of exemplars and events. But whether the apes literally knew the *meaning* of what they were signing (e.g., what the symbol represented) was not, at least initially, a focal question. The emphasis was on production, that is, on the use of signs by the apes. Attention to "meaning" was to be either nonsystematically inferred (e.g., through the analyses of fortuitous instances of signing in novel situations) or deferred, if for no other reason than that methods for measuring meaning in nonverbal subjects called for brand new research tactics. Additionally, the impact of behaviorism remained sufficiently strong to make one hesitate to ask, "Does the chimpanzee *know* what he/she is saying?"

Terrace's research with Nim started with high optimism that the operant methods to be employed might result in Nim reporting on certain aspects of even his private life and views. With the help of dozens of enthusiastic students of Columbia University, and even the use of what had at one time been the university president's mansion, Nim was taught via standard reinforcement techniques how to sign. But Nim became known not so much for his precise signing in association with various exemplars as for his penchant for long strings of signs, such as "give orange me give eat orange me eat orange give me eat orange give me you." Clearly, this was Nim's best effort to get an orange, but whether it was a seemingly interminable string of signs, or one that, from Nim's perspective, only needed punctuation here and there as he reiterated that he wanted "you" to "give orange me" so that he might "eat," cannot, of course, be known. Terrace's interpretation was that Nim's stringing of signs in this manner did not provide any more information than the use of just one or two signs would have, and that such strings were laced with "wild cards" (e.g., more, hug, give, Nim, etc.) that were always good to use—whatever the goal! Thus, at best, Nim was not a chimp of "few words," that is, not succinct; at worst, he suggested incompetence from a language perspective. Perhaps from his perspective he was just trying hard to find the golden sign, the one that would get him what he wanted (Seidenberg & Petitto, 1987; Terrace, 1985).

More serious, however, was Terrace's conclusion that a great deal of Nim's signs were either partial or complete imitations of what he had seen others about him use in the recent past. It was this conclusion that led Terrace and his colleagues (Terrace, Petitto, Sanders, & Bever, 1979) to reverse their earlier view that Nim was acquiring language and, instead, to conclude just the opposite: Nim had no language!

Even more serious was this development: when segments of tapes of chimpanzee signing in other laboratories were subjected to the same kind of analysis as that applied to Nim's tapes, the same conclusion was justified.

Both Washoe and Koko (a gorilla, studied by Patterson, 1990), as well as Nim, were said primarily to be just imitating on a deferred basis the signs they saw others use! A furor understandably resulted, with those with a vested interest in Washoe and Koko ardently defending the competence of their subjects. But the view of Terrace and his associates prevailed, to the extent that it soon became widely accepted that: (1) because Terrace's Nim did not have language and (2) because analyses at Terrace's laboratory of taped materials from the laboratories of other signing projects indicated that others' apes were also imitating, there was the strong implication that (3) no ape had demonstrated *any* language competence whatsoever and that (4) language was beyond the competence of apes! Even if (1) and (2) were true, (4) certainly did not follow, and (3) *might* not.

Although Lana's computerized-keyboard data were not included in Terrace's conclusion that imitative behavior accounted for the manual-signing "language" of other chimpanzees, her productions were simplistically attributed to "rote memory" by Gardner and Gardner (1978) and Terrace (1982, 1985). A comprehensive characterization of Lana's data as rote memory, however, notably fails to take into account her constructive variations of basic stock sentences when she coped with novel problems (Rumbaugh, 1977). Neither does it take into account the Pate and Rumbaugh (1983) evaluation of the Thompson and Church (1980) stock-sentence model of Lana's productions, based on a larger corpus obtained from a later stage in research with Lana, discussed earlier. Thus, while Lana's initial stock-sentence mastery was essentially by rote, a rote memory model failed to account for the variations and constructions that she made of those sentences.

Nonetheless, that was the tone of the day. None were to survive. The proverbial babies, even if extant, had to be thrown out with the bath waters! Nothing else mattered much, given that Nim was concluded to be a senseless, sentenceless, repetitive signer. Shortly after Terrace's analyses, Premack concluded that his attempts to teach language did not confer human language upon the chimpanzee, though it appeared to upgrade its ability to solve problems that entailed representation (Premack, 1983, p. 354). For Premack, there was " . . . the lack of any degree of language among nonhuman," be it natural or learned (1986, p. 149).

From the view of many, it was a sad day in the history of science—sad, not because of a challenge, but because of the stifling effect that it infused into all research with chimpanzees that addressed language. It was as though the null hypothesis had, indeed, been proved: Apes don't have language and never will. Such a conclusion is, of course, scientifically impossible. (Although the null hypothesis may be either retained or rejected on probabilistic grounds, it can never be proved.)

As a direct consequence, Project Nim was terminated and Nim was given a one-way fare to the University of Oklahoma, from whence he had come. There, it would seem, according to reports published by O'Sullivan and Yeager (1989), Nim "got hold" of himself and his language skills and used them impressively, appropriately, and without evincing the problems of imitating and interrupting the signing of others about him as he had while at Columbia University. It is unfortunate that those reports have yet to have their deserved effect. Nim could not redeem himself, given the negative judgments of him by Terrace and his colleagues in *Science* regarding his competence.

In retrospect, the truth was probably somewhere in the midfield. Given what is now known about chimpanzee competence for language, Nim was probably not as incompetent as he was judged to be. And, it is worth noting that although Nim was raised in a socially rich environment in which people spoke both to him and to each other to achieve communicative ends, no test of Nim's speech comprehension was ever given. Such tests might have revealed at least some speech comprehension. Additionally, Nim's signing competence was never subjected to appropriate blind tests, which might have served to insulate him from others' signing and thus stimulate him to "use what he knew." Several critics have suggested that there was also too rapid a turnover in the students who worked with Nim as technicians and that this might have discouraged the prospects that Nim would ever learn the communicative use of *meaningful symbols*. If subjected to controlled tests, which might have served to reduce social pressures and expectancies, Nim might have manifested a competence otherwise clouded by his best efforts to "find a way" (Terrace, 1985) to do what he must to get what he wanted. Such tests certainly would have precluded imitation of his teachers, a penchant that he had strongly developed.

In support of this conclusion, not all analyses agree with the interpretation of pervasive imitation. For example, Chantek's signs were not attributable to imitation, as were Nim's, and were appreciably more spontaneous (Miles, 1990). Similarly, Lana only infrequently (<10%) used lexigrams that had been used by her experimenters within the 10-min time frame preceding her own utterances, hence she rarely imitated (Betz, 1981).

Quite independently, the whole issue of imitation in language acquisition was addressed by scientists of child language acquisition (Greenfield & Savage-Rumbaugh, 1991; Nelson, 1985, 1986) who concluded that imitation could, indeed, be both a normal and effective means whereby children learn and affirm the appropriate use of words! Thus, from this perspective, to the degree that Nim did imitate in Terrace's laboratory, it might have served to instate skills that, although unused or unmeasured in controlled tests at Columbia University, became accessible and useful to him on the plains of Oklahoma. There, as suggested (O'Sullivan & Yeager, 1989), he became selective and efficient in his signs.

III. CATEGORIZATION SKILLS AND SEMANTICS

Concurrent with the pall cast upon ape manual-signing research in the wake of the final days of Project Nim was a study reported by Sue Savage-Rumbaugh and her associates (Savage-Rumbaugh, 1986; Savage-Rumbaugh, Rumbaugh, Smith, & Lawson, 1980) that concluded that Sherman and Austin, but not Lana, had manifested basic categorization skills that evinced *semanticity*. Semantics, that is, *word meaning,* is the most basic building block of languages. Unless words have meaning, it is unlikely that their use will have rational effects.

Through use of a series of training steps, Sherman and Austin first learned to classify three kinds of foods and three tools, for which they previously had learned lexigrams, by placing them in separate bins. Subsequent steps required them to classify not the foods and tools *per se,* but rather photographs and then lexigrams *in lieu* of the items that comprised this limited set of only six exemplars. The final training step entailed their learning to use two newly introduced lexigrams, one glossed "tool" and the other "food," enabling them to label each of the three tool lexigrams and three food lexigrams categorically.

In the final controlled tests that followed, Sherman and Austin were presented with 17 *other* previously learned lexigrams that stood for a variety of other foods and drinks and other tools. The question was whether they could accurately label the correct category for each one, in turn, through use of the food and tool lexigrams. They did so quite accurately. Only one error was made between the two chimps—Sherman called a *sponge* a food, rather than a tool. This might not have been an error from his perspective, for he literally consumed sponges as he sucked avidly on them when soaked with favorite juices.

It was noted above that Lana, in contrast to Sherman and Austin, did *not* learn to categorize symbols, other than those used in the initial training phases, as "food" and "tool." Why did she fail? A probable answer is that her training *emphasized production*. It was, in retrospect, naïve to assume that just because an ape *used* a symbol (be it a manual sign, plastic token, or lexigram), it must also "know" its meaning. Nothing could have been more untenable. Production and comprehension can be independent systems when apes learn language via formal instruction.

Thus, a "word" used by an ape as an initial consequence of basic operant-conditioning procedures probably is nothing more than a simple operant behavior. It does not necessarily ever acquire meaning to the user. Meaningfulness is instated not as a conditioned response, but rather as the result of learning to comprehend—to "listen."

Thus, Lana was doomed to failure in the categorization study, for she had not learned to comprehend. Although she could respond accurately to familiar "stock" requests, she could not respond appropriately to new ones. Although from time to time we wondered whether she really "knew" what she was saying as she answered "stock" questions (e.g., "What color of this box?" and "What name of this that-is red?") and used strings of lexigrams to activate delivery systems for milk, juice, a variety of foods, music, slides, special views, and so on, we never took the next step. Frankly, at the time we saw no clear way of asking Lana whether she really "knew" what she and we were saying with lexigrams.

But, as stated, with Sherman and Austin, Savage-Rumbaugh did work out a scheme to test whether they knew the meaning of each lexigram. Only if they literally knew the meanings of the 17 test lexigrams could they have labeled them correctly as "food" or "tool."

This categorization study, a benchmark effort in that it served to sustain our research apes and language skills, received little attention in the wake of Terrace's declaration that Nim was language*less*. Nevertheless, the findings were recognized by a sufficient number of peers, as well as by ourselves, as justification sufficient to continue research—and we did.

A. Cross-Modal Tests and Statements of Imminent Actions

Project Sherman and Austin also demonstrated impressive symbol-based, cross-modal matching. Without specific training to do so, Sherman and Austin were able to look at a lexigram and then reach into a box, into which they could not see, and select the appropriate object (Savage-Rumbaugh, Sevcik, & Hopkins, 1988). They also declared which of several food items they would retrieve from a return visit to the room in which they had inspected a limited array of randomly selected foods and drinks (Savage-Rumbaugh, Pate, Lawson, Smith, & Rosenbaum, 1983). On their first visit to the room, they individually surveyed the assortment of foods and drinks available for that trial and covertly made "a choice." Next, they returned to the room where the keyboard and the experimenter were located. There they declared via lexigram the food or drink that they would select on their return visit. Finally, they revisited the room where the food/drink array was located and retrieved an item. If that item was the one that had been announced via lexigram, the chimpanzee was praised and permitted to ingest the food or drink. Thus, on the basis of visually surveying the specific set of foods/drinks available across trials, a choice was made, announced, and verified by retrieval. Both Sherman and Austin did so with about 90% concurrence between what they declared via lexigram and what they retrieved from the array on their second visit. Their ability to use symbols to declare a future course of action had been documented.

B. Simulation Studies

Epstein, Lanza, and Skinner (1980) had earlier attempted to reduce claims by Savage-Rumbaugh, Rumbaugh, and Boysen (1978) regarding Sherman and Austin's symbolic skills to the behavioristic principles of shaping, chaining, and discrimination learning, through use of a simulation study that used pigeons (*Columba livia domestica*).

Consistent with the general tactics of simulation studies, they conditioned pigeons in certain key-pressing operants and chained responses and then argued that the resulting performances were essentially analogous to those of Sherman and Austin. Epstein and his colleagues stopped well short of simulating important critical tests, in this instance ones in which both Sherman and Austin spontaneously used an alternative symbol system (food container labels) for communication once their keyboard was deactivated (Savage-Rumbaugh & Rumbaugh, 1980, for a response to Epstein et al.). Because the results of such transfer tests are critical to the argument of symbolic function, their simulation tests fail in providing a meaningful alternative explanation. That such simulation studies serve to remind everyone about the ubiquitous power of reinforcers is no substitute for comprehensive replication of procedures *and* results. We find it interesting that there have been no simulation studies with pigeons or other animals of Sherman and Austin's categorization skills or their ability to announce imminent courses of actions regarding their choice of a specific food/drink from a random, changing array as discussed above. It is probable that the accomplishments of Sherman and Austin are so qualitatively different from what one could hope to condition in an pigeon, for instance, that attempts at simulation are viewed as infeasible.

A frequently overlooked truism is that science can be no better than its methods. Inadequate methods produce whatever they do—but they can never be definitive. Adequate methods are, at any given time, not always easy to differentiate from inadequate ones. But, given time, the methods of science will sort them out and lead the quest to better, though perhaps never perfect, methods and perspectives of phenomena.

C. Training of Comprehension Skills

How and why were Sherman and Austin able to categorize lexigrams in controlled test virtually without error? The answer probably rests in the fact that their training *emphasized* comprehension. Their comprehension skills were cultivated by tasks that required their coordinated use of lexigrams to convey a message of request or information to one another. They did so both in food sharing and in tool-use tasks. In the tool tasks, for example, typically one chimpanzee requested a tool needed to open a puzzle box.

That chimpanzee could obtain the needed tool only if the second chimpanzee comprehended, and then complied with, the first chimpanzee's lexigram-encoded requests. In a similar manner, they requested specific foods/drinks of one another through use of lexigrams.

Thus, Savage-Rumbaugh argues that Sherman and Austin succeeded in manifesting knowledge regarding the meanings of lexigrams, where Lana had failed to do so, because their language history was symmetrical and balanced. Both *comprehension* and *production* were emphasized.

Whether Washoe, Sarah, Lana, Nim, and other apes literally knew the meanings of their lexical token, rather than more simply knowing which token was associated with a given exemplar, remains unclear and unknown. Certainly they all knew a great deal about "how" and "when" to use manual signs or other nonspeech systems, but that is not reason enough to conclude that they also necessarily understood the meanings of those language tokens.

Their language-training protocols were, from the perspective of language acquisition, "wrong side up." Human parents do not teach their children language by having them talk. Rather, they talk to their children about their environments, about what has just happened, what is just about to happen, and so on. It is through listening that the child comes to understand, to comprehend language. Only later does it come to speak, to produce language. Even if an adult does not talk, we do not say that he/she is without language—as long as it is clear that they comprehend in detail what is said by others. Rather, it is merely said that they do not talk.

The lesson for science is, once again, that the fruits of research can be no better than methods employed. The methods of researching apes' language skills through the mid-1980s were as tactically compromised as the building of a house would be if it were built upside down, with the roof on the ground and the foundation in the air. The methods of reference emphasized production, but it is comprehension, not production, that is the foundation of language.

D. Fostering Comprehension Skills through Early Rearing

Project Sherman and Austin, then, clearly demonstrated that apes can come to comprehend the symbols or words of language if they are given special experiences and training (Savage-Rumbaugh, 1986). By cultivating the appropriate use of each word-lexigram in a variety of tasks and situations, the lexigrams become decontextualized and acquire meanings as symbols. On the other hand, and of particular significance, is the recent finding that language comprehension in apes also can be achieved by rearing them from birth in an environment where language is used to coordinate social activities (Savage-Rumbaugh et al., 1993). As discussed below, such rearing provides the apes with the opportunity to both hear words and see their respective lexigram equivalents used in a variety of contexts, where each word serves to predict events of the immediate future. As a consequence, the words acquire meaning and come to serve as symbols for events or for objects not necessarily present.

IV. STUDIES WITH THE BONOBO

Lana, Sherman, and Austin are so-called "common" chimpanzees (*Pan troglodytes*). The bonobo (*P. paniscus*) is the other species of chimpanzee, which by tradition has been called the "pygmy" chimpanzee, a misnomer, for it is *not* a pygmoid version of another form.

Bonobos are even more rare and endangered than are chimpanzees and more closely resemble humans (Kano, 1980): For example, bonobos can walk erect more readily and competently than *P. troglodytes,* and use eye contact to initiate joint attention; they use iconic gestures to entice others to assume physical orientations and actions (Savage & Wilkerson, 1978); and they can modulate their vocalizations much more than can the common chimpanzee. Their capacity to learn language observationally was discovered quite fortuitously in efforts to teach lexigrams to a bonobo, Matata.

Matata was a wild-born female bonobo who dwelled in the forest until an estimated age of 6 years. Her several talents did not include a facility in learning lexigrams (Savage-Rumbaugh, Sevcik, Brakke, & Rumbaugh, 1990). Even after years of effort, Matata failed to use even a small number of lexigrams reliably and gave no evidence that they served as symbolic representations of referents in her world. Nevertheless, these unavailing efforts to teach Matata had an unexpected and striking consequence. Throughout Matata's training sessions, her adopted son, Kanzi, was with her, but no effort was made to teach him. The one and only subject was Matata. Kanzi, however, always had the unintended opportunity to *observe.*

A. Kanzi: Observational Learning and Single-Word Speech Comprehension

When Kanzi was about 2 1/2 years old, he became a research subject when Matata was separated from him so that she might be bred at another site. It was then that a number of surprises occurred in rapid succession, the most remarkable being that Kanzi knew the symbols that others had tried to teach to Matata. Kanzi needed no special training to request that things be given to him, to name things, or even to announce what he was about to do through use of lexigrams. Quite evidently, Kanzi had learned by watching people trying to teach Matata, because he had not been an active participant in that instruction.

As a consequence, Savage-Rumbaugh instituted a major alteration in her research tactics. Attempts to teach Kanzi through the use of structured training protocols were stopped. In their stead, Kanzi was immersed in a life in which throughout his waking hours he was a full partner in the social scene that took place in the laboratory and its surrounding 55-acre forest. People talked to Kanzi as though he understood what was being said, and any lexigram on his keyboard that coincided with a word being spoken was activated. Thus, the use of lexigrams coupled with spoken words served to demonstrate to Kanzi the concordance between what he heard in speech and the functions of his keyboard. And because several of his keyboards electronically "spoke" the English equivalent of the lexigram activated, he had opportunity to learn that, somewhat literally, he could talk to others through appropriate use of his keyboards. Project Kanzi was, thus, begun.

All activities and events that were imminent were announced to Kanzi in this manner. Although Kanzi was encouraged to observe others' use of the keyboard and even to use it himself, he was never denied objects or participation in activities by failing to do so. Kanzi thus could learn of things that were of particular interest to him, rather than being constrained by topics chosen by his caregivers.

Kanzi readily learned the names of the several food sites located throughout the forest and how to go to them. Whenever he would activate a lexigram that was the name of one of those sites, it was assumed that he had "asked" to go to it. Accordingly, travel would commence to that site. Along the way, people reminded him, through speech, through use of his keyboard, and even through the use of pictures, of the place being visited. Through social activities en route, he also learned to ask to play a number of games and to get special foods and materials from backpacks carried by others. Back at the laboratory he learned how to ask to visit other chimps through use of their lexigram-names, to get and even cook specific foods, to invite the playing of various games, to watch videotapes of his own choice, and so on.

Through these experiences, Kanzi's lexigram vocabulary increased to 149 words by the time he was 5 1/2 years old, and it was about then that Kanzi appeared to be comprehending human speech—not just single words, but sentences as well. Because no other chimpanzee or bonobo in our laboratory had evinced comprehension of speech in formal tests, the prospect that he could do so was viewed with conservatism. But even the initial tests of his speech comprehension (see Figure 2) of single words were convincing, particularly when contrasted with the essential absence thereof in Sherman and Austin (Savage-Rumbaugh, McDonald, Sevcik, Hopkins, & Rubert, 1986).

Additional studies were undertaken to determine whether Kanzi's skills could be replicated. In one such study, Kanzi's sister, Mulika, was exposed

FIGURE 2   Kanzi, a bonobo (*Pan paniscus*), wears his headphones to hear spoken words and sentences. His comprehension approximates that of a 2 1/2-year-old child.

to the same kind of linguistic environment as was Kanzi. She, too, first developed comprehension of spoken English and lexigrams as used by others and then began to use them productively. At the age of about 2 1/2 years, although Mulika could use fewer than 10 lexigrams with competence for social communication, multiple and varied tests revealed that she understood dozens of others when spoken to her (whereupon she had to select the appropriate lexigram or picture of the item named) or announced by lexigrams (whereupon she had to select a picture of its exemplar; Savage-Rumbaugh et al., 1990; Sevcik, 1989). Mulika scored 100% in all test conditions for 22 different words, two of which she had never used at the keyboard.

B. Chimpanzee and Bonobo Compared

To determine whether bonobos differed from chimpanzees in their prowess for language, a bonobo (Panbanisha) and a chimpanzee (Panzee) were co-reared from shortly after birth in a language-saturated environment (Savage-Rumbaugh, Brakke, & Hutchins, 1992). When the subjects were about two years old, the bonobo was clearly superior to the chimpanzee both in comprehension of speech and in the use of the keyboard. Early rearing clearly fostered the spontaneous emergence of language skills, perhaps more readily in the bonobo than in the chimpanzee.

C. Ape Speech Comprehension Compared to a Human Child

Another experiment assessed Kanzi's sentence comprehension in direct comparison with a human child, Alia (Savage-Rumbaugh et al., 1993). Alia's mother worked with Kanzi and other chimpanzees for half of each day, then worked with Alia, her daughter, and a lexigram keyboard in a manner that approximated Kanzi's post-Matata experience with it. Thus, it became possible to give both Kanzi and Alia a set of novel sentences (all requests) for purposes of comparative assessment. Controls for the test trials (415 for Kanzi; 407 for Alia) entailed the experimenter sitting behind a one-way vision mirror and attendants listening to loud music through earphones so they could not hear the test sentences. Throughout the testing, there were varied arrays of objects on the floor in front of the subjects with which they might work to fulfill the requests. An important control is that they were asked to fulfill a variety of requests with the items of any given set to avoid the possibility that they might do only the "obvious" action and be speciously correct.

Seven different sentence types were given. Type 1 entailed asking that object X be placed in/on a transportable/nontransportable object Y (e.g., Put the milk in the water; Put the backpack on the Fourtrax). Type 2 sentence types asked, "Give (or show) object X to animate A" (e.g., Give Rose a carrot); "Give object X and object Y to animate A" (e.g., Give the peas and the beans to Kelly); "Do action A on animate A" (e.g., Hide the gorilla), and "Do action A on animate A with object X" (e.g., Make the snake bite Linda). Type 3 asked, "Do action A on object X (with object Y)" (e.g., Bite the picture of the oil; Stab your ball with the sparklers). Type 4 announced impending events in which the subjects were to engage, such as "Alia is going to chase Mommy," whereupon she was to do so. Type 5 sentences asked that object X be taken to location Y (e.g., Take the can opener to the bedroom), to go to location Y and get object X (e.g., Go to the microwave and get the tomato), and to go get object X that's in location Y (e.g., Get the telephone that's outdoors). Type 6 asked to make pretend that animate A do action A on recipient Y (e.g., "Can you make the bug bite the doggie?"). Type 7 consisted of other variations on the previous categories (e.g., Take the potato outdoors and get the apple). Generally, there were duplicate items so that, for instance, Kanzi had to get item X from location Y, even though there was another identical item X in front of him. On the controlled trials, Kanzi was 74% correct (307/415 sentences; age 8 years) and Alia was 66% correct (267/407 sentences; age 2 1/2 years).

Because many requests could be fulfilled solely on the basis of processing syntactic relations, other studies addressed the issue of word order directly though the use of three kinds of reversals (e.g., "Take the potato outdoors" vs. "Go outdoors and get the potato"; "Put the doggie in the refrigerator" vs. "Go get the dog that's in the refrigerator"; "Put the shoe in the raisins" vs. "Put the raisins in the shoe". Kanzi and Alia's performances were more similar than they were different and did not differ significantly across these three types of reversals given. Overall, Kanzi was correct on 81% (71/88), and Alia was correct on 64% (53/83) of the reversal sentences. These data support the interpretation that both Kanzi and Alia were sensitive to word order as well as to semantic and syntactic cues.

This study served to corroborate conclusions argued earlier by Savage-Rumbaugh (Savage-Rumbaugh et al., 1986) and her refutation (Savage-Rumbaugh, 1987) of Seidenberg and Petitto's (1987) argument that neither chimpanzees nor bonobos nor any other ape could learn about the symbolic, representational nature of words, let alone sentences. Their perspective was similar to the one they had advanced with Terrace et al. in 1979, namely, that, because of their advanced cognition, apes were able to learn and use "words"—but only as tools with which to get incentives, not as names, symbols, or representations of referents. Nelson's (1987) analysis of the justification for their argument led to the conclusion ". . . all of the evidence that Seidenberg and Petitto allege . . . fails to make the case that Kanzi does not use words to name or does not have mental representations" and "Kanzi . . . appears to use those symbols much as very young children do" (1987, p. 295).

Other studies (Greenfield & Savage-Rumbaugh, 1991, 1993) revealed that Kanzi's productive competence, with lexigrams in combination with gestures, was comparable to a 1 1/2 year-old child. Although some of Kanzi's grammatical rules appeared to be modeled after those of his human caretakers, he also invented his own (Greenfield & Savage-Rumbaugh, 1991). For example, Kanzi made up his own rule for combining agent *gesture* with an action *lexigram:* place lexigrams first (e.g., Kanzi would ask a caregiver to go hide by using the lexigram, LIZ, and then gesture, HIDE, by placing his hand over his eyes). Methodology accepted for studying child language (spoken or sign) indicated that Kanzi's two-element combinations (lexigram–lexigram and lexigram–gesture), where contextual information was available, were classified according to their semantic relations—such as agent–action and action–object. He also frequently combined two action lexigrams with regularities that reflected both natural action categories and preferred action orders in social play. Although Kanzi's use of grammar designed to alter meaning was limited, it occurred in social games where he would designate the agent and the social object of action (e.g., in playing CHASE, Kanzi would alter his selection of agents who were to do the chasing and who was to be chased). Many of Kanzi's statements were declarations of what it was that he was about to do.

Thus, with his comprehension of spoken English commensurate with that of a 2 1/2-year-old child and his production commensurate with that of a 1 1/2-year-old child, it was concluded that Kanzi was the first ape that had acquired language skills *without* formal training programs (Savage-Rumbaugh et al., 1993). Bates assessed the study as ". . . a fair test of the hypothesis that apes are capable of at least some language comprehension, at both the lexical and the structural levels" and concluded that at least one bonobo ". . . is capable of language comprehension that approximates (in level if not detail) the abilities of a human 2-year-old on the threshold of full-blown sentence processing" (Bates, 1993, pp. 222–223). Given that the keyboard is, at best, a poor substitute for speech, it is interesting to consider how Kanzi's utterances might have been increased in grammar and syntax if he had had speech rather than a lexigram keyboard with which to talk (Hopkins & Savage-Rumbaugh, 1991).

The importance of Kanzi's achievement is at least two fold: (1) an ape had acquired language spontaneously and without formal training in a manner that parallels the course whereby children acquire language; and (2) an ape that had acquired language in this manner manifested comprehension skills that were much more advanced than its production skills, a sequence which, once again, parallels the pattern whereby children acquire language. In contrast, all other apes had acquired such language as they did as a result of protracted formal training, with the atypical consequence that their production skills were in advance of their comprehension skills.

V. A SENSITIVE AGE FOR LANGUAGE ACQUISITION

The contrast in findings obtained with Kanzi, Panbanisha, and Panzee with those obtained with all other apes in language research over the past 30 years strongly suggests that during infancy a sensitive period exists for language acquisition both by children (Greenough, Black, & Wallace, 1987; Greenough, Wither, & Wallace, 1990) and by apes (Rumbaugh et al., 1991). Especially during the first few months of life, exposure to language, including speech, is critical if the acquisition of language is to be optimal (Savage-Rumbaugh, 1991). Observations of language use in the social contexts of everyday life appear to be far more effective for the acquisition of language than are the event-specific reinforcements of language productions (Lock, 1978, 1980), be they speech or use of a keyboard. Comprehension of language by the human child develops long before the speech musculature has matured enough for vocal control permitting language production (Golinkoff, Hirsh-Pasek, Cauley, & Gordon, 1987).

It is believed likely that use of the lexigram keyboard might help structure speech comprehension by the ape (Savage-Rumbaugh et al., 1993). By having the attention brought to the lexigram symbol as the word is spoken, the ape might be better able to parse out spoken words from the sound stream. In turn, this parsing might serve to enhance the learning of individual word meaning.

VI. SUMMARY

The twentieth century will be noted for a wide variety of scientific and technological advancements, including powered flight, antibiotics, space travel, and the breaking of the genetic code. It also should be noted as the century in which major psychological, as well as biological, continuities between animal and human have been defined.

Charles Darwin (1859) was quite right when he anticipated continuity in mental processes, some of which provide for language. Though none will argue that any animal has the full capacity of humans for language, none should deny that at least some animals have quite impressive competencies for language skills, including speech comprehension.

The finding that the language skills in the bonobo and the chimpanzee are likely more fully and efficiently developed as a result of early rearing than by formal training at a later age declares a continuity even stronger than that defined by the language acquisition potential of the ape. To clarify, because early rearing facilitates the emergence of language in ape as well as in child, a naturalness to the familiar course of language acquisition, whereby comprehension precedes production, is also corroborated. In turn, the continuity and the shared naturalness of language acquisition serve jointly to define an advanced and critical point of linkage between the genera *Pan* and *Homo*—and, as concluded by Domjan (1993), one worthy of contributing to the series of reconceptions of ourselves as anticipated by Ploog and Melnechuk (1971).

Acknowledgments

Preparation of this chapter was supported by Grant HD-06016 from the National Institute of Child Health and Human Development, by Grant NAG2-438 from the National Aeronautics and Space Administration, and by additional support from the College of Arts and Sciences, Georgia State University. The authors thank David A. Washburn and Sally Coxe for critical comments on this manuscript.

References

Andrews, P., & Martin, L. (1987). Cladistic relationships of extant and fossil hominoids. *Journal of Human Evolution, 16,* 101–108.

Bates, E. (1993). Comprehension and production in early language environment: A commentary on Savage-Rumbaugh, Murphy, Sevcik, Brakke, Williams, and Rumbaugh, "Language comprehension in ape and child." *Monographs of the Society for Research in Child Development,* Vol. 58 Nos. 3 & 4, 222–242.

Bates, E., Thal, D., & Marchman, V. (1991). Symbols and syntax: A Darwinian approach to language development. In N. A. Krasnegor, D. M. Rumbaugh, R. L. Schiefelbusch, & M. Studdert-Kennedy (Eds.), *Biological and behavioral determinants of language development* (pp. 29–65). Hillsdale, NJ: Erlbaum.

Bellugi, U., Bihrle, A., & Corina, D. (1991). Linguistic and spatial development: Dissociations between cognitive domains. In N. A. Krasnegor, D. M. Rumbaugh, R. L. Schiefelbusch, & M. Studdert-Kennedy (Eds.), *Biological and behavioral determinants of language development* (pp. 363–393). Hillsdale, NJ: Erlbaum.

Betz, S. (1981). *Sentence expansion by LANA chimpanzee.* Unpublished master's thesis, Georgia State University, Atlanta.

Chomsky, N. (1965). *Aspects of a theory of syntax.* Cambridge, MA: MIT Press.

Chomsky, N. (1988). *Language and problems of knowledge; The Managua.* Cambridge, MA: MIT Press.

Darwin, C. (1859). *The origin of species.* London: Murray.

Domjan, M. (1993). *Domjan and Burkhard's The Principles of Learning and Behavior.* Pacific Grove, CA: Brooks/Cole Pub.

Epstein, R., Lanza, R. P., & Skinner, B. F. (1980). Symbolic communication between two pigeons. *Science, 207,* 543–545.

Essock, S. M. (1977). Color perception and color classification. In D. M. Rumbaugh (Ed.), *Language learning by a chimpanzee* (pp. 207–224). New York: Academic Press.

Fouts, R. S. (1972). Use of guidance in teaching sign language to a chimpanzee *(Pan). Journal of Comparative Psychology, 80,* 515–522.

Fouts, R. S., & Fouts, D. H. (1989). Loulis in conversation with the cross-fostered chimpanzees. In R. G. Gardner, B. T. Gardner, & T. E. Van Cantfort (Eds.), *Teaching sign language to chimpanzees* (pp. 293–307). New York: State University of New York Press.

Furness, W. H. (1916). Observations on the mentality of chimpanzees and orang-utans. *Proceedings of the American Philosophical Society, 55,* 281–290.

Gardner, R. A., & Gardner, B. T. (1969). Teaching sign language to a chimpanzee. *Science, 165,* 664–672.

Gardner, R. A., & Gardner, B. T. (1978). Comparative psychology and language acquisition. *Annals of the New York Academy of Sciences, 309,* 37–76.

Gardner, R. G., Gardner, B. T., & Van Cantfort, T. E. (Eds.). (1989). *Teaching sign language to chimpanzees* (pp. 293–307). New York: State University of New York Press.

Golinkoff, R. M., Hirsh-Pasek, K., Cauley, K. M., & Gordon, L. (1987). The eyes have it: Lexical and syntactic comprehension in a new paradigm. *Journal of Child Language, 14,* 23–45.

Gory, J., & Xitco, M. (1993). *Dolphins learn to communicate through use of a computer-monitored keyboard.* Paper presented at the Marine Biology Convention, Orlando Fl.

Greenfield, P., & Savage-Rumbaugh, E. S. (1991). Imitation, grammatical development, and the invention of protogrammar by an ape. In Krasnegor, N. A., Rumbaugh, D. M., Schiefelbusch, R. L., & Studdert-Kennedy, M. (Eds.), *Biological and behavioral determinants of language development* (pp. 235–258). Hillsdale, NJ: Erlbaum.

Greenfield, P., & Savage-Rumbaugh, E. S. (1993). Comparing communicative competence in child and chimp: The pragmatics of repetition. *Journal of Child Language, 20,* 1–26.

Greenough, W. T., Black, J. E., & Wallace, C. S. (1987). Experience and brain development. *Child Development, 58,* 539–559.

Greenough, W. T., Wither, G. S., and Wallace, C. S. (1990). Morphological changes in the nervous system arising from behavioral experience: What is the evidence that they are involved in learning and memory? In L. R. Squire & E. Lindenlaub (Eds.), *The biology of memory Symposia Medica Hoechst, 23,* (pp. 159–185). Schattauer: Stuttgart & New York.

Harlow, H. F. (1977). Review of *Intelligence in ape and man* by David Premack, 1976. *American Scientist, 65,* 639–640.

Hayes, C. (1951). *The ape in our house.* Harper: New York.

Hayes, K. J., & Hayes, C. (1950). *Vocalization and speech in chimpanzees* (16mm. sound film). State College, PA: Psychological Cinema Register.

Hayes, K. J., & Nissen, C. (1971). Higher mental functions of a home-raised chimpanzee. In A. M. Schrier & F. Stollnitz (Eds.), *Behavior of nonhuman primates: Modern research trends.* Vol 4. (pp. 59–115). New York: Academic Press.

Herman, L. M. (1987). Receptive competencies of language-trained animals. In J. S. Rosenblatt, C. Beer, M. C. Busnel, & P. J. B. Slater (Eds.), *Advances in the Study of Behavior, Vol. 17* (pp. 1–60). Orlando, FL: Academic Press.

Herman, L. M., Pack, A. A., & Morrel-Samuels, P. (1993). Representational and conceptual skills in dolphins. In H. L. Roitblat, L. M. Herman, & P. E. Nachtigall (Eds.), *Language and communication: Comparative perspectives* (pp. 403–442). Hillsdale, NJ: Erlbaum.

Hopkins, W. D., & Savage-Rumbaugh, E. S. (1991). Vocal communication as a function of a differential rearing experiences in *Pan paniscus:* A preliminary report. *International Journal of Primatology, 12*(6), 559–583.

Jerison, H. J. (1985). On the evolution of mind. In D. A. Oakley (Ed.), *Brain and mind* (pp. 1–31). London & New York: Methuen.

Kano, T. (1980). Social behavior of wild pygmy chimpanzee (*Pan paniscus*) of Wamba: A preliminary report. *Journal of Human Evolution, 9,* 243–260.

Kellogg, W. N., & Kellogg, L. A. (1933). *The ape and the child.* New York: McGraw-Hill.

Krasnegor, N. A., Rumbaugh, W. M. Schiefelbusch, R. L., & Studdert-Kennedy, A. M. (Eds.). (1991). *Biological and behavioral determinants of language development.* Hillsdale, NJ: Erlbaum.

Lieberman, P. (1968). Primate vocalizations and human linguistic ability. *Journal of the Acoustical Society of America, 44,* 1157–1164.

Lock, A. (1978). *Action, gesture and symbol: The emergence of language.* London: Academic Press.

Lock, A. (1980). *The guided reinvention of language.* New York: Academic Press.

Matsuzawa, T. (1985a). Color naming and classification in a chimpanzee (*Pan troglodytes*). *Journal of Human Evolution, 14,* 283–291.

Matsuzawa, T. (1985b). Use of numbers by a chimpanzee. *Nature (London), 315,* 57–59.

Miles, H. L. W. (1990). The cognitive foundations for reference in a signing orangutan. In S. T. Parker & K. R. Gibson (Eds.), *"Language" and intelligence in monkeys and apes: Comparative developmental perspectives* (pp. 511–539). New York: Cambridge University Press.

Napier, J. R., & Napier, P. H. (1967). *A handbook of living primates.* New York: Academic Press.

Nelson, K. (1985). *Making sense: The acquisition of shared meaning.* New York: Academic Press.

Nelson, K. (1986). *Event knowledge: Structure and function in development.* Hillsdale, NJ: Erlbaum.

Nelson, K. (1987). What's in a name? Reply to Seidenberg and Petitto? *Journal of Experimental Psychology: General, 116*(3), 293–296.

Oden, D. L., Thompson, R. K. R., & Premack, D. (1990). Infant chimpanzees spontaneously perceive both concrete and abstract same/different relations. *Child Development, 61,* 621–631.

O'Sullivan, C., & Yeager, C. (1989). Communicative context and linguistic competence. In R. G. Gardner, B. T. Gardner, & T. E. Van Cantfort (Eds.), *Teaching sign language to chimpanzees* (pp. 269–279). New York: State University of New York Press.

Parker, S. T., & Gibson, K. R. (Eds.), (1990). *"Language" and intelligence in monkeys and apes: Comparative developmental perspectives*. New York: Cambridge University Press.

Pate, J. L., & Rumbaugh, D. M. (1983). The language-like behavior of Lana chimpanzee: Is it merely discrimination and paired associate learning? *Animal Learning and Behavior, 11,* 134–138.

Patterson, F. L. (1990). Language acquisition by a lowland gorilla: Koko's first ten years of vocabulary development. *Word, 41,*(2), 97–143.

Patterson, F. L., & Linden, E. (1981). *The education of Koko.* New York: Holt, Rinehart, & Winston.

Pepperberg, I. M. (1993). Cognition and communication in an African grey parrot (*Psittacus erithacus*): Studies on a nonhuman, nonprimate, nonmammalian subject. In H. L. Roitblat, L. M. Herman, & P. E. Nachtigall (Eds.), *Language and communication: Comparative perspectives.* (pp. 221–292). Hillsdale, NJ: Erlbaum.

Ploog, D., & Melnechuk, T. (1971). Are apes capable of language? *Neurosciences Research Program Bulletin, 9*(5), 600–700.

Premack, D. (1971). On the assessment of language competence in the chimpanzee. In A. M. Schrier & F. Stollnitz (Eds.), *Behavior of nonhuman primates (Vol 4,* pp. 185–228). Academic Press: New York.

Premack, D. (1983). Animal cognition. *Annual Review of Psychology, 34,* 351–362.

Premack, D. (1986). *Gavagai!* Cambridge, MA: MIT Press.

Premack, D., & Premack, A. J. (1983). *The mind of an ape.* New York: Norton.

Roitblat, H. L., Herman, L. M., & Nachtigall, P. E. (Eds.). (1993). *Language and communication: Comparative perspectives.* Hillsdale, NJ: Erlbaum.

Rumbaugh, D. M. (Ed.). (1977). *Language learning by a chimpanzee: The LANA project.* New York: Academic Press.

Rumbaugh, D. M., & Gill, T. V. (1976). The mastery of language-type skills by the chimpanzee (*Pan*). *Annals of the New York Academy of Sciences, 280,* 562–578.

Rumbaugh, D. M., & Gill, T. V. (1977). Lana's acquisition of language skills. In D. M. Rumbaugh (Ed.), *Language learning by a chimpanzee* (pp. 207–224). New York: Academic Press.

Rumbaugh, D. M., Hopkins, W. D., Washburn, D. A., & Savage-Rumbaugh, E. S. (1991). Comparative perspectives of brain, cognition, and language. In N. A. Krasnegor, D. M. Rumbaugh, R. L. Schiefelbusch, & M. Studdert-Kennedy (Eds.), *Biological and behavioral determinants of language development.* (pp. 145–164). Hillsdale, NJ: Erlbaum.

Rumbaugh, D. M., & Pate, J. L. (1984). The evolution of primate cognition: A comparative perspective. In H. L. Roitblat, T. G. Bever, & H. S. Terrace (Eds.), *Animal cognition* (pp. 569–587). Hillsdale, NJ: Erlbaum.

Rumbaugh, D. M., Savage-Rumbaugh, E. S., & Scanlon, J. L. (1982). The relationship between language in apes and human beings. In L. Rosenblum (Ed.), *Primate behavior* (pp. 361–384). New York: Academic Press.

Sarich, V. M. (1983). Retrospective on hominoid macromolecular systematics. In R. L. Ciochon & R. S. Corruccini (Eds.), *New interpretations of ape and human ancestry* (pp. 137–150). New York: Plenum.

Savage, E. S., & Wilkerson, B. J. (1978). Socio-sexual behavior in *Pan paniscus* and *Pan troglodytes:* A comparative study. *Journal of Human Evolution, 1,* 327–334.

Savage-Rumbaugh, E. S. (1986). *Ape language: from conditioned response to symbol.* New York: Columbia University Press.

Savage-Rumbaugh, E. S. (1987). Communication, symbolic communication, and language: A reply to Seidenberg and Petitto. *Journal of Experimental Psychology: General, 116,* 288–292.

Savage-Rumbaugh, E. S. (1991). Language learning in the bonobo: How and why they learn.

In N. A. Krasnegor, D. M. Rumbaugh, R. L. Schiefelbusch, & M. Studdert-Kennedy (Eds.), *Biological and behavioral determinants of language development* (pp. 209–233). Hillsdale, NJ: Erlbaum.

Savage-Rumbaugh, E. S., Brakke, K. E., & Hutchins, S. S. (1992). Linguistic development: Contrasts between co-reared *Pan troglodytes* and *Pan paniscus*. In T. Nishida, W. C. McGrew, P. Marler, M. Pickford, & F. B. M. de Waal (Eds.), *Topics in primatology* (pp. 51–66). Tokyo: University of Tokyo Press.

Savage-Rumbaugh, E. S., & Lewin, R. (1994). *Kanzi: At the brink of the human mind.* New York: Wiley.

Savage-Rumbaugh, E. S., McDonald, K., Sevcik, R. A., Hopkins, W. D., & Rubert, E. (1986). Spontaneous symbol acquisition and communicative use by a pygmy chimpanzee *(Pan paniscus).* *Journal of Experimental Psychology: General, 115,* 211–235.

Savage-Rumbaugh, E. S., Murphy, J., Sevcik, R. A., Brakke, K., Williams, S., & Rumbaugh, D. M. (1993). Language comprehension in ape and child. *Monographs of the Society for Research in Child Development,* Vol. 58, Nos. 3 & 4, 1–221.

Savage-Rumbaugh, E. S., Pate, J. L., Lawson, J., Smith, S. T., and Rosenbaum, S. (1983). Can a chimpanzee make a statement? *Journal of Experimental Psychology: General, 112,* 457–492.

Savage-Rumbaugh, E. S., & Rumbaugh, D. M. (1980). Requisites of symbolic communication—Or, are words for birds? *Psychological Record, 30,* 305–318.

Savage-Rumbaugh, E. S., Rumbaugh, D. M., & Boysen, S. (1978). Symbolic communication between two chimpanzees *(Pan troglodytes).* *Science, 201,* 641–644.

Savage-Rumbaugh, E. S., Rumbaugh, D. M., Smith, S. T., & Lawson, J. (1980). Reference: The linguistic essential. *Science, 210,* 922–924.

Savage-Rumbaugh, E. S., Sevcik, R. A., & Hopkins, W. D. (1988). Symbolic cross-modal transfer in two species of chimpanzees. *Child Development, 59,* 617–625.

Savage-Rumbaugh, E. S., Sevcik, R. A., Brakke, K. E., & Rumbaugh, D. M. (1990). Symbols: Their communicative use, comprehension, and combination by bonobos *(Pan paniscus.* In C. Rovee-Collier & L. P. Lipsitt (Eds.), *Advances in infancy research* (Vol. 9, pp. 221–278. Norwood, NJ: Ablex.

Schusterman, R. L., Gisiner, R., Grimm, B. K., & Hanggi, E. B. (1993). Behavior control by exclusion and attempts at establishing semanticity in marine mammals using match-to sample paradigms. In H. L. Roitblat, L. M. Herman, & P. E. Nachtigall (Eds.), *Language and communication: Comparative perspectives.* (pp. 249–274). Hillsdale, NJ: Erlbaum.

Seidenberg, M. S., & Petitto, L. A. (1987). Communication, symbolic communication, and language: Comment on Savage-Rumbaugh, McDonald, Sevcik, Hopkins, and Rubert (1986). *Journal of Experimental Psychology: General, 116,* 279–287.

Sevcik, R. A. (1989). *A comprehensive analysis of graphic symbol acquisition and use: Evidence from an infant bonobo (Pan paniscus).* Unpublished doctoral dissertation, Georgia State University, Atlanta.

Sevcik, R. A., Romski, M. A., & Wilkinson, K. M. (1991). Roles of graphic symbols in the language acquisition process for persons with severe cognitive disabilities. *Augmentative and Alternative Communication, 7,* 161–170.

Sibley, C. G., & Ahlquist, J. E. (1987). DNA hybridization evidence of hominoid phylogeny: Results from an expanded data set. *Journal of Molecular Evolution, 26,* 99–121.

Skinner, B. F. (1957). *Verbal behavior.* New York: Appleton-Century-Crofts.

Terrace, H. S. (1979a). Is problem-solving language? *Journal of the Experimental Analysis of Language, 31,* 161–175.

Terrace, H. S. (1979b). *Nim.* New York: Knopf.

Terrace, H. S. (1982). Why Koko can't talk. *Sciences (New York), 22*(9), 8–9.

Terrace, H. S. (1985). In the beginning was the "name." *American Psychologist, 40*(9), 1011–1028.

Terrace, H. S., Petitto, L. A., Sanders, R. J., & Bever, T. G. (1979). Can an ape create a sentence? *Science, 206,* 891–900.

Thompson, C. R., & Church, R. M. (1980). An explanation of the language of a chimpanzee. *Science, 208,* 313–314.

von Glasersfeld, E. (1977). The Yerkish language and its automatic parser. In D. M. Rumbaugh, (Ed.), *Language learning by a chimpanzee: The LANA project.* (pp. 55–71). New York: Academic Press.

Yerkes, R. M., & Learned, B. W. (1925). *Chimpanzee intelligence and its vocal expressions.* Baltimore, MD: Williams & Wilkins.

CHAPTER 12

Human Associative Learning

David R. Shanks

I. INTRODUCTION

There is a widespread feeling that the best place to study the elementary processes of learning and memory is in the animal rather than the human laboratory. Since it is possible to have greater control over factors such as motivation and prior learning history with nonhumans than with humans, and because it is easier to present events of genuine motivational significance to animals than to humans, there is a natural temptation to think that theoretical and empirical progress comes chiefly from studies of animal learning and only later filters through to researchers interested in human learning, who get carried along on the rear seat, as it were. One of the principal aims of this chapter is to argue that, contrary to this view, research conducted with human subjects during the last decade has provided genuine insights into associative learning, and that these insights should be of special interest to animal learning researchers.

There are at least two ways in which research using human subjects has been particularly fruitful. First, there are a number of issues relevant to learning that can be studied with human subjects but which would be extremely difficult, if not impossible, to investigate in animals. Unlike nonhuman organisms, amenable human subjects can be coaxed into performing almost any task given appropriate instructions: tasks requiring

Animal Learning and Cognition

subjects to make similarity or probability judgments, from which a wealth of interesting results have emerged, are obvious examples. Second, data can be obtained from humans that are orders of magnitude more complex, and therefore more theoretically challenging, than those obtainable from non-humans. The clearest case is language acquisition, where even data from circumscribed situations like the learning of past tenses of verbs can be extremely complex (see Pinker, 1991). The increasingly successful quantitative modeling of large data sets by human learning researchers (e.g., Nosofsky & Kruschke, 1992; Plunkett & Marchman, 1991; Seidenberg & McClelland, 1989) represents a considerable achievement.

The research reviewed in this chapter comes from a subset of human learning studies, namely those examining *associative* learning. In an associative learning task, the experimenter arranges a contingent relationship between events, which can be of two types, either external stimuli or the subject's actions. The term "associative learning" provides a contrast with "nonassociative learning," in which no explicit contingency between events is programmed, but where learning can nevertheless be observed. The prototypical examples of the latter cited by animal learning researchers are habituation and perceptual learning, while in human learning the major class of nonassociative tasks is probably that concerned with various priming phenomena (e.g., Benzing & Squire, 1989). I do not review data from nonassociative studies here, but it is worth noting that at the explanatory level the associative/nonassociative distinction may not be especially meaningful. Even when only a single stimulus is presented, the experimenter is implicitly arranging contingencies among the elements or features of that stimulus, and therefore so-called nonassociative learning may, like associative learning, be grounded in knowledge of those contingencies (Hall, 1991; McLaren, Kaye, & Mackintosh, 1989). The principles of associative learning may be perfectly applicable to nonassociative learning as well. Recent reviews of human learning and memory, which include discussions of priming, have been provided by Johnson and Hasher (1987) and Hintzman (1990).

Theories of learning have to address two fundamental questions. First, they must address the issue of what—at the representational level—is actually learned. For instance, learning may involve the encoding of relatively unanalyzed experiences in a multiple-trace memory store, the construction of complex hypotheses, or the acquisition of associative connections. For each theory, a different representational assumption is made concerning the sort of data structure that is created. The second question, given an adequate answer to the first question, is "what is the mechanism of learning?" How are the representations cited in the answer to the first question actually acquired? For example, what rules govern the growth of associative connections? What are the stimulus conditions that affect this process? I shall term the answers to these two questions the representational and learning assumptions, respectively, of a given theory. This is an important distinction, because theories of learning are sometimes rejected in their entirety when only one of their component assumptions is actually inadequate.

Dependent Measures

Associative learning tasks utilize a variety of dependent measures, including rate of responding in action–outcome tasks, numerical judgments concerning cue–outcome or action–outcome relationships, and responding to transfer stimuli in cue-category learning. Given this variety of measures, it is important to consider whether conclusions drawn on the basis of one type of measure also extend to other measures. In this regard, a variety of studies of human *implicit learning* have advanced the interesting conclusion that dissociations between different indices of learning can be obtained, specifically between verbally reported, conscious judgments on one hand and performance measures such as rate of responding on the other (e.g., Berry & Broadbent, 1984; Hartman, Knopman, & Nissen, 1989; Hayes & Broadbent, 1988; Lewicki, Czyzewska, & Hoffman, 1987; Reber, 1989; Willingham, Nissen, & Bullemer, 1989). Such dissociations, if they are genuine, suggest that instead of there being a single knowledge source that can be examined by any of a variety of tests, there are multiple sources some of which can only influence certain response measures. Multiple knowledge systems would in turn seem to imply multiple learning systems.

As an illustration of an apparent dissociation, consider an experiment by Willingham, Greeley, and Bardone (1993). On each trial, an asterisk appeared in one of four locations on a computer screen and subjects simply had to press as fast as possible the response key corresponding to that location. Subjects were given instructions appropriate for a typical choice RT task, but in fact for one group of subjects there was a sequence underlying the selection of the stimulus on each trial. For these subjects, a 16-trial stimulus sequence was repeated many times over, while for control subjects the order of stimuli within each of the 16-trial sets was randomized.

Willingham et al. (1993) observed that RTs fell across 420 training trials, and that this decrease was greater for the sequence than for the control subjects. Thus a performance measure—in this case reaction time—indicated sequence learning. However, some of the sequence subjects were not only unable to verbally report any of the predictive relationships in the stimulus sequence, but were also unaware that there had even been a sequence. Yet, at the same time, these subjects still showed more RT speedup than the random subjects. Thus, in this study, as in others (e.g., Hartman et al., 1989; Stadler, 1989; Svartdal, 1989; Willingham et al., 1989) we appear to have evidence of a dissociation between different dependent measures used to index learning: reaction times indicate sequence learning, whereas verbal judgments do not. Since a verbal judgment reports a conscious state of knowledge, many authors have been keen to interpret these dissociations as evidence for unconscious ("implicit") learning. Specifically, it has been claimed that measures indicative of conscious knowledge (e.g., verbal reports) dissociate from those that do not require conscious knowledge (performance measures) and which instead may reveal unconscious sources of knowledge.

I shall argue that researchers interested in associative learning need not be unduly concerned by such results, because in fact the evidence for a dissociable implicit learning system is fragile (for reviews, see Perruchet & Amorim, 1992; Shanks, Green & Kolodny, 1994; Shanks & St.John, 1994). Two points need to be emphasized. First, there have been many clear examples of *associations* rather than dissociations between performance and report measures (e.g., Perruchet & Amorim, 1992; Sanderson, 1989), suggesting that special conditions may be required to obtain dissociations. Secondly, there has been an extended debate about whether observed dissociations are indicative of a psychologically significant distinction between different knowledge sources, or merely reflect different degrees of sensitivity in different tests of knowledge (e.g., Dulany, Carlson, & Dewey, 1984; Perruchet & Amorim, 1992). Certainly, different tests are well known to vary in sensitivity, with uncued verbal reports coming near the bottom of the league table.

As an illustration of the extent to which sensitivity may account for these dissociations, consider a further piece of data from Willingham, Greeley, and Bardone's experiment. After the RT phase of their experiment, subjects who had been exposed to the sequence were not only questioned about that sequence, but also asked to undertake a recognition memory test. They were first instructed that there had been a sequence during the RT phase, and then shown either fragments of the sequence they had seen or novel fragments that had not been part of the sequence. Willingham et al. found that only two out of an original group of 45 subjects could still be classified as unaware of the sequence given their performance on the verbal report *and* recognition tests. Thus a recognition test, which provides many more retrieval cues than a test of verbally reportable knowledge, seems to reveal evidence of conscious sequence knowledge in subjects who may otherwise be classed as unaware, suggesting that in these and other experiments appearing to show dissociations between different measures of learning, variations in test sensitivity are likely to be responsible (see Dienes, Broadbent, & Berry, 1991, for further evidence).

II. EFFECTS OF CONTINGENCY ON ASSOCIATIVE LEARNING

In the remainder of this review, I organize the various theoretical approaches around a single empirical phenomenon, well known in animal conditioning experiments, namely, the effect of contingency on learning. This simple effect, which any theory of learning must be able to account for, can in fact be explained in a variety of highly contrasting ways, and so serves to illustrate the key differences that exist between the various theories. The different sorts of theories that are discussed include statistical, instance, associative, and rule-based theories.

Objectively, temporal correlation or contingency is probably the principal informational clue that the environment provides about whether a pair of events is causally related or not, and so if we view the associative learning mechanism as one that has evolved to detect causal relations in the world (Dickinson & Shanks, 1994), then sensitivity to contingency should be readily observable. The Chapters by Hall, Dickinson and Gallistel (2, 3, and 8, respectively) in this volume discuss results that prove that this prediction is true of animal learning, and despite some findings to the contrary (e.g., Allan & Jenkins, 1983), in the last 15 years there has been a good deal of evidence showing that associative learning in humans is also finely tuned to detect variations in the degree of contingency between a cue or action and an outcome. One of the first such demonstrations was reported in 1979 by Alloy and Abramson in an action–outcome learning task. Alloy and Abramson defined the actual degree of contingency by the metric ΔP:

$$\Delta P = P(O/A) - P(O/-A), \tag{1}$$

where $P(O/A)$ is the probability of the outcome given the action, and $P(O/-A)$ is the probability of the outcome in the absence of the action. As Allan (1980) has shown, the ΔP rule is closely related to normative statistical measures of the association of two stimuli, such as the ϕ^2 and χ^2 coefficients, except that these coefficients are bidirectional and take into account $P(A/O)$ and $P(A/-O)$. Since prediction and causation are unidirectional, from the cue or action to the outcome, one can argue that $P(A/O)$ and $P(A/-O)$ are not relevant and that therefore the ΔP rule is indeed normative (see Anderson, 1990).

We can illustrate subjects' sensitivity to contingency by considering an experiment by Wasserman, Elek, Chatlosh, and Baker (1993, Experiment 3). In that experiment, subjects were required to make judgments of the extent to which pressing a telegraph key caused a white light to flash. They were presented with 25 different problems constructed by taking all possible pairings of $P(O/A)$ and $P(O/-A)$ with the values of 1.0, 0.75, 0.5, 0.25, and 0.0 per s. In situations where $P(O/A)$ is greater than $P(O/-A)$, the action to some degree causes the outcome, whereas in situations where $P(O/A)$ is less than $P(O/-A)$, the action is a preventive cause that makes the outcome *less* likely to occur. When $P(O/A)$ and $P(O/-A)$ are equal, there is no objective contingency between the action and outcome.

Each condition lasted for 1 min and was divided into 60 1-s intervals. If the subject responded during a given interval by pressing the telegraph key, then the white light flashed for 0.1 s at the end of that interval with a probability $P(O/A)$, and if the subject did not respond during the 1-s interval, the light flashed with probability $P(O/-A)$. At the end of each problem, the subjects rated the response–outcome relation on a scale from -100 ("prevents the light from occurring") to $+100$ ("causes the light to occur").

In general, the results shown in Table 1 indicated an impressive degree of sensitivity to the actual degree of contingency, so much so that 96.7% of the variance in the judgments is accounted for by the actual contingency. As $P(O/A)$ is held constant, judgments decrease as $P(O/-A)$ is raised from 0.0 to 1.0. Conversely, judgments increase (become less negative) when $P(O/-A)$ is held constant as $P(O/A)$ is raised from 0.0 to 1.0. Judgments are close to zero when $P(O/A)$ and $P(O/-A)$ are equal. In sum, in Wasserman et al.'s procedure subjects were remarkably sensitive to the actual contingency between an action and an outcome when judging their causal relation.

The sensitivity of action–outcome judgments to event contingency has been replicated in a large number of studies (e.g., Alloy & Abramson, 1979; Chatlosh, Neunaber, & Wasserman, 1985; Dickinson, Shanks, & Evenden, 1984; Wasserman, Chatlosh, & Neunaber, 1983). The effect also emerges, unsurprisingly, in cue–outcome learning tasks. Table 2 shows the design and results of a category learning experiment (Shanks, 1991a, Experiment 1) in which subjects saw hypothetical patients presenting with certain symptoms (the cues). For each patient, the subject had to diagnose what illness

TABLE 1   Mean Judgments of Contingency, Divided by 100.0, as a Function of $P(O/A)$ and $P(O/-A)$

$P(O/A)$	$P(O/-A)$				
	0.00	0.25	0.50	0.75	1.00
1.00	0.85	0.52	0.37	0.13	-0.03
0.75	0.65	0.43	0.16	0.06	-0.12
0.50	0.37	0.19	0.01	-0.19	-0.34
0.25	0.12	-0.10	-0.14	-0.37	-0.58
0.00	-0.08	-0.45	-0.51	-0.66	-0.76

Source: From Wasserman, Elek, Chatlosh, and Baker (1993, Experiment 3).

TABLE 2 Design and Results of Experiment

Condition	Trial type	Test symptom	Mean rating
Contingent	AB → 1[a]	A	62.3
	B → 0		
Noncontingent	CD → 2	C	41.8
	D → 2		

Source: From Shanks (1991a, Experiment 1).
[a] A–D are the cues (symptoms) and 1, 2 are the outcomes (diseases). 0 indicates no outcome.

(the category) they thought that patient had, and feedback was provided on each trial. Some patients had symptoms A and B, and the correct disease was disease 1 (AB→1), while others had just symptom B and no disease (B→0). Thus, there is a positive contingency between symptom A and disease 1, and this was reflected in subjects' judgments of the A→1 relationship (mean = 62.3 on a scale from 0 to 100). Some patients had symptoms C and D and disease 2 (CD→2) and others symptom D and disease 2 (D→2). The D→2 trials increase the probability of disease 2 in the absence of cue C, relative to the situation for cue A, and hence reduce the C→2 contingency, a fact that was reflected in the lower judgments for the C→2 relationship (mean = 41.8) than for the A→1 relationship.

Finally, there is also evidence that sensitivity to contingency is found when the dependent measure is response rate rather than a numerical judgment of causality. In an action–outcome learning task, Chatlosh et al. (1985; see also Shanks & Dickinson, 1991) found that response rate attained a higher level when there was a positive contingency than when there was a zero contingency, which in turn was higher than when there was a negative contingency between key pressing and a light flashing. Each flash of the light earned the subject a point.

The fact that the human learning system is sensitive to the contingency between events means that it is well adapted to detect at least one piece of useful information about causality provided by the environment. We can now examine how various theories of learning attempt to explain this fundamental learning phenomenon. One approach is to assume that the subject behaves like a statistician and computes a measure of the degree of covariation between a given cue and the outcome. A second proposes that each stimulus configuration is encoded as a separate trace in a multiple-trace memory store, with responding to novel stimuli being a function of their similarity to stored exemplars. A third approach assumes that a mental association or connection is formed between the cue and the outcome, and a final theory is that the subject constructs a hypothesis or rule that picks out those cues or combinations of cues whose presence is necessary and sufficient for the occurrence of the outcome. We will consider each of these approaches in turn. Although it is not possible at present to draw definitive conclusions concerning the viability of each of these theories, I shall try to highlight the main strengths and weaknesses of each approach.

III. STATISTICAL MODELS

There has been a long tradition in research on associative learning of trying to determine whether organisms actually use a mentally represented version of a contingency metric such as the ΔP measure to make associative judgments. Typically, the ΔP rule has emerged as a better predictor of judgments than simpler rules relying, for example, just on the total number of outcomes (e.g., Wasserman et al., 1983; but see Allan & Jenkins, 1983). From a normative perspective, of course, there is a strong a priori case for thinking that subjects might indeed derive their judgments by applying the ΔP rule to the events they witness, and that learning therefore consists of *acquiring representations of the conditional probabilities that figure in the rule*. In a cue (C)–outcome task, the learning assumption would presumably cite some computational strategy for extracting running mean conditional probabilities $P(O/C)$ and $P(O/-C)$ from the training trials.

How does the rule explain contingency effects? The answer is obvious: increasing the number of outcomes that occur in the absence of the action, for instance, has the effect of reducing judgments, since an increase in $P(O/-A)$ leads to a reduction in ΔP, the statistic on which judgments are assumed to be based. However, although this account appears at face value to be straightforward, there is an interesting complication that can be illustrated by reference to a simple replication of the experiment shown in Table 2. In that replication (Shanks, 1991a, Experiment 2), subjects were presented with six trial types: AB→1, B→0, C→1, DE→2, E→2, F→0. As before, a contingency effect was obtained in that judgments were lower for cue D than for cue A. But if contingency is calculated across *all* trials, the inclusion of the C and F trials means that ΔP is equal for cues A and D and therefore the difference in judgments is unexplained. In this design, $P(1/A) = P(2/D)$, while $P(1/-A) = P(2/-D)$, so ΔP must be equal for these two cues. Many previous attempts to challenge statistical theories have rested on the assumption that contingency is calculated across the entire set of experimental trials (e.g., Chapman, 1991; Shanks & Dickinson, 1987).

However, it is reasonable to suggest that the assumption that ΔP is computed across all trial types is incorrect: instead, it should be calculated only across a subset (the "focal" set) of all trials. Specifically, as Cheng and Holyoak (in press) and Melz, Cheng, Holyoak, and Waldmann (1993) have emphasized, a more appropriate way of measuring contingency is to calculate $P(O/A)$ for a target cue A across all A trials, and to calculate $P(O/-A)$

across all trials that are identical to the A trials with the exception that A is absent. With respect to the Shanks (1991a, Experiment 2) study, the procedure is to calculate $P(1/A)$ across all relevant trials, which in this case means the AB trials, yielding a conditional probability of 1.0. For $P(1/-A)$, we focus only on trials that are identical to the trials that contribute to the computation of $P(1/A)$, except that A is absent; in this case, that means we consider only the B trials, from which we obtain $P(1/-A) = 0.0$ and hence $\Delta P = 1.0$. A comparable computation for cue D yields a value of $\Delta P = 0.0$, and hence the effect of contingency is accounted for. The idea is simply that the causal efficacy of a cue has to be determined by contrasting what happens when the cue is present versus what happens when it is absent, everything else being held constant. Note that we normally restrict the trials that contribute to the computation of ΔP anyway: we ignore trials occurring outside the experimental setting.

Thus qualified, there are several learning phenomena besides the basic effect of contingency that the ΔP theory can accommodate (see Cheng & Holyoak, in press). One such phenomenon is evident in Wasserman et al.'s (1993) results shown in Table 1: despite the close concordance between subjective and objective contingencies, subjects tend to underestimate the magnitude of extreme contingencies. To try to explain this finding, Wasserman et al. asked their subjects both to give contingency ratings and to estimate the two conditional probabilities $P(O/A)$ and $P(O/-A)$ at the end of each problem, and then compared the observed contingency judgments with the values of ΔP computed from the subjects' own probability estimates. The results showed that the tendency to underestimate extreme contingencies could be explained on the basis that subjects were computing ΔP over inaccurate estimates of $P(O/A)$ and $P(O/-A)$. For instance, in the condition where $P(O/A)$ was 1.0 and $P(O/-A)$ was 0.0, subjects' means estimates of these probabilities were 0.91 and 0.05, respectively, from which we derive a ΔP of 0.86. In fact, as Table 1 shows, the mean contingency judgment (0.85) was much closer to this than to the objective contingency (1.0).

Another phenomenon that is consistent with the ΔP model has been investigated in an elegant series of experiments by Baker, Mercier, Vallee-Tournageau, Frank, and Pan (1993). They used a video game task in which subjects were able to fire shells at tanks traversing the screen. By firing at the tanks, subjects could change their color, which might in turn allow them to avoid being blown up by color-sensitive mines in a minefield the tanks had to traverse. Subjects were asked to judge the relationship between the action (firing at the tanks) and the outcome (avoiding destruction).

Figure 1A gives a schematic description of the experiment (Baker et al., 1993, Experiment 1, .5/0 and .5/1 conditions). In the control condition, $P(O/A)$ was 0.75 and $P(O/-A)$ was 0.25, and subjects accurately estimated

A
O   -O

	O	-O
A	15	5
-A	5	15

B
O   -O

	O	-O
A	0	5
-A	0	15

FIGURE 1   Schematic description of Baker, Mercier, Vallee-Tourangeau, Frank, and Pan's (1993) study. (A) Overall contingency table showing that $P(O/A) = .75$, $P(O/-A) = 0.25$, and $\Delta P = 0.5$. In the experimental condition an additional cue was presented on all occasions that the outcome occurred, that is, on the 20 trials in the O column. (B) According to the focal set contingency theory, trials with the additional cue are removed from the focal set, leaving a contingency table in which $\Delta P = 0.0$. A, Action; $-$A, no action; O, outcome: $-$O, no outcome.

the action–outcome relationship: they gave ratings close to 50 on a rating scale from 0 to 100. In the experimental condition, each time the outcome was programmed to occur, an additional cue was present. Specifically, a plane flew over the tank, and subjects were instructed that this was a spotter plane that might relay to the tanks information about the mines and hence allow them to avoid destruction. Because this cue appeared on all trials on which the outcome occurred, it was a perfect predictor of the outcome, and the results indicated that this cue had a profound effect on ratings of the action–outcome relationship: judgments were now close to zero. While the presence of the additional cue did not alter the gross relationship between the action and the outcome—the total number of occasions on which the action was paired with the outcome, for instance, was identical in the experimental and control conditions—judgments were substantially affected by it.

This sort of cue selection effect has traditionally been taken as being problematic for statistical theories, but the "focal set" interpretation of contingency theory provides a ready explanation of Baker et al.'s results. Briefly, the subject has to compute the probability of the outcome in the presence and in the absence of the action, which requires contrasting trials on which the action plus the causal background are present versus trials where just the causal background is present. Thus, all trials on which the additional cue occurs must be removed from the focal set. The effect of this is to change the contingency table to that shown in Figure 1B, which gives the residual frequencies when trials with the additional cue have been removed. It is obvious from this figure that $P(O/A)$ and $P(O/-A)$ are now both zero and hence the action–outcome contingency is zero, which concords with subjects' judgments. In sum, contingency theory can provide a straightforward explanation for the sort of cue selection effect Baker et al. obtained.

Despite these encouraging results, there have been some further findings challenging the idea that learning is mediated by internal representations of conditional probabilities (see Shanks, 1993). I shall concentrate on one particular problem well illustrated in an experiment by Chapman (1991, Experiment 4). Statistical accounts predict that the order in which trial types are witnessed should make little difference in the observed judgments, since probabilities are calculated—either on-line or subsequently—across the entire set of trials and are unaffected by order. However, there is evidence suggesting that trial order *does* have an affect on associative learning. In Chapman's study, subjects were exposed to a training procedure designed to establish one cue as having a negative contingency with the outcome. Using a simulated medical diagnosis task, Chapman gave subjects 12 trials in the first stage in which symptom A was associated with a fictitious disease (see Table 3). In the second stage, 12 other patients had symptoms A and B but did not have the disease, and 12 patients had symptoms C and D and also did not have the disease. These AB and CD trial types were intermixed. Finally, in the third stage, 12 patients had both symptom C and the disease.

The A→1 and AB→0 trials should establish cue B as having a negative relationship with the disease. According to the probabilistic contrast model, ΔP for cue B will be calculated just across the A and AB trials, yielding a value of -1.0, and in accordance with this prediction, cue B was given a negative rating. Turning to the CD→0 and C→1 trials, it is clear that apart from the order in which the trials are witnessed, the evidence presented to the subjects concerning the relationship between D and the disease is exactly comparable to that concerning the relationship between B and the disease, and the value of ΔP is therefore the same for B and D.

Contrary to this prediction, Chapman found that subjects gave a less negative rating for cue D than they had for cue B, and she therefore concluded that trial order is an important factor in learning. Note that Chapman's experiment makes it unlikely that the effect is due to differential

TABLE 3   Design of Experiment

Stage 1	Stage 2	Stage 3	Test trials
A → 1[a]	AB → 0		B
	CD → 0	C → 1	D
E → 1	E → 1	E → 1	

Source: From Chapman (1991, Experiment 4).
[a] A–E are the cues (symptoms), 1 is the outcome (disease), and 0 indicates no disease. The E → 1 trials are fillers.

forgetting concerning B and D, since the three-stage procedure ensured that these cues occurred contemporaneously and hence should have suffered from forgetting—if at all—to equal degrees. If the result cannot be explained by a statistical theory, is there an alternative explanation? Consider a subject observing A→1 trials followed by AB→0 trials. Given that cue A has been established as a predictor of the disease, the absence of the disease on the subsequent AB trials is surprising and should lead the subject to reason that the new cue, B, must have an effect that cancels out that of cue A, and hence must be negatively related to the disease. For the CD→0, C→1 trials, the subject has no prior expectations on the CD trials and so the absence of the disease should be neither surprising nor unsurprising. According to some theories, this might mean that no relationship between D and the disease will be formed. In Section V we will see that connectionist theories posit a close relationship between prior expectancy, and therefore surprise, and learning. Such theories can explain Chapman's data on the basis that the difference in trial order alters the degree to which the outcome on CD trials, relative to AB trials, is surprising.

IV. INSTANCE THEORIES

One of the principal challenges to any theory of learning is to provide an account of generalization. How will knowledge of previously encountered stimuli or events determine responding to some further stimulus or event? Statistical models assume that subjects can determine how to respond to a new stimulus because they have prior knowledge of the probabilities of various outcomes given the presence of the various elements of which the stimulus is composed. In contrast, instance theories (also known as "exemplar" or "multiple-trace" theories) propose that responding to a novel stimulus is a function of its similarity in psychological space to other stimuli that have been stored in memory. The idea—which has a venerable history going back to the work of Richard Semon around the turn of the century (see Schacter, Eich, & Tulving, 1978)—is that every stimulus configuration (together with the outcome or category with which it is associated) that the subject encounters is stored in a relatively unanalyzed form in a multiple-trace memory system (see Pearce, Chapter 5, this volume).

The explanation of contingency effects that emerges from instance theories is extremely simple, but I shall postpone discussion of this account for the moment. First, it is necessary to consider the evidence for the crucial process of instance memorization. Suppose a subject witnesses a large number of slightly different stimuli such as dot patterns, and learns that half are associated with category 1 and half with category 2. Why should we imagine that each of these patterns is separately memorized, as proposed by instance theories? Is it not more likely that some degree of abstraction takes place? According to *prototype* theories, for example, learning involves abstracting the underlying prototype from a set of varying stimuli. Knowledge of the category *bird,* for example, might be represented by a prototypical bird that has been mentally abstracted from our experience of a large number of different actual birds.

On this rather simpler account, responding to a new stimulus is a function of its similarity to the prototype rather than to a collection of different memorized instances. As test stimuli get closer to the prototype, they should therefore become easier to categorize, an effect that is readily demonstrated in the laboratory (e.g., Rosch, Simpson, & Miller, 1976). Moreover, there is abundant evidence that the prototype stimulus itself will be classified accurately and rapidly, even when it was never presented in the training stage. For instance, Homa, Sterling, and Trepel (1981) trained subjects to classify various geometrical patterns generated from three prototype patterns into three corresponding categories. When tested on new and old patterns, subjects were in some cases more likely to correctly classify the prototype, which they had never seen, than any of the specific training instances that they had seen. This was particularly the case when a long interval (1 week) intervened between training and testing, and when the category contained a large number of instances (20). Such results seem to imply that the prototype, at least in some cases, is mentally represented.

It turns out on closer examination, however, that the prototype view of associative learning is untenable. In the formation of a prototype, a large amount of information is discarded that people can be shown to be sensitive to (see Medin, 1989, for a review). Perhaps the clearest evidence comes from experiments showing that information is retained about specific training items (e.g., Brooks, 1978; Malt, 1989; Shin & Nosofsky, 1992; Whittlesea, 1987; Whittlesea, & Brooks, 1988). Such "exemplar" effects are easy to demonstrate: in the Homa et al. (1981) experiment (see also Homa, Dunbar, & Nohre, 1991), for example, a set of test instances were arranged to be equidistant from the prototype of a category but to vary in terms of their similarity to specific training instances of that category. Homa et al. found that subjects became more accurate in classifying the test stimuli the closer they were to training instances. If information about the specific training items is discarded in the formation of the prototype, it is difficult to see why such a result should emerge.

Direct evidence that representations of training instances are retrieved in the process of classification comes from a series of studies by Malt (1989). She presented subjects in the first stage with pictures of animals for classification into two categories. Once these had been learned to a criterion, a test stage was presented. On some trials, new items had to be classified, and Malt reasoned that if the classification of a new item involved the explicit retrieval of an earlier similar training instance, then if that training instance was presented for classification on the immediately following trial, it should be classified—as a result of priming— more rapidly than it would otherwise have been. This is exactly what she found: classification of old exemplars was faster if new but similar items had occurred on the preceding trial.

A multitude of studies has now shown that performance in associative learning tasks can be well understood in terms of instance storage alone (see Medin & Florian, 1992). For example, it is not necessary to cite prototype abstraction in order to explain the accurate and rapid classification of the prototype stimulus: Hintzman (1986), Medin and Schaffer (1978), and Nosofsky (1986) have all shown that multiple-trace theories can account for the prototype effect. As a test item gets closer to the prototype, its summed similarity to the training instances also increases. Thus, classification decisions based on summed similarity will be maximal for the prototype pattern. Although there remain advocates of prototype extraction (e.g., Homa et al., 1991) as well as skeptics about the distinguishability of prototype and instance theories (e.g., Barsalou, 1990), the evidence seems overwhelmingly to go against the notion that category learning is based on prototype abstraction. Instead, the instance view proposes that subjects encode in a multiple-trace memory array representations of the actual instances seen during training, and base their classifications on the similarity between a test item and stored instances (Estes, 1986a, b; Hintzman, 1986; Nosofsky, 1986). The power of such an instance-storage mechanism can be illustrated by considering the results of a study by Shin and Nosofsky (1992) who examined category learning with dot patterns. Shin and Nosofsky first obtained judgments of pairwise similarity for the patterns from one group of subjects. On the assumption that stimuli are represented in terms of their component dimensions as points in a psychological space, similarity judgments are assumed to be a direct function of distance in the space. Other subjects were then trained to classify some of the patterns into three categories, following which they were tested on their classification decisions for the remaining patterns.

Since Shin and Nosofsky had obtained independent similarity data, they were therefore able to compare subjects' test pattern classification decisions with the predictions of an instance theory, specifically, the context model of Medin and Schaffer (1978) and Nosofsky (1986). The basic idea is that the probability of assigning a stimulus S_i to category J is a function of the summed similarity of S_i to each of the j members of category J that have been stored in memory, divided by the summed similarity of S_i to all k exemplars of all K categories:

$$P(J/S_i) = \frac{\sum\limits_{j} s_{ij}}{\sum\limits_{K} \sum\limits_{k} s_{ik}} \qquad (2)$$

where s_{ij} is the similarity between stimuli i and j. One additional component to the model is the assumption that selective attention can alter perceived similarities. Thus, learning may lead subjects to differentially attend to the dimensions along which the stimuli vary as a function of their usefulness in classifying the training stimuli, in turn leading to the psychological space being shrunk or expanded in certain directions.

Shin and Nosofsky found a remarkable degree of concordance between predicted and observed classifications, with over 95% of the variance in the observed classifications being accounted for in a one-parameter model. A prototype theory, assuming that the training instances formed the basis for an abstracted prototype with which the test stimulus was then compared, performed much more poorly in predicting responses.

A. Contingency Effects

We know from the results described in Section II that the extent to which a given element of a stimulus will come to control responding will be affected by the degree of contingency between that element and the outcome. In a categorization experiment, for instance, the to-be-classified object can be conceived of as a collection of elements, some of which are highly predictive because in the past they have covaried with the category, and some of which are less predictive. Having found an impressive degree of support for the multiple-trace view of associative learning, it is now appropriate to return to the effect of contingency on associative learning and ask how it is explained by instance theories. The basic idea, which is in marked contrast to the explanation given by statistical theories, is illustrated in Figure 2. The figure shows stimuli as points in a psychological space, with the dimensions of the space corresponding to the dimensions of variation among the stimuli. Each stimulus is represented by an independent binary dimension on which the values are present and absent. In a design in which two types of trials (AB and B) are presented, the relevant stimuli occupy the left (B) and right (AB) upper vertices of the space.

The explanation of contingency effects is straightforward. Suppose AB is paired with the outcome (1), while B on its own does not predict the outcome (0). According to the instance account, attention will be selectively directed to the A dimension, since it is presence versus absence of this cue that predicts the outcome. In terms of the psychological space depicted in Figure 2, the net effect of this is to stretch the space horizontally and shrink it vertically. In contrast, when AB and B trials are both accompanied by the outcome, the A dimension is irrelevant and it is now the B dimension to which attention should be directed. Hence, the space is stretched in the vertical direction and shrunk in the horizontal direction. Note that because attention is assumed to be limited, stretching in one dimension is accompanied by shrinking in the other (Nosofsky, 1986).

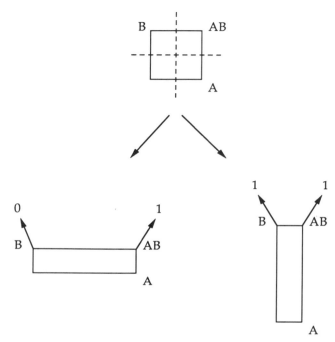

FIGURE 2   Explanation of contingency effects provided by instance theory. Two independent cues, A and B, are represented as points in a psychological space where the end points
of each dimension are the presence and absence of the cues (*top*). When the outcome (1) occurs
given the compound cue AB but not (0) given B alone (*lower left*), the space is stretched along
the relevant A dimension and shrunk along the irrelevant B dimension, making A more similar
to AB. In contrast, when both AB and B are accompanied by the outcome (*lower right*), A
becomes less similar to AB. The dotted lines refer to the possible classification rules discussed
in Section VI.

The outcome is that presentation of A on its own will elicit a greater
response tendency in the contingent than in the noncontingent case, since,
as the figure shows, A is more similar to AB for the situation shown on the
lower left than for that shown on the lower right. Since AB is associated
with the outcome, this greater degree of similarity translates into a greater
response tendency.

B. Artificial Grammar Learning

It turns out that the conception of associative learning adopted by instance
theories can encompass a truly impressive amount of experimental data,
often at an exquisite degree of quantitative detail (see Nosofsky, 1992, for a
review). For example, such theories have been able to account for data from

experiments in which the stimuli were Munsell color patches (Nosofsky, 1987), dot patterns (Shin & Nosofsky, 1992), schematic faces (Nosofsky, 1991), slides of skin disorders (Brooks, Norman, & Allen, 1991), pseudowords (Whittlesea, 1987), and phrases (Whittlesea & Brooks, 1988). In the present section, I shall try to illustrate the power of the instance approach by considering recent examinations of a learning task involving complex stimuli, and which is usually thought to require something altogether more sophisticated than the memorization of training exemplars. Studies of artificial grammar learning (e.g., Reber, 1989) are modeled in a highly simplified way on language learning. During the study phase of a typical experiment, subjects read strings of letters (e.g., MXRVXT), which have been generated from a finite-state grammar. The task is to memorize the strings. Prior to the test phase, subjects are instructed that the strings were generated according to a complex set of rules and that their task in the test phase is to decide which new items are also constructed according to those rules. Subjects then make grammaticalness decisions for grammatical and nongrammatical test items. The standard result is that subjects are able to perform above chance on this grammaticalness test (the typical level is about 60–70% correct classifications).

Much of the interest in artificial grammar learning has been driven by the question of whether it is possible to learn about the rules of a grammar unconsciously (see Perruchet & Pacteau, 1990; Reber, 1989; Shanks & St.John, 1994), but that is an issue that I shall not address here. Instead, I focus on the equally interesting question of what is learned. Do subjects learn abstract grammatical rules or do they memorize the training strings? Reber's (1989) view has been that abstract rule learning is the principal determinant of classification performance, and the evidence for this has come from a variety of sources, but one of the most persuasive is that subjects can transfer what they have learned to a completely new set of test strings containing letters not used in the study phase, provided the test strings are generated by a grammar structurally identical to the one that generated the study strings (e.g., Mathews et al., 1989; Reber, 1969). On the notion that learning consists of abstracting the underlying composition rules of the grammar, transfer of this sort should indeed be expected, for the same reason that a listener can judge sentences of a natural language containing unfamiliar words to be grammatical.

In contrast to this view, recent examinations have revealed that much of the data from artificial grammar-learning experiments—including transfer—can be explained on the assumption that subjects merely store the training strings as whole items in memory and then respond to test strings on the basis of similarity to the memorized items (Brooks & Vokey, 1991; Vokey & Brooks, 1992). For example, Brooks and Vokey (1991) trained subjects on strings from one grammar and tested them either on new strings from that grammar or on new strings from a structurally identical grammar that used a different letter set. In addition, they divided the test strings into ones that were similar to study items (differing by only one letter) and ones that were dissimilar (differing by more than one letter). Consider the training string MXTRRRX. The test string VXTRRRX differs by one letter, while the string MXRTMXR differs by 4 letters, yet both are grammatical. For subjects tested with the changed letter set, the test item BDGHHHD is structurally similar to the original test string, while the string CBHGBDG is dissimilar.

There were two key findings, both of which challenge the view that learning is based solely on abstraction of grammatical rules. First, for subjects trained and tested on items using the same letter set, Brooks and Vokey found a large effect of similarity on grammaticalness decisions. Indeed, while most (64%) similar grammatical strings were called grammatical, the minority (45%) of dissimilar grammatical strings were classified as grammatical. This result is consistent with the idea that comparison to memorized training strings is the basis of classification, but is at variance with the claim that classifications are based simply on abstracted grammatical rules: similarity does not objectively affect grammaticalness.

Second, although Brooks and Vokey replicated the finding that above-chance grammaticalness performance was possible when subjects were trained on one letter set and tested on another, they also found that similarity influenced these decisions just as it influenced responding in subjects tested on the same letter set. Brooks and Vokey did find that grammaticalness per se had an additional influence on grammaticalness decisions independently of similarity, which suggests that another source of knowledge exists in addition to the stored instances, but the overall implication is that similarity to memorized instances can account for much of the data obtained in grammar-learning experiments. I return in Sections V and VI to consider whether this account is adequate, or whether we must appeal to an additional abstract rule-learning mechanism.

In sum, a wealth of data can be interpreted in terms of instance memorization. In addition to the concept-learning phenomena discussed here, instance theories have been shown to be able to account for many other observations, such as the fact that learning curves usually conform to a power law (Logan, 1988).

C. Representation and Learning in Instance Theories

I suggested earlier that an adequate theory of learning needs to state what it is that is learned as well as what sort of mechanism governs the learning process. Instance theories concentrate on the former question: the general thrust of the findings reviewed in this section is that learning in many contexts can be interpreted as the simple memorization of relatively un-analyzed training instances, with responding to a test item being based on its computed similarity to those stored items. On this account, the learning *mechanism* is an elementary process in which an instance is stored in memory together with its category label or associated outcome. The evidence certainly argues that the representational and learning claims of instance theories allow a great deal of data to be explained, but I end this section by considering some evidence that is difficult to explain with such a simple learning rule.

Recent data have suggested that it is not sufficient to assume that each instance is automatically stored in its entirety. The evidence comes from influential experiments originally conducted by Gluck and Bower (1988), and subsequently replicated and extended by Estes, Campbell, Hatsopoulos, and Hurwitz (1989), Nosofsky, Kruschke, and McKinley (1992), and Shanks (1990b, 1991b). There has been some controversy concerning the design originally adopted by Gluck and Bower (see Shanks, 1990a, b; Gluck & Bower, 1990), but the critical result is quite well established. Before describing one of the actual experiments, I illustrate the basic effect. The design is shown in Figure 3.

The figure shows a psychological space corresponding to the presence and absence of two stimuli, A and B, with the top right-hand corner corresponding to the stimulus combination AB. Suppose subjects are presented with a series of AB trials, on half of which this compound cue is associated with outcome or category 1, and on half with category 2. Suppose also that on a number of additional trials, B is associated with category 1. The question is, what will the subject do on a test trial with cue A? Instance theories, and perhaps common sense, predict that subjects will either be indifferent between categories 1 and 2 for this cue, or will associate it with category 1. The reason is straightforward: there are an equal number of AB→1 and AB→2 instances, so A is equally similar to each subset of those instances. There are also some B→1 instances, and if A is at all similar to B, then this will tip the balance toward category 1 as the choice response, but if there is no similarity between A and B, then A will be assigned to categories 1 and 2 with equal probability. Selective attention to the dimensions will clearly not alter these conclusions, as the lower left- and right-hand diagrams of Figure 3 show. However the space is stretched or shrunk, cue A will never be more similar to the stored instances of category 2 than to those of category 1.

What happens in an actual experiment? Tests of this prediction have used the simulated medical diagnosis procedure described earlier. In one such study (Shanks, 1991b, Experiment 2), subjects made one of two diagnoses on each trial for patients presenting with between 0 and 4 different symptoms such as swollen glands. One disease occurred three times as often as

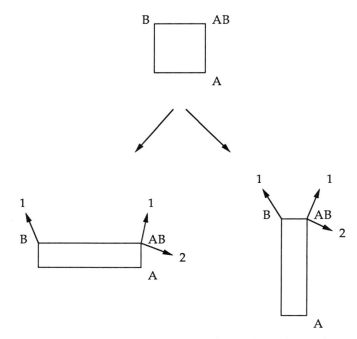

FIGURE 3   Schematic illustration of Gluck and Bower's (1988) experiment. Cue compound AB is paired with outcomes 1 and 2 on equal numbers of trials, while B is paired with outcome 1. Possible ways in which the space might be distorted as a result of selective attention are shown on the lower left and lower right. Note that A is always at least as similar to the instances of category 1 as it is to those of category 2.

the other, and corrective feedback was given on each trial. A target symptom, corresponding to cue A, was paired with diseases 1 and 2 on an equal number of trials. On many of these trials other symptoms were present too, but the equality $P(1/A) = P(2/A) = 0.5$ held whether all trials with cue A were considered or just those consisting of A on its own. Thus, objectively, on an A trial the subjects should have predicted diseases 1 and 2 with equal probability. However, in the absence of cue A, disease 1 was more common than disease 2. These trials, on which other symptoms were present, correspond to the B→1 trials.

The result was clear-cut. After a series of learning trials, subjects were more likely ($p = 0.71$) to predict disease 2 than disease 1 on test trials with symptom A, contradicting the prediction of instance theories. Nosofsky et al. (1992) have confirmed at the quantitative level that the context model is unable to predict this outcome. The reason why the effect occurs is presumably as follows. Because of the B→1 trials, cue B is a good predictor of category 1. On the AB trials the subject must learn to expect categories 1 and 2 equally, and hence A needs to become associated more with category 2 than with category 1 in order to offset the false expectation, given the presence of cue B, that category 1 is more likely to be correct than category 2.

The implication of this kind of experiment, initiated by Gluck and Bower (1988), is that the assumptions underlying instance theories are inadequate. Instances are not necessarily encoded in their entirety together with their category labels. Instead of being stored in memory as relatively unanalyzed wholes, stimuli are analyzed into their component parts, with some elements able to gain more control than others over the category response. Note also that the bias observed in these experiments is hard to reconcile with normative accounts of learning (e.g., Anderson, 1990): the probabilities of outcomes 1 and 2 given cue A are equal.

There are at least two further reasons to question the underlying assumptions of instance theories. The first comes from function learning experiments demonstrating that subjects can make accurate interpolations and extrapolations (e.g., DeLosh, 1992; Koh & Meyer, 1991). Suppose a subject learns to make a range of responses, $R_1 \ldots R_n$, to a range of stimuli, $S_1 \ldots S_n$. The stimuli might be lines of varying length and the responses button presses of varying durations. If learning consists solely of instance memorization, then when subjects are tested on an extrapolation trial with stimulus S_{n+1}, we would have to predict that they will make the response appropriate for the training stimulus that is most similar to S_{n+1}, namely R_n. In fact, subjects are quite good at making novel responses (i.e., R_{n+1}) to such test stimuli (e.g., DeLosh, 1992), but this is a phenomenon that lies outside the scope of instance theories. The implication is that it is inadequate to claim that associative learning is in general based on the memorization of instances, with responding being governed by similarity to those instances. Instead, the evidence suggests that subjects can learn the abstract functions that relate stimuli to responses (Koh & Meyer, 1991).

The second reason to question the assumptions on which instance theories are based comes from a different source. Watkins and Kerkar (1985) obtained evidence from an ingenious set of experiments that when an item is presented twice, recollection cannot be explained by reference to the retrieval of two separate memorized instances. In their studies, subjects were presented with a list of words such as "umbrella," with some words being repeated on the list. In one of the experiments (Experiment 1), words were written in different colors such that each of the once-presented words and each occurrence of a twice-presented word was in a different color. At the end of the list subjects were first given a brief distracting task and then required to remember as many of the words as possible.

As would be expected, Watkins and Kerkar observed better recall of twice-presented than of once-presented words, but also observed *superad-*

ditivity. Let us assume that the probability of remembering either presentation of a twice-presented word is just the probability of remembering a once-presented word. If the probability, P2, of remembering a twice-presented word is just the probability of remembering each separate occurrence, then it should be readily predictable from the probability, P1, of remembering once-presented words:

$$P2 = P1 + P1 - P1 \cdot P1. \tag{3}$$

That is, the probability of remembering a twice-presented word is just the probability of remembering its first presentation plus that of remembering its second presentation minus their product. The latter is subtracted because recall probability will not be increased if the subject remembers both presentations rather than just one. Watkins and Kerkar found that the probability of recalling a twice-presented word, P2, was much greater (0.46) than the value predicted (0.32) from Equation 3 given the value of P1, which was 0.18. This indicates that the above assumption is incorrect: the probability of remembering, say, the first presentation of a twice-presented word is greater than that of remembering a word presented only once, which in turn means that the second presentation makes the first more recallable, and vice versa.

This result may not, on its own, be too problematic for instance theories. Although they provide no mechanism whereby memorized items may affect each other's recallability, it is perfectly possible that such a process exists and that subjects in Watkins and Kerkar's experiment were retrieving stored instances. But a further aspect of the results makes this look extremely unlikely. If each stored trace of a twice-presented item is enhanced in its memorability compared to that of a once-presented item, then we would expect this to extend to *all* aspects of the trace: specifically, subjects should be better able to remember the color of a twice-presented word on its first or second presentation than that of a once-presented word. To test this, after subjects had attempted to recall the words, Watkins and Kerkar presented a complete list of the words with the twice-presented words marked, and asked subjects to say what color (or colors in the case of the marked words) each had appeared in. Watkins and Kerkar observed exactly the opposite of the predicted result. Subjects were significantly *poorer* at remembering the colors associated with twice-presented words than those associated with once-presented ones. After a correction for guessing, the probability of recalling the color of a once-presented word was 0.37, while the probability of recalling one of the colors associated with a twice-presented word was 0.25.

Watkins and Kerkar interpreted this result in terms of the formation of generic memory traces, a process not readily accommodated with instance theories. Subjects are assumed to abstract from multiple presentations of a stimulus only those aspects that are invariant, and discard trial-by-trial fluctuations. Of course, a range of other findings, such as the memory deficits of patients with anterograde amnesia, have been used to argue for this distinction between episodic (instance-based) and semantic (generic) knowledge (see Humphreys, Bain, & Burt, 1989). The abstraction process requires a mechanism whereby a stimulus is analyzed into its constituent elements rather than stored in a holistic form, a process that is not provided by instance theories but which has been central to the sorts of network models that are considered in the next section.

V. CONNECTIONIST MODELS

Connectionist models take the view that associative knowledge is represented in the form of mental associations between cues or actions and outcomes, with these associations being incremented or decremented on a trial-by-trial basis according to an adaptive learning rule. Many animal learning theorists, of course, will be sympathetic to such a view, since it has been the dominant way of analyzing nonhuman learning for several decades, and as Pearce argues in Chapter 5 of this volume, the dispute between connectionist and instance theories is mirrored in animal learning research by that between elemental and configural theories. Connectionist models have had an enormous impact in the last decade across the whole field of human learning, being successfully applied to tasks as diverse as perceptual-motor learning (e.g., Cleeremans & McClelland, 1991), concept learning (e.g., McClelland & Rumelhart, 1985), and language acquisition (e.g., Elman, 1990).

The connectionist approach answers the question of "what is learned" by positing connections between units that represent features or combinations of features, and answers the question of "what is the mechanism of learning" by positing one of a variety of learning rules or algorithms. While there exist a number of such learning rules, I shall focus on one of the simplest, namely the LMS or "delta" rule of Widrow and Hoff (1960), which has played a major role in several recent connectionist models of human learning (e.g., Gluck & Bower, 1988; Kruschke, 1992; McClelland & Rumelhart, 1985). As Hall notes in Chapter 2 of this volume, this learning rule is formally equivalent—given certain assumptions—to the Rescorla and Wagner (1972) theory of conditioning. Suppose we have a large set of potential cues or actions, denoted c_i, and a large number of possible outcomes or categories denoted o_j. In a feedforward network (also called a pattern associator), each of the cues is represented by a unit in a homogeneous input layer of a large, highly interconnected network. Each output is also represented by a unit in a separate output layer, and each input unit is connected to each output unit with a modifiable connection. On every trial, some set of cues is present and some outcome occurs. We calculate the output o_j on each output unit:

$$o_j = \sum_i a_i w_{ij}, \tag{4}$$

where a_i is the activation of input unit i (1 if that cue is present, 0 if it is not), and w_{ij} is the current weight from cue unit i to output unit j. Next, the "error" (or extent to which the outcome is surprising) $d_j = t_j - o_j$ is computed as the difference between the obtained output, o_j, and the desired output, t_j (1 if that output event actually occurs, 0 if it does not). Finally, we change each of the weights in proportion to the error:

$$\Delta w_{ij} = \alpha\, a_i\, d_j, \tag{5}$$

where Δw_{ij} is the weight change and α is a learning-rate parameter.

At the end of a series of training trials, there are a variety of different ways to look at the system's behavior. If we are modeling a situation in which subjects make cue–outcome association judgments, then the appropriate measure from the network is simply the activation o_j of the appropriate outcome unit when the input unit corresponding to that cue is turned on. On the other hand, if we are interested in classification performance, we can assume that the probability of classifying the stimulus as a member of category j will simply be the output produced on the category j output node divided by the total output across all n output units.

We have seen two previous explanations of the effect of contingency on associative learning—the fact that increasing the probability of the outcome in the absence of the cue, $P(O/-C)$, reduces the degree to which the cue and outcome are judged to be related. How do connectionist models account for this phenomenon? As $P(O/-C)$ is increased, more and more outcomes occur in the absence of the target cause. Present on these trials will be a number of background or contextual cues that will receive increments in their weights. On trials where the cue occurs, d_j will be reduced because of the larger weight for the background cues. Hence, there will be less increment [Equation (5)] in the weight of the target cue, and at asymptote the weight for cue C will be inversely related to $P(O/-C)$.

A. Experimental Tests

Connectionist models have been able to provide excellent fits to data from associative learning experiments varying the degree of contingency between events. For example, Dickinson et al. (1984) reported good fits to data from an action–outcome learning task, and Wasserman et al. (1993) obtained good fits to the data presented in Table 1, even reproducing the underestimation of extreme contingencies. Further, the typical negatively accelerated learning curves obtained in action–outcome learning tasks (e.g., Baker, Berbrier, & Vallee-Tourangeau, 1989; Shanks, 1985a, 1987) also can be well reproduced. Since the w_{ij} start at zero, the initial error d_j will be large and increments to the weights will in turn be large, but successive increments will get smaller and smaller as d_j decreases. The net effect will be negatively accelerated acquisition curves (see Shanks, 1993).

Perhaps more important, these models can explain many of the selective effects that are observed in associative learning studies. The ability to account for effects of the sort observed by Baker et al. (1993) (see Figure 1) is well known. Essentially, the additional cue that is present on all trials where the outcome occurs will acquire a positive weight for the outcome because it is an excellent predictor of it. On trials where the action and additional cue occur together, the weight of the additional cue will make the term o_j large in Equation (4). Hence, the mismatch d_j will be smaller than it would be in the control group, and the weight change in Equation (5) for the action will in turn be small. Thus, w_{ij} for the action will not attain the value that it will in the control group.

In the last section we saw that an instance theory can account for much of the data obtained in concept learning tasks, Recall that there are two principal phenomena that a categorization model must account for, prototype and instance effects. McClelland and Rumelhart (1985) and Knapp and Anderson (1984) have shown that connectionist networks are able to reproduce prototype effects. The effect comes about simply because the prototypical pattern activates most of the input units that are most strongly associated with the target category, and few of the units that are strongly associated with other categories. The matrix of weights a network acquires during training represents an abstraction from the training exemplars corresponding (in a loose way) to a prototype.

There is now emerging evidence that connectionist models can also show instance effects, and that these models in many respects may approximate the performance of instance theories (Kruschke, 1992; McClelland & Rumelhart, 1985; Nosofsky & Kruschke, 1992; Shanks & Gluck, 1994). As an illustration, remember that one of the clearest examples came from the experiment by Homa et al. (1981) in which subjects' classification responses to test stimuli equidistant from the prototype were biased by similarity to specific training items. To try to reproduce such effects, I (Shanks, 1991c) presented a network with four hypothetical training stimuli from each of two categories, and then tested the network with new stimuli equidistant from the prototypes but differing in similarity to specific training items. The network responded to the test stimuli in accordance with their similarity to training stimuli. The reason for this is that although the weights in a network are abstractions in the sense that they combine information across multiple presentations of different stimuli, they nevertheless do implicitly retain information about specific training instances. Each instance possesses some idiosyncratic features, and although such features will be seen only infrequently, the network will nevertheless be able to learn that such features are weakly predictive of the category. Two stimuli equidistant from the prototype may then differ in terms of possession of such weakly-predictive idiosyncratic features.

What about Gluck and Bower's (1988) categorization data that were so problematic for instance theories? Remember that subjects were presented with AB→1, AB→2, and B→1 trials. Despite being equally paired with categories 1 and 2, subjects tended to classify cue A in category 2. Estes et al. (1989), Gluck and Bower (1988), Nosofsky et al. (1992), and Shanks (1990b, 1991b) have all shown that this bias can be reproduced in a connectionist network. Shanks (1991b) describes a connectionist network, which, when trained on the same set of learning trials as the subjects, chose category 2 on exactly the same proportion (0.71) of trials as they did (Shanks, 1991b, Experiment 2). The reason is basically the same as the explanation of simple contingency effects: cue B becomes strongly associated with category 1, and this means that on the AB→1 and AB→2 trials, cue A must acquire a greater weight for category 2 in order to offset cue B's bias.

In the previous section, I argued that the learning of artificial grammars can be interpreted in terms of instance memorization. Dienes (1992) has recently shown that a simple associationist learning model can also reproduce much of the observed data from grammar-learning experiments. Because the task involves learning associations among a set of simultaneously presented cues (the letters), the architecture of Dienes's model was that of an autoassociator rather than that of a feedforward network (McClelland & Rumelhart, 1988); strings of letters are presented to a homogeneous layer of units, each of which represents a given letter in a given position, with each unit being connected to every other unit (but not itself). These connections have modifiable weights, which can be updated by Equation (5). The network's task is to reproduce across the units the pattern that was actually presented.

Dienes (1992) found that such a network could provide an excellent account of artificial grammar learning, not only being able to discriminate between new grammatical and nongrammatical strings, but also performing at the same overall level of accuracy as subjects (60–70% correct), and producing approximately the same rank ordering of difficulty of the test strings. Like other connectionist models, it learned in a memorizationlike fashion by encoding individual stimuli and modifying its weights in response to each stimulus to process that stimulus better.

In addition to being able to explain the basic phenomena of artificial grammar learning, the connectionist approach has an advantage over instance theories. Remember that in Brooks and Vokey's (1991) study, both similarity to studied instances and grammaticalness per se had effects on grammaticalness performance. Perruchet (1993) has argued that both of these factors can be reduced to something simpler, namely, fragmentary information about the bigrams and trigrams that composed the study strings. Perruchet was able to show that similar test items tend to contain more of the bigrams and trigrams that composed the study strings than do dissimilar items, and the same is independently true for grammatical versus nongrammatical test strings. Thus, substring knowledge provides a more parsimonious account than the dual-process account of Brooks and Vokey. In addition, Perruchet and Pacteau (1990) found that subjects recognized frequent bigrams better than infrequent ones. These results require some representation of the frequency of bigrams and trigrams in the study strings, and an important feature of distributed memories (Cleeremans & McClelland, 1991; Dienes, 1992), is that they produce exactly such frequency statistics, in the form of strengths of encodings, as a by-product of the learning process. Such a process, unlike whole-string memorization, can account for Perruchet's results as well as those of Brooks and Vokey.

B. Internal Representations

Several results from associative learning studies suggest that simple pattern associator or autoassociator networks of the sort considered so far—in which only a single layer of connections mediates between input and output—are computationally insufficient as general models of human learning. For instance, people can learn nonlinear classifications in which it is impossible to construct a linear boundary separating the members of two categories (e.g., Medin & Schwanenflugel, 1981; Nosofsky, 1987), yet single-layer networks are constrained to learn only linear classifications. For reasons such as this, more powerful models have been studied in which a layer of hidden units mediates between input and output.

The best known such multilayer system uses an adapted version of the delta rule (the "generalized delta rule") as its learning algorithm (see McClelland & Rumelhart, 1988) and utilizes hidden units with sigmoidal activation functions. Such "backpropagation" networks have no difficulty learning nonlinear classifications (indeed, they can learn essentially any input–output mapping), yet despite their widespread application to human learning tasks (e.g., Seidenberg & McClelland, 1989), there are good reasons to question whether such networks are adequate as general models of human associative learning. The biggest problem is that multilayer networks trained with the generalized delta rule suffer from catastrophic retroactive interference where associations tend to be almost totally overwritten and unlearned as a result of later learning (McCloskey & Cohen, 1989). Although the same can be said to some extent of human learning, humans do not suffer from this sort of interference to anything like the degree that networks governed by the generalized delta rule do (Kruschke, 1993). For this reason, alternative networks have been developed (e.g., Kruschke, 1992; Shanks & Gluck, 1994) in which hidden units again mediate between input and output—in order that nonlinear classifications may be learned—but in which different learning procedures and internal representations allow knowledge to be better protected from retroactive interference.

I mentioned earlier that one of the principal achievements of recent studies of human learning has been to provide extremely good quantitative fits to large sets of data, something rarely attempted in animal learning research. As an illustration of the power of current connectionist learning models in this regard, consider an experiment and simulation by Nosofsky et al. (1992), which again used the medical diagnosis procedure. For each of 240 hypothetical patients, some combination of 4 binary symptoms (e.g., high vs. low blood pressure) was present and the subject had to predict the accompanying disease (either burlosis or midrosis). One disease was more common than the other and occurred on 75% of trials, and corrective feedback was supplied on each trial. Because there were four symptom dimensions, there were 16 different symptom patterns: if we denote the two values of each symptom dimension by 1 and 2, we obtain patterns such as 2122, where the first number refers to low blood pressure, the second to high muscle tension, and so on. The relationship between the symptoms and diseases was probabilistic. For a given symptom pattern, the outcome was uncertain: for pattern 2122, for instance, the rare disease occurred on 34% of trials and the common disease on 66%. Overall, the probability of the rare disease varied from 3% for pattern 2211 to 76% for pattern 1122.

Nosofsky et al. obtained choice predictions in a transfer stage in which there were 81 test patterns, including the 16 training patterns, plus a further 65 incomplete patterns in which some dimensions were unspecified. They then compared the transfer performance of their subjects on these test trials with that of Kruschke's (1992) ALCOVE network. In this network, 8 input units coded the input symptoms: each symptom dimension was represented by 2 input units, with one of these being turned on to indicate one value of a dimension (e.g., high blood pressure), and the other turned on to indicate the other value (low blood pressure). Incomplete patterns were represented by turning on neither of the nodes for the relevant dimension. Two output units represented the diseases. The network also had an intermediate layer of 16 hidden units each of which was maximally activated by one of the 16 training patterns and was activated by other patterns to the extent that they were similar to the pattern that caused maximal activation. These hidden units can be considered as having graded receptive fields centered at specific

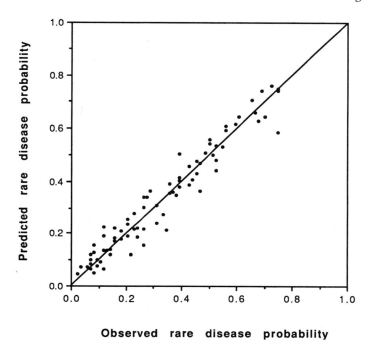

Observed   rare   disease   probability

FIGURE 4   Observed and predicted (ALCOVE) probabilities with which the rare disease was chosen on each of 81 test trials in Nosofsky, Kruschke, and McKinley's (1992) Experiment 1.

locations in the input space. The network was trained with exactly the same number and type of trials as the subjects, and the weights from the hidden to the output units were updated using the delta rule of Equation (5).

The results are shown in Figure 4, which plots the predicted versus observed probability of choosing the rarer disease for each of the 81 test patterns. Overall, the model accounted for an impressive 94.0% of the variance. In short, the model was able to predict with considerable accuracy the probability that subjects would choose the rare disease for a given transfer trial. In addition to providing this sort of impressive fit to categorization data, Kruschke (1993) has shown that the major drawbacks of standard backpropagation networks, such as their susceptibility to catastrophic interference, do not apply to the ALCOVE model.

C. Current Challenges to Connectionist Models

One major limitation of most connectionist models is that they typically fail to incorporate any mechanism for selective attention. At first sight, this may not seem a serious problem; after all, these models can explain contingency effects without the selective attention mechanism that instance theories need to appeal to. But there are a number of demonstrations of attentional effects that are outside the scope of standard single- or multi-layer networks. A simple illustration that will be familiar to animal learning researchers is latent inhibition. Learning of a cue–outcome relationship is retarded if the cue has previously been presented on its own in the absence of the outcome (e.g., Lipp, Siddle, & Vaitl, 1992). This is hard to explain in terms of simple connectionist processes, since the absence of any outcome in the pre-exposure stage means that the teaching signal t_j is zero and hence no learning should occur. On the other hand, the effect can be accounted for if an attentional mechanism serves to reduce the processing that the cue receives in the second stage of the experiment.[1] Kruschke (1992, 1993) discusses several other phenomena that appear to demonstrate selective attention, and describes one potential adaptive attention-learning component that is incorporated into the ALCOVE model.

A second challenge comes from the phenomenon of "backward blocking" (Shanks, 1985b). While normal blocking effects (e.g., Chapman & Robbins, 1990) can be explained by standard connectionist models, a further blocking effect is highly problematic. To illustrate, Chapman (1991, Experiment 1) found that learning of the relationship between cue A and outcome 1 was moderate when subjects were presented with a set of AB→1 trials, but was reduced if cue B had previously been paired on its own with the outcome (B→1). This blocking result is readily accounted for, like the results of Baker et al.'s (1993) studies, on the basis of a reduction in the error term d_j on the AB→1 trials when B has been pretrained, and hence a reduction in the weight changes in Equation (5). However, Chapman (1991, Experiment 3) also found that blocking occurred if the B→1 trials *followed* the AB→1 trials. In this case, the two stages of the blocking experiment are reversed, and hence the effect is called "backward blocking."

The effect is inconsistent with connectionist accounts because while the B→1 trials should affect the weight connecting cue B and the outcome, they should not affect the weight for cue A. In fact, almost all learning rules make the assumption—which Chapman's results question—that an input unit has to be activated on a trial in order for its weights to be modified as in Equation (5). Modifying weights for inputs when they are absent introduces some unpalatable side effects: learning something new about one cue would

---

[1] This effect, taken in isolation, can also be accommodated by a contingency theory, as Cheng and Holyoak (in press) have pointed out: the cue-alone trials serve to reduce P(outcome/cue) and hence reduce ΔP. However, this explanation fails to explain the well-known finding that latent inhibition retards inhibitory as well as excitatory learning. If ΔP is negative in the second stage of the experiment, then the reduction in P(O/C) brought about by the cue-alone trials should hasten rather than retard learning.

lead to (un)learning about every other cue that has a representation in memory. Busemeyer, Myung, and McDaniel (1993) discuss another cue-competition effect that appears to be at variance with the general class of connectionist models.

VI. RULE LEARNING

The picture of associative learning that emerges from the previous sections is of a rather passive process, in which instances are separately encoded in memory or weights in a network are adjusted. But there is evidence that in some circumstances a different, more active process can operate whereby a person considers various hypotheses that might determine category membership and, in short, tries to induce a rule. The basic idea is that associative learning can give rise to some internal hypothesis specifying how stimuli are associated with the outcome or category. The rule can then be applied to novel stimuli in order to determine an appropriate response.

Psychologists and philosophers have been divided on what sort of behavior is characteristic of rule-based knowledge, but a widely adopted instrumental definition is that performance is based on a rule if no differences are observable in performance to trained and to untrained stimuli. To see why this definition is relevant, we must first remind ourselves of the characteristics of non-rule-based behavior. Let us suppose that subjects in some category-learning task improve their classification performance in the study phase by memorizing the training stimuli just in the way the context model proposes. Thus, if the stimuli fall into two categories, subjects respond "category 1" if a stimulus is more similar to the exemplars of category 1 than to the exemplars of category 2. Clearly, on this account responding to test stimuli is going to be graded: some new test stimuli will be highly similar to trained stimuli from one category, and hence will evoke rapid and accurate responses, whereas others will be more equally similar to training stimuli from the two categories and hence will be classified more slowly and less accurately.

In contrast to the behavior expected if subjects memorize the training instances, we would expect to observe none of these differences if subjects learn and respond on the basis of a rule. If a decision boundary perfectly divides members of one category from those of the other, then it constitutes a rule for classifying the stimuli: a stimulus on one side of the boundary is in category 1, one on the other side is in category 2. We would be strongly motivated to conclude that subjects have learned this rule if the probability and latency of making a correct response is the same to all stimuli falling on one side of the boundary, for such a result would suggest that the subject is merely analyzing the stimulus to decide on which side of the boundary it falls and is not concerned in the least with comparing it to previously seen stimuli. Thus, suppose the stimuli are rectangles varying in width and height, and suppose the two categories are defined by a rule that unequivocally assigns a stimulus to category 1 if its width is greater than its height, and to category 2 if its width is less than its height. If subjects are able to learn this rule from exposure to some learning procedure, then when they make a classification decision they should merely be interested in whether the stimulus is wider or not than it is high; its similarity to training items is immaterial.

Laboratory demonstrations of rule learning, revealing no differences in performance to trained and novel stimuli, have been provided in a number of studies (e.g., Allen & Brooks, 1991; Kemler Nelson, 1984; Nosofsky, Clark, & Shin, 1989; Regehr & Brooks, 1993; Smith & Shapiro, 1989; Ward & Scott, 1987). For example, the rationale of Allen and Brooks's (1991) and Regehr and Brooks's (1993) experiments was as follows: suppose that subjects learn to classify stimuli in a situation where a simple, perfectly predictive classification rule exists, and are then tested on transfer items that vary in similarity to the training stimuli. Observed behavior to the transfer items can be of two contrasting types: (1) transfer items similar to old items from the opposite category (bad transfer items) may be classified as quickly and as accurately as items similar to old items from the same category (good transfer items); or (2) the bad transfer items may be classified much less rapidly and accurately than the good transfer items. The first case would be consistent with classification being determined by the speeded application of a rule: in this case all that matters is whether the rule assigns the transfer item to one category or the other. On the other hand, the second outcome described above would be consistent with categorization on the basis of similarity to training instances, and there would be no need to cite a rule as being part of the classification process.

Allen and Brooks (1991) and Regehr and Brooks (1993) obtained evidence that both types of outcome can occur, depending on the type of stimuli used and the precise nature of the task. Evidence for rule learning— no difference in latency or accuracy in classifying a new item similar to a training item and in the same category versus a new item similar to a training item but in the opposite category—was related to a number of factors. For instance, rule learning was more likely to be the controlling process when subjects were actually told the rule prior to the task, although this was not a necessary condition. It was also dependent on the nature of the stimuli used: highly individual stimuli tended to elicit more similarity-based classification, and stimuli composed of interchangeable features tended to elicit more rule-based classification behavior.

As another example, consider again the learning of artificial grammars. Reber (1989) has argued that incidental learning of strings generated from a finite-state grammar is abstract, a view I questioned in Section IV. There is evidence, though, that under some circumstances abstract rule learning in this domain can occur. For instance, Mathews et al. (1989) trained subjects on strings from a standard finite-state grammar under either incidental or rule-searching instructions, and observed that grammaticalness decisions to new transfer strings were equally accurate in the two cases. However, when the strings were derived from a bidirectional grammar, in which strings consisted of two halves with rules mapping a letter in one half (e.g., X) onto a letter in the corresponding position in the other half (e.g., T), no evidence of learning emerged in the incidental condition, although performance on transfer strings was good following rule-searching instructions. There was a considerable difference in performance to trained and novel stimuli under incidental learning conditions, but not under rule-searching instructions. The implication is that while instance memorization may be sufficient for learning about one sort of grammar, rule learning is required for another. The nature of the relationships among the study items (i.e., the sort of grammar used) appears to determine the role these different processes play.

How do rule-based theories explain the effects of variations in contingency on associative learning? When presented with a contingent relationship between cue A and outcome 1 in the form of a set of intermixed AB→1, B→0 trials, the subject is assumed to search for a predictive rule. The obvious hypothesis is that cue A is the significant cue, which leads to a rule corresponding to the vertical dotted line in the top panel of Figure 2. On a test trial with A alone, the outcome will thus be expected. In contrast, a set of noncontingent AB→1, B→1 trials licenses the hypothesis that cue B is the significant feature (the horizontal line in Figure 2). In this case, the category will not be expected on a test trial with cue A alone. Although this account of contingency effect has yet to be tested in detail, some recent investigations have shown that rule-based models may be able to explain a large number of associative learning phenomena (Nosofsky et al., 1989; Nosofsky, Palmeri, & McKinley, 1994).

Several factors determining the balance between rule learning and other bases of classification have been investigated. Thus Smith and Shapiro (1989) found that rule learning was less likely when subjects had to perform a secondary task during the learning stage. Smith and Kemler Nelson (1984) found that rule learning was less likely in a speeded than in an unspeeded learning task, and also that there seems to be a developmental trend in rule learning: in situations where adults classify according to a rule, children often do so on the basis of similarity to training instances. In sum, laboratory studies have established the reality of rule- or hypothesis-based concept learning, and have begun to identify the circumstances that determine when it predominates.

VII. CONCLUSIONS

It has been possible to discuss only a fraction of the human associative learning literature in this chapter. Nevertheless, an impressive amount of data can be interpreted in terms of the acquisition of contingency information, the encoding of instances, or the accumulation of weighted connections in an associative network. Considerable progress has been made in understanding the conditions under which these different types of learning may occur. In addition, research is beginning to identify the circumstances in which rule learning may occur. Each of these theories is able to provide a different explanation of the basic empirical phenomenon on which this chapter has focused: that a subject's behavior in the presence of a certain stimulus (which can be conceived of as a configuration of elements) depends on the prior statistical relationship or contingency between each of those elements and various outcomes.

As these theoretical developments continue, it is likely that processing models will be applied in a formal way to more and more complex learning tasks. Already, it is possible to account in fine detail for learning data obtained in laboratory concept-learning and artificial grammar-learning tasks; perhaps it will not be long before substantial progress is also made in developing computational models of even more complex abilities such as language learning.

Acknowledgments

The preparation of this chapter was supported by the UK Medical Research Council. I am very grateful to Anthony Dickinson for his helpful comments.

References

Allan, L. G. (1980). A note on measurement of contingency between two binary variables in judgment tasks. *Bulletin of the Psychonomic Society, 15,* 147–149.

Allan, L. G., & Jenkins, H. M. (1983). The effect of representations of binary variables on judgment of influence. *Learning and Motivation, 14,* 381–405.

Allen, S. W., & Brooks, L. R. (1991). Specializing the operation of an explicit rule. *Journal of Experimental Psychology: General, 120,* 3–19.

Alloy, L. B., & Abramson, L. Y. (1979). Judgment of contingency in depressed and non-depressed students: Sadder but wiser? *Journal of Experimental Psychology: General, 108,* 441–485.

Anderson, J. R. (1990). *The adaptive character of thought.* Hillsdale, NJ: Erlbaum.

Baker, A. G., Berbrier, M. & Vallee-Tourangeau, F. (1989). Judgments of a 2 × 2 contingency table: Sequential processing and the learning cure. *Quarterly Journal of Experimental Psychology, 41B,* 65–97.

Baker, A. G., Mercier, P., Vallee-Tourangeau, F., Frank, R., & Pan, M. (1993). Selective associations and causality judgments: The presence of a strong causal factor may reduce judgments of a weaker one. *Journal of Experimental Psychology: Learning, Memory, and Cognition, 19,* 414–432.

Barsalou, L. W. (1990). On the indistinguishability of exemplar memory and abstraction in category representation. In T. K. Srull & R. S. Wyer (Eds.), *Advances in social cognition* (Vol. 3, pp. 61–88). Hillsdale, NJ: Erlbaum.

Benzing, W. C., & Squire, L. R. (1989). Preserved learning and memory in amnesia: Intact adaptation-level effects and learning of stereoscopic depth. *Behavioral Neuroscience, 103,* 538–547.

Berry, D. C., & Broadbent, D. E. (1984). On the relationship between task performance and associated verbalizable knowledge. *Quarterly Journal of Experimental Psychology, 36A,* 209–231.

Brooks, L. (1978). Nonanalytic concept formation and memory for instances. In E. Rosch & B. B. Lloyd (Eds.), *Cognition and categorization* (pp. 169–211). Hillsdale, NJ: Erlbaum.

Brooks, L. R., Norman, G. R., & Allen, S. W. (1991). The role of specific similarity in a medical diagnostic task. *Journal of Experimental Psychology: General, 120,* 278–287.

Brooks, L. R., & Vokey, J. R. (1991). Abstract analogies and abstracted grammars: Comments on Reber (1989) and Mathews et al. (1989). *Journal of Experimental Psychology: General, 120,* 316–323.

Busemeyer, J. R., Myung, I. J., & McDaniel, M. A. (1993). Cue competition effects: Theoretical implications for adaptive network learning models. *Psychological Science, 4,* 196–202.

Chapman, G. B. (1991). Trial order affects cue interaction in contingency judgment. *Journal of Experimental Psychology: Learning, Memory, and Cognition, 17,* 837–854.

Chapman, G. B., & Robbins, S. J. (1990). Cue interaction in human contingency judgment. *Memory & Cognition, 18,* 537–545.

Chatlosh, D. L., Neunaber, D. J., & Wasserman, E. A. (1985). Response-outcome contingency: Behavioural and judgmental effects of appetitive and aversive outcomes with college students. *Learning and Motivation, 16,* 1–34.

Cheng, P. W., & Holyoak, K. J. (in press). Adaptive systems as intuitive statisticians: Causality, contingency, and prediction. In J.-A. Meyer & H. L. Roitblat (Eds.), *Comparative approaches to cognition.* Cambridge, MA: MIT Press.

Cleeremans, A. & McClelland, J. L. (1991). Learning the structure of event sequences. *Journal of Experimental Psychology: General, 120,* 235–253.

DeLosh, E. L. (1992). *Interpolation and extrapolation in a functional learning paradigm* (Technical Report). West Lafayette, IN: Purdue University, Purdue Mathematical Psychology Program.

Dickinson, A., & Shanks, D. R. (1994). Instrumental action and causal representation. In G. Lewis, D. Premack, & D. Sperber (Eds.), *Causal understandings in cognition and culture.* Oxford: Oxford University Press.

Dickinson, A., Shanks, D. R., & Evenden, J. L. (1984). Judgment of act-outcome contingency: The role of selective attribution. *Quarterly Journal of Experimental Psychology, 36A,* 29–50.

Dienes, Z. (1992). Connectionist and memory-array models of artificial grammar learning. *Cognitive Science, 16,* 41–80.

Dienes, Z., Broadbent, D. E., & Berry, D. (1991). Implicit and explicit knowledge bases in artificial grammar learning. *Journal of Experimental Psychology: Learning, Memory, and Cognition, 17,* 875–887.

Dulany, D. E., Carlson, R. A., & Dewey, G. I. (1984). A case of syntactical learning and judgment: How conscious and how abstract? *Journal of Experimental Psychology: General, 113,* 541–555.

Elman, J. L. (1990). Representation and structure in connectionist models. In G. T. M. Altmann (Ed.), *Cognitive models of speech processing: Psycholinguistic and computational perspectives* (pp. 345–382). Cambridge, MA: MIT Press.

370   David R. Shanks

Estes, W. K. (1986a). Array models for category learning. *Cognitive Psychology, 18*, 500–549.
Estes, W. K. (1986b). Memory storage and retrieval processes in category learning. *Journal of Experimental Psychology: General, 115*, 155–174.
Estes, W. K., Campbell, J. A., Hatsopoulos, N., & Hurwitz, J. B. (1989). Base-rate effects in category learning: A comparison of parallel network and memory storage-retrieval models. *Journal of Experimental Psychology: Learning, Memory, and Cognition, 15*, 556–571.
Gluck, M. A., & Bower, G. H. (1988). From conditioning to category learning: An adaptive network model. *Journal of Experimental Psychology: General, 117*, 225–244.
Gluck, M. A., & Bower, G. H. (1990). Component and pattern information in adaptive networks. *Journal of Experimental Psychology: General, 119*, 105–109.
Hall, G. (1991). *Perceptual and associative learning*. Oxford: Clarendon Press.
Hartman, M., Knopman, D. S., & Nissen, M. J. (1989). Implicit learning of new verbal associations. *Journal of Experimental Psychology: Learning, Memory, and Cognition, 15*, 1070–1082.
Hayes, N. A., & Broadbent, D. E. (1988). Two modes of learning for interactive tasks. *Cognition, 28*, 249–276.
Hintzman, D. L. (1986). "Schema abstraction" in a multiple-trace memory model. *Psychological Review, 93*, 411–428.
Hintzman, D. L. (1990). Human learning and memory: Connections and dissociations. *Annual Review of Psychology, 41*, 109–139.
Homa, D., Dunbar, S., & Nohre, L. (1991). Instance frequency, categorization, and the modulating effect of experience. *Journal of Experimental Psychology: Learning, Memory, and Cognition, 17*, 444–458.
Homa, D. Sterling, S., & Trepel, L. (1981). Limitations of exemplar-based generalization and the abstraction of categorical information. *Journal of Experimental Psychology: Human Learning and Memory, 7*, 418–439.
Humphreys, M. S., Bain, J. D., & Burt, J. S. (1989). Episodically unique and generalized memories: Applications to human and animal amnesics. In S. Lewandowsky, J. C. Dunn, & K.Kirsner (Eds.), *Implicit memory: Theoretical issues* (pp. 139–156). Hillsdale, NJ: Erlbaum.
Johnson, M. K., & Hasher, L. (1987). Human learning and memory. *Annual Review of Psychology, 38*, 631–668.
Kemler Nelson, D. G. (1984). The effect of intention on what concepts are acquired. *Journal of Verbal Learning and Verbal Behavior, 23*, 734–759.
Knapp, A. G., & Anderson, J. A. (1984). Theory of categorization based on distributed memory storage. *Journal of Experimental Psychology: Learning, Memory, and Cognition, 10*, 616–637.
Koh, K., & Meyer, D. E. (1991). Function learning: Induction of continuous stimulus-response relations. *Journal of Experimental Psychology: Learning, Memory, and Cognition, 17*, 811–836.
Kruschke, J. K. (1992). ALCOVE: An exemplar-based connectionist model of category learning. *Psychological Review, 99*, 22–44.
Kruschke, J. K. (1993). Human category learning: Implications for backpropagation models. *Connection Science, 5*, 3–36.
Lewicki, P., Czyzewska, M., & Hoffman, H. (1987). Unconscious acquisition of complex procedural knowledge. *Journal of Experimental Psychology: Learning, Memory, and Cognition, 13*, 523–530.
Lipp, O. V., Siddle, D. A. T., & Vaitl, D. (1992). Latent inhibition in humans: Single-cue conditioning revisited. *Journal of Experimental Psychology: Animal Behavior Processes, 18*, 115–125.

Logan, G. D. (1988). Toward an instance theory of automatization. *Psychological Review, 95,* 492–527.

Malt, B. (1989). An on-line investigation of prototype and exemplar strategies in classification. *Journal of Experimental Psychology: Learning, Memory, and Cognition, 15,* 539–555.

Mathews, R. C., Buss, R. R., Stanley, W. B., Blanchard-Fields, F., Cho, J. R., & Druhan, B. (1989). Role of implicit and explicit processes in learning from examples: A synergistic effect. *Journal of Experimental Psychology: Learning, Memory, and Cognition, 15,* 1083–1100.

McClelland, J. L., & Rumelhart, D. E. (1985). Distributed memory and the representation of general and specific information. *Journal of Experimental Psychology: General, 114,* 159–188.

McClelland, J. L., & Rumelhart, D. E. (1988). *Explorations in parallel distributed processing.* Cambridge, MA: MIT Press.

McCloskey, M., & Cohen, N. J. (1989). Catastrophic interference in connectionist networks: The sequential learning problem. In G. H. Bower (Ed.), *The psychology of learning and motivation* (Vol. 24, pp. 109–165). San Diego: Academic Press.

McLaren, I. P. L., Kaye, H. & Mackintosh, N. J. (1989). An associative theory of the representation of stimuli: Applications to perceptual learning and latent inhibition. In R. G. M. Morris (Ed.), *Parallel distributed processing: Implications for psychology and neurobiology* (pp. 102–130). Oxford: Oxford University Press.

Medin, D. L. (1989). Concepts and conceptual structure *American Psychologist, 44,* 1469–1481.

Medin, D. L., & Florian, J. E. (1992). Abstraction and selective coding in exemplar-based models of categorization. In A. F. Healy, S. M. Kosslyn, & R. M. Shiffrin (Eds.), *From learning processes to cognitive processes: Essays in honor of William K. Estes* (Vol. 2, pp. 207–234). Hillsdale, NJ: Erlbaum.

Medin, D. L., & Schaffer, M. M. (1978). A context theory of classification learning. *Psychological Review, 85,* 207–238.

Medin, D. L., & Schwanenflugel, P. J. (1981). Linear separability in classification learning. *Journal of Experimental Psychology: Human Learning and Memory, 7,* 355–368.

Melz, E. R., Cheng, P. W., Holyoak, K. J., & Waldmann, M. R. (1993). Cue competition in human categorization: Contingency or the Rescorla-Wagner learning rule? Comments on Shanks (1991). *Journal of Experimental Psychology: Learning, Memory, and Cognition, 19,* 1398–1410.

Nosofsky, R. M. (1986). Attention, similarity and the identification–categorization relationship. *Journal of Experimental Psychology: General, 115,* 39–57.

Nosofsky, R. M. (1987). Attention and learning processes in the identification and categorization of integral stimuli. *Journal of Experimental Psychology: Learning, Memory, and Cognition, 13,* 87–108.

Nosofsky, R. M. (1991). Tests of an exemplar model for relating perceptual classification and recognition memory. *Journal of Experimental Psychology: Learning, Memory, and Cognition, 17,* 3–27.

Nosofsky, R. M., (1992). Exemplar-based approach to relating categorization, identification, and recognition. In F. G. Ashby (Ed.), *Multidimensional models of perception and cognition.* (pp. 363–393). Hillsdale, NJ: Erlbaum.

Nosofsky, R. M., Clark, S. E., & Shin, H. J. (1989). Rules and exemplars in categorization, identification, and recognition. *Journal of Experimental Psychology: Learning, Memory, and Cognition, 15,* 282–304.

Nosofsky, R. M., & Kruschke, J. K. (1992). Investigations of a exemplar-based connectionist model of category learning. In D. L. Medin (Ed.), *The psychology of learning and motivation.* Vol. 28, pp. 207–250. Academic Press.

Nosofsky, R. M., Kruschke, J. K., & McKinley, S. C. (1992). Combining exemplar-based category representations and connectionist learning rules. *Journal of Experimental Psychology: Learning, Memory, and Cognition, 18,* 211–233.

Nosofsky, R. M., Palmeri, T. J., & McKinley, S. C. (1994). Rule-plus-exception model of classification learning. *Psychological Review, 101,* 53–79.

Perruchet, P. (1993). Defining the knowledge units of a synthetic language: Comment on Vokey and Brooks (1992). *Journal of Experimental Psychology: Learning, Memory, and Cognition, 20,* 223–228.

Perruchet, P., & Amorim, M.-A. (1992). Conscious knowledge and changes in performance in sequence learning: Evidence against dissociation. *Journal of Experimental Psychology: Learning, Memory, and Cognition, 18,* 785–800.

Perruchet, P., & Pacteau, C. (1990). Synthetic grammar learning: Implicit rule abstraction or explicit fragmentary knowledge? *Journal of Experimental Psychology: General, 119,* 264–275.

Pinker, S. (1991). Rules of language. *Science, 253,* 530–535.

Plunkett, K., & Marchman, V. (1991). U-shaped learning and frequency effects in a multi-layered perceptron: Implications for child language acquisition. *Cognition, 38,* 43–102.

Reber, A. S. (1969). Transfer of syntactic structure in synthetic languages. *Journal of Experimental Psychology, 81,* 115–119.

Reber, A. S. (1989). Implicit learning and tacit knowledge. *Journal of Experimental Psychology: General, 118,* 219–235.

Regehr, G., & Brooks, L. R. (1993). Perceptual manifestations of an analytic structure: The priority of holistic individuation. *Journal of Experimental Psychology: General, 122,* 92–114.

Rescorla, R. A., & Wagner, A. R. (1972). A theory of Pavlovian conditioning: Variations in the effectiveness of reinforcement and nonreinforcement. In A. H. Black & W. F. Prokasy (Eds.), *Classical conditioning II: Current research and theory* (pp. 64–99). New York: Appleton-Century-Crofts.

Rosch, E., Simpson, C., & Miller, R. S. (1976). Structural bases of typicality effects. *Journal of Experimental Psychology: Human Perception and Performance, 2,* 491–502.

Sanderson, P. M. (1989). Verbalizable knowledge and skilled task performance: Association, dissociation, and mental models. *Journal of Experimental Psychology: Learning, Memory, and Cognition, 15,* 729–747.

Schacter, D. L., Eich, J. E., & Tulving, E. (1978). Richard Semon's theory of memory. *Journal of Verbal Learning and Verbal Behavior, 17,* 721–743.

Seidenberg, M. S., & McClelland, J. L. (1989). A distributed, developmental model of word recognition and naming. *Psychological Review, 96,* 523–568.

Shanks, D. R. (1985a). Continuous monitoring of human contingency judgment across trials. *Memory & Cognition, 13,* 158–167.

Shanks, D. R. (1985b). Forward and backward blocking in human contingency judgment. *Quarterly Journal of Experimental Psychology, 37B,* 1–21.

Shanks, D. R. (1987). Acquisition functions in causality judgment. *Learning and Motivation, 18,* 147–166.

Shanks, D. R. (1990a). Connectionism and human learning: Critique of Gluck and Bower (1988). *Journal of Experimental Psychology: General, 119,* 101–104.

Shanks, D. R. (1990b). Connectionism and the learning of probabilistic concepts. *Quarterly Journal of Experimental Psychology, 42A,* 209–237.

Shanks, D. R. (1991a). Categorization by a connectionist network. *Journal of Experimental Psychology: Learning, Memory, and Cognition, 17,* 433–443.

Shanks, D. R. (1991b). A connectionist account of base-rate biases in categorization. *Connection Science, 3,* 143–162.

Shanks, D. R. (1991c). Some parallels between object recognition and associative learning. In J.-A. Meyer & S. Wilson (Eds.), *Simulation of behaviour: From animals to animats* (pp. 337–343). Cambridge, MA: MIT Press.

Shanks, D. R. (1993). Human instrumental learning: A critical review of data and theory. *British Journal of Psychology, 84,* 319–354.

Shanks, D. R., & Dickinson, A. (1987). Associative accounts of causality judgment. In G. H. Bower (Ed.). *The Psychology of Learning and Motivation* (Vol. 21, pp. 229–261). Orlando, FL: Academic Press.

Shanks, D. R., & Dickinson, A. (1991). Instrumental judgment and performance under variations in action-outcome contingency and contiguity. *Memory & Cognition, 19,* 353–360.

Shanks, D. R., & Gluck, M. A. (1994). Tests of an adaptive network model for the identification and categorization of continuous-dimension stimuli. *Connection Science, 6,* 69–99.

Shanks, D. R., Green, R. E. A., & Kolodny, J. A. (1994). A critical examination of the evidence for unconscious (implicit) learning. In C. Umilta & M. Moscovitch (Eds.), *Attention and performance XV: Conscious and nonconscious information processing.* Cambridge, MA: MIT Press.

Shanks, D. R. & St.John, M. F. (1994). Characteristics of dissociable human learning systems. *Behavioral and Brain Sciences, 17.*

Shin, H. J., & Nosofsky, R. M. (1992). Similarity-scaling studies of "dot-pattern" classification and recognition. *Journal of Experimental Psychology: General, 121,* 278–304.

Smith, J. D., & Kemler Nelson, D. G. (1984). Overall similarity in adults' classification: The child in all of us. *Journal of Experimental Psychology: General, 113,* 137–159.

Smith, J. D., & Shapiro, J. H. (1989). The occurrence of holistic categorization. *Journal of Memory and Language, 28,* 386–399.

Stadler, M. A. (1989). On learning complex procedural knowledge. *Journal of Experimental Psychology: Learning, Memory, and Cognition, 15,* 1061–1069.

Svartdal, F. (1989). Shaping of rule-governed behaviour. *Scandinavian Journal of Psychology, 30,* 304–314.

Vokey, J. R., & Brooks, L. R. (1992). Salience of item knowledge in learning artificial grammars. *Journal of Experimental Psychology: Learning, Memory, and Cognition, 18,* 328–344.

Ward, T. B., & Scott, J. (1987). Analytic and holistic modes of learning family-resemblance concepts. *Memory & Cognition, 15,* 42–54.

Wasserman, E. A., Chatlosh, D. L., & Neunaber, D. J. (1983). Perception of causal relations in humans: Factors affecting judgments of response-outcome contingencies under free-operant procedures. *Learning and Motivation, 14,* 406–432.

Wasserman, E. A., Elek, S. M., Chatlosh, D. L., & Baker, A. G. (1993). Rating causal relations: The role of probability in judgments of response-outcome contingency. *Journal of Experimental Psychology: Learning, Memory, and Cognition, 19,* 174–188.

Watkins, M. J., & Kerkar, S. P. (1985). Recall of a twice-presented item without recall of either presentation: Generic memory for events. *Journal of Memory and Language, 24,* 666–678.

Whittlesea, B. W. A. (1987). Preservation of specific experiences in the representation of general knowledge. *Journal of Experimental Psychology: Learning, Memory, and Cognition, 13,* 3–17.

Whittlesea, B. W. A., & Brooks, L. R. (1988). Critical influence of particular experiences in the perception of letters, words, and phrases. *Memory & Cognition, 16,* 387–399.

Widrow, B., & Hoff, M. E. (1960). Adaptive switching circuits. IRE WESCON Convention Record, Pt. 4, 96–104.

Willingham, D. B., Greeley, T., & Bardone, A. M. (1993). Dissociation in a serial response time task using a recognition measure: Comment on Perruchet and Amorim (1992). *Journal of Experimental Psychology: Learning, Memory, and Cognition, 19,* 1424–1430.

Willingham, D. B., Nissen, M. J., & Bullemer, P. (1989). On the development of procedural knowledge. *Journal of Experimental Psychology: Learning, Memory, and Cognition, 15,* 1047–1060.

Index